Second Supplements to the 2nd Edition of

RODD'S CHEMISTRY OF CARBON COMPOUNDS

ELSEVIER SCIENCE B.V.
Sara Burgerhartstraat 25
P.O. Box 211, 1000 AE Amsterdam, The Netherlands

First edition 2000

Library of Congress Cataloging in Publication Data
A catalog record from the Library of Congress has been applied for.

ISBN: 0 444 82980 6

⊗ The paper used in this publication meets the requirements of ANSI/NISO Z39.48-1992 (Permanence of Paper).
Printed in The Netherlands.

Second Supplements to the 2nd Edition of

RODD'S CHEMISTRY OF CARBON COMPOUNDS

VOLUME I

ALIPHATIC COMPOUNDS

★

VOLUME II

ALICYCLIC COMPOUNDS

★

VOLUME III

AROMATIC COMPOUNDS

★

VOLUME IV

HETEROCYCLIC COMPOUNDS

★

VOLUME V

MISCELLANEOUS

GENERAL INDEX

Second Supplements to the 2nd Edition of

RODD'S CHEMISTRY OF CARBON COMPOUNDS

A modern comprehensive treatise

Edited by
MALCOLM SAINSBURY
School of Chemistry, The University of Bath,
Claverton Down, Bath BA2 7AY, England

Second Supplement to

VOLUME IV HETEROCYCLIC COMPOUNDS

Part I: Six-membered Heterocyclic Compounds with Two Hetero-Atoms from Group V of the Periodic Table: the Pyridazine and Pyrimidine Groups.

Part J: Six-membered Heterocyclic Compounds with Two Hetero-Atoms from Group V of the Periodic Table: the Pyrazine Group, Phenoxazine, Phenothiazine, Phenazine and Sulphur Dyes. Six-membered Heterocyclic Compounds with Three or More Hetero-Atoms.

2000
ELSEVIER
Amsterdam–Lausanne–New York–Oxford–Shannon–Singapore-Tokyo

Contributors to this Volume

R. BOLTON
Department of Chemistry, University of Surrey, Guildford,
Surrey GU2 5XH, England

C.J. GRANVILLE SHAW
Department of Chemistry, Brunel University, Uxbridge UB8 3PH, England

D. HURST
Kingston University, Penrhyn Road, Kingston upon Thames KT1 2EE,
Surrey, England

K.J. McCULLOUGH
Department of Chemistry, Heriot Watt University, Edinburgh EH14 4AS,
Scotland, U.K.

L.K. MEHTA
Department of Chemistry, Brunel University, Uxbridge UB8 3PH, England

S. JOHNE
Friedensstrasse 60, D-06749 Bitterfeld, Germany

D.F. O'SHEA
Department of Chemistry, University College Dublin,
Belfield, Dublin 4, Republic of Ireland

P.J. PARRICK
Department of Chemistry, Brunel University, Uxbridge UB8 3PH, England

Preface to Volume IV I/J

The current volume comprises chapters 42 to 47 of the 2nd Supplement of *Rodd's Chemistry of the Carbon Compounds*. I have been extremely fortunate that most of the contributors to this volume were also responsible for similar chapters in the 1st Supplement, published in 1995. As a result they were able to compile this new supplement exactly on time and with minimal editorial input. In only 5 years, however, the chemistry previously covered in Volume I/J has burgeoned to the extent that my colleagues found it necessary to read several thousand references in order to present the balanced account assembled here.

In this volume Drs Parrick and Shaw have upgraded their earlier chapter on pyridazines, cinnolines and their allies, and Dr Hurst and Dr McCullough have done the same for their chapters on the subjects of pyrimidines and quinazolines, and pyrazines, respectively. Dr Bolton has contributed to this series several times before, but not on the subject of phenazine, oxazine and thiazine sulfur dyes. However, his wide experience in heterocyclic chemistry enabled him to analyse recent progress in this area without difficulty and with his customary skill. Professor Johne returns to one of his research interests, quinazoline alkaloids, while Professor O'Shea completes the volume with a timely review of six-membered rings with three or more heteroatoms: triazines, tetrazines pentazines and hexazines.

I wish to thank all these industrious chemists for their help and forbearance in bringing this work to so successful a conclusion.

Malcolm Sainsbury

Bath
May 1999

Contents
Volume IV I/J

Six-membered Heterocyclic Compounds with Two Hetero-Atoms from Group V of the Periodic Table: the Pyridazine and Pyrimidine Groups, the Pyrazine Group, Phenoxazine, Phenothiazine, Phenazine and Sulphur Dyes. Six-membered Heterocyclic Compounds with Three or More Hetero-Atoms

List of contributors v
Preface vii
List of common abbreviations and symbols used xiii

Chapter 42. Pyridazines, Cinnolines, Benzocinnolines and Phthalazines
by J. PARRICK, C.J. GRANVILLE SHAW AND L.K. MEHTA

1. Pyridazines 1
 (a) Methods of synthesis 1
  (i) From arylhydrazones, 1 – (ii) Cyclisations using hydrazines and hydrazones, 3 – (iii) Cycloadditions reactions, 10 – (iv) From other heterocyclic compounds, 12 – (v) Microwave techniques, 14 –
 (b) Reactions 15
 (c) Complexes with metal ions 28
 (d) Physical properties and spectra 30
 (e) Biological activity 32
 (f) Herbicides 40
 (g) Miscellaneous applications 41
2. Cinnoline and its derivatives 42
 (a) Methods of synthesis 42
  (i) From diazonium salts, 42 – (ii) From phenylhydrazones, 42 – (iii) Replacement of fluorine, 43 – (iv) Addition reactions, 44 – (v) By formation of the N–N bond, 45 – (vi) From γ-ketoesters or cyclohexan-1,3-diones, 45 –
 (b) Physical properties 46
 (c) Reactions 47
3. The benzocinnolines 49
 (a) Benzo[*c*]cinnoline 49
  (i) Methods of preparation, 49 – (ii) Physical properties, 50 – (iii) Reactions, 51 -
 (b) Benzo[*f*]cinnolines 52
 (c) Benzo[*h*]cinnolines 52
4. Phthalazine and its derivatives 53
 (a) Methods of synthesis 53
  (i) From phthalic acid derivatives, 53 – (ii) From cycloaddition reactions, 54 – (iii) From amidines, 57 –
 (b) Physical properties 57
 (c) Reactions of the heterocycle 59

(d) Reactions of substituted phthalazines 60
(e) Phthalazine ylides 63
(f) Phthalazin-1(2*H*)-ones 65
(g) Phthalazin-1,4(2*H*,3*H*)-diones and phthalazine quinones 66
(h) Complexes with metal ions 67
(i) Analytical and environmental aspects 67
(j) Pharmaceuticals 68

Chapter 43. Pyrimidines and Quinazolines
by D.T. HURST

1. Introduction 71
2. Pyrimidines 72
(a) Properties and structural aspects 72
(b) Synthesis 72
(c) Reactions of pyrimidines 79
(i) Substitution, 79 – (ii) Metallation reactions, 84 – (iii) Ring transformations and miscellaneous reactions of pyrimidines, 87 –
3. Quinazolines 91
(a) Properties and structural aspects 91
(b) Synthesis 91
(c) Reactions 95

Chapter 44. Pyrazines and Related Ring Structures
by K.J. McCULLOUGH

1. Pyrazines, 1,4-diazines 99
(a) Introduction 99
(b) Physical and spectroscopic properties 101
(c) General methods of synthesis 101
(d) Pyrazine, its homologues and derivatives 102
(i) Pyrazine, alkyl-, alkenyl-, alkynyl- and aryl-pyrazines, 102 – (ii) Halopyrazines, 106 – (iii) Aminopyrazines, 110 – (iv) Pyrazinols (pyrazinones), alkoxypyrazines, alkylthiopyrazines and related derivatives, 111 – (v) Pyrazine *N*-oxides, 113 – (vi) Pyrazinoic acids and related derivatives, 115 – (vii) Cyanopyrazines, 116 – (viii) Pyrazine *C*-nucleosides, 119 –
(e) Reduced pyrazines 121
(i) Dihydropyrazines, 121 – (ii) Tetrahydropyrazines, 127 – (iii) Hexahydropyrazines (piperazines), 128 – (iv) Ketopiperazines, 133 – (v) Miscellaneous fused pyrazines, 137 –
2. Quinoxalines, benzopyrazines 137
(a) Introduction 137
(b) Physical and spectroscopic properties 138
(c) General methods of synthesis 139

(d) Quinoxaline, its homologues and derivatives 142
(i) Quinoxaline, alkyl-, alkenyl-, alkynyl- and aryl quinoxalines, 142 – (ii) Haloquinoxalines, 145 – (iii) Aminoquinoxalines, 147 – (iv) Quinoxalin-2(1*H*)-ones and related compounds, 148 – (v) Quinoxaline *N*-oxides, 151 – (vi) Miscellaneous quinoxaline derivatives, 156 – (vii) Reduced quinoxalines, 158 –
3. Phenazines, dibenzopyrazines 162
(a) Introduction 162
(b) Physical and spectroscopic properties 164
(c) General methods of synthesis 164
(d) Phenazine, its homologues and derivatives 166
(i) Phenazine and its derivatives, 166 – (ii) Reduced phenazines, 168 – (iii) Miscellaneous phenazine derivatives, 170 –

Chapter 45. Phenazine, Oxazine, Thiazine and Sulfur Dyes
by R. BOLTON

1. Dye-mediated electron-transference 174
(a) Application in polymerisation 174
(b) Application to biological systems 175
2. Analytical applications 178
(a) Biosensors 179
(b) Kinetic/catalytic application in analysis 183
(c) Spectroscopic applications to analysis 187
(d) Analysis using luminescence/fluorescence 190
3. Application in synthesis 191
4. Oxazine dyes 193
5. Spiro-oxazines and spiropyrans 193
6. Triphenodioxazine dyes 197
7. Sulfur dyes 200

Chapter 46. Quinazoline Alkaloids
by S. JOHNE

1. Bicyclic quinazolines 203
2. Pyrroloquinazolines 210
3. Indolo-, pyrido- and indolopyrido-quinazolines 214
4. Other quinazoline alkaloids 223

Chapter 47. Six-membered Rings with Three or More Hetero-Atoms
by D.F. O'SHEA

(I) Nitrogen rings 233
1. Triazines 233
(a) 1,2,3-Triazines 233
(i) Synthesis and structure, 233 – (ii) Reactions of 1,2,3-triazines, 235 –
(b) 1,2,4-Triazines (as-triazines) 238
(i) Synthesis and structure, 238 – (ii) Reactions of 1,2,4-triazines, 242 –
(c) 1,3,5-Triazines (s-triazines) 248
(i) Synthesis and structure, 248 – (ii) Reactions of 1,3,5-triazines, 253 –

2. Tetrazines 254
(a) 1, 2, 3, 4- and 1, 2, 3, 5-Tetrazines 254
(i) Synthesis and structure, 254 –
(b) 1,2,4,5-Tetrazines (s-tetrazines) 256
(i) Synthesis and structure, 256 – (ii) Reactions of 1,2,4,5-tetrazines, 258 –
(c) Verdazyls 262
3. Pentazine and hexazine 263
(II) Nitrogen with Sulphur and/or Oxygen 265
1. Sulphur–nitrogen rings 265
2. Oxygen–nitrogen rings 268
3. Oxygen–nitrogen–sulphur rings 271
(III) Oxygen, Sulphur Rings 272
1. Trioxanes and oxygen–sulphur rings 272
2. Tri- and Tetrathianes 273

Guide to the Index 277
Index 279

List of Common Abbreviations and Symbols Used

A	acid
Å	Ångström units
Ac	acetyl
a	axial
as, asymm.	asymmetrical
at.	atmosphere
B	base
Bu	butyl
b.p.	boiling point
c, C	concentration
CD	circular dichroism
conc.	concentrated
crit.	critical
D	Debye unit, 1×10^{-18} e.s.u.
D	dissociation energy
D	dextro-rotatory; dextro configuration
d	density
dec., decomp	with decomposition
deriv.	derivative
E	energy; extinction; electromeric effect
E	entgegen (opposite) configuration
*E*1, *E*2	uni- and bi-molecular elimination mechanisms
E1cB	unimolecular elimination in conjugate base
ESR	electron spin resonance
Et	ethyl
e	nuclear charge; equatorial
f.p.	freezing point
G	free energy
GLC	gas liquid chromatography
g	spectroscopic splitting factor, 2.0023
H	applied magnetic field; heat content
h	Planck's constant
Hz	hertz
I	spin quantum number; intensity; inductive effect
IR	infrared
J	coupling constant in NMR spectra
J	Joule
K	dissociation constant
k	Boltzmann constant; velocity constant
kcal	kilocalories
L	laevorotatory, laevo configuration
M	molecular weight; molar; mesomeric effect
Me	methyl

m	mass; mole; molecule; *meta*-
m.p.	melting point
Ms	mesyl (methanesulphonyl)
[M]	molecular rotation
N	Avogadro number; normal
NMR	nuclear magnetic resonance
NOE	Nuclear Overhauser Effect
n	normal; refractive index; principal quantum number
o	*ortho*-
ORD	optical rotatory dispersion
P	polarisation; probability; orbital state
Pr	propyl
Ph	phenyl
p	*para*-; orbital
PMR	proton magnetic resonance
R	clockwise configuration
S	counterclockwise configuration; entropy; net spin of incompleted electronic shells; orbital state
S_N1, S_N2	uni- and bi-molecular nucleophilic substitution mechanism
S_Ni	internal nucleophilic substitution mechanism
s	symmetrical; orbital
sec	secondary
soln.	solution
symm.	symmetrical
T	absolute temperature
Tosyl	*p*-toluenesulphonyl
Trityl	triphenylmethyl
t	time
temp.	temperature (in degrees centrigrade)
tert	tertiary
UV	ultraviolet
v	velocity
Z	zusammen (together) configuration
α	optical rotation (in water unless otherwise stated)
$[\alpha]$	specific optical rotation
ϵ	dielectric constant; extinction coefficient
μ	dipole moment; magnetic moment
μ_B	Bohr magneton
μg	microgram
μm	micrometer
λ	wavelength
ν	frequency; wave number
χ, χ_d, χ_μ	magnetic; diamagnetic and paramagnetic susceptibilities
~	about
(+)	dextrorotatory
(–)	laevorotatory
–	negative charge
+	positive charge

Second Supplements to the 2nd Edition of Rodd's Chemistry of Carbon Compounds, Vol.IV IJ, edited by M. Sainsbury

Chapter 42

PYRIDAZINES, CINNOLINES, BENZOCINNOLINES AND PHTHALAZINES

JOHN PARRICK, C. J. GRANVILLE SHAW AND LINA K. MEHTA

This chapter covers the chemistry of pyridazine and its benzo derivatives since publication of the first supplement. A useful background review of the chemistry in this area in the period 1982-95 is available (W.J. Coates, Comprehensive Heterocyclic Chemistry II, 1996, 6, 1).

1. Pyridazines

a. Methods of Synthesis

(i) From arylhydrazones

Arylhydrazones are accessible intermediates from which *N*-arylpyridazines can be formed. Self-cyclisation in acidic conditions of cinnamoylacetonitrile aryl hydrazones ArNHN=C(CN)COCH=CHPh gives cyanopyridazines (1) (G.A.M. Nawwar, R.M. Swellem and L.M. Chabaka, Coll. Czech. chem. Commun., 1994, 59, 186).

The coupling of 1-amino-2,6,6-tricyanocyclohexadiene derivative (2) with a variety of aryldiazonium salts produces the hydrazones (3) which, in the presence of weak base in ethanol at elevated temperatures, furnish 2-aryl-4-cyanopyridazine-3(2*H*)-ones (4) (M.J. Mokrosz, Arch. Pharm., 1993, 326, 39).

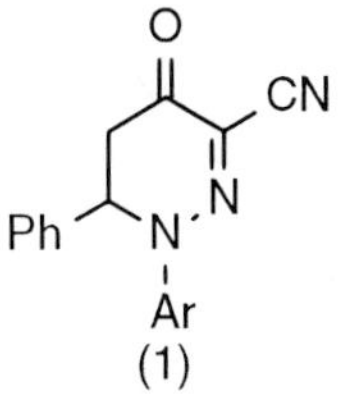

O CN Ph N N Ar

(1)

R^2 R^3 NC R^1 NC H_2N CH_2R CN

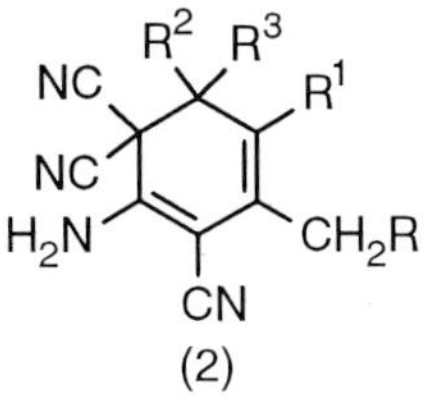

(2)

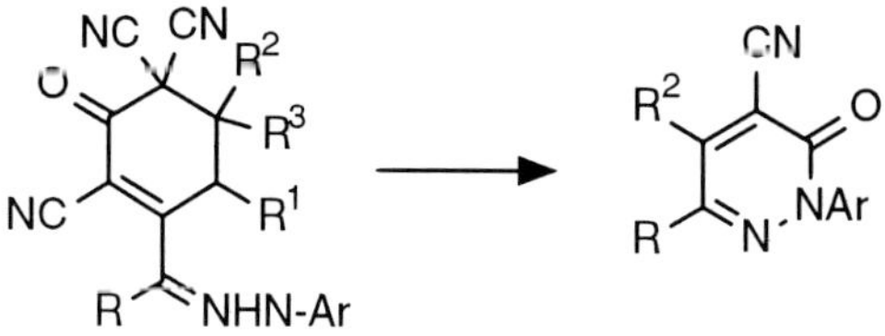

Treatment of benzaldehyde *N*-(2-propenyl)-*N*-(4-methylsulfonylphenyl)hydrazone (5) (R^1= 4-$MeSO_2$, R^2 = H) with 45% H_2SO_4 gave 2-(4-methylsulfonylphenyl)-6-phenylhexahydro-4-pyridazinol (6) in 35% yield rather than the expected 1-(4-methylsulfonylphenyl)-1-(2-propenyl)-hydrazine. The halogen substituted hydrazones (5, R^1 = 2-Cl, 3-Cl, 4-F, R^2 = H) and benzaldehyde 1-(2-methyl-2-propenyl)-phenylhydrazone (5, R^1 = H, R^2 = Me) gave similar results (R. Bakthavatchalam, E. Ciganek and J.C. Calabrese, J. heterocyclic Chem., 1996, 33, 213).

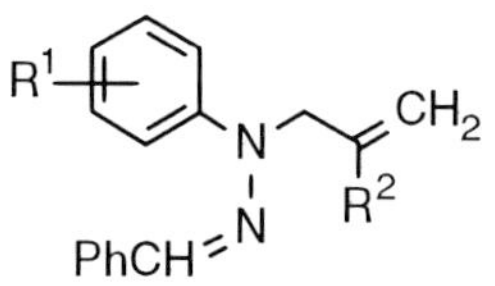

Cyclisation of the halohydrazones (7, R^1 = H, Me, Ph, R^2 = CO_2Et, Ph, X = H) prepared by the reaction of 5-hydrazinopyridazinones (8) with ethyl bromopyruvate and phenacyl bromide in ethanol, produced pyrrolo[2,3-*d*]pyridazinones (9). On the other hand, heating ethyl bromopyruvate pyridazinylhydrazone (7, R^1 = H, R^2 = CO_2Et, X = H) in dioxane afforded 1,4-dihydropyridazino[4,5-*c*]pyridazine (10) in low yield. Compounds 10 (R^1 = H, Me, Ph, CH_2Ph, R^2 = CO_2Et, R^1 = H, R^2 = Ph) were also obtained in good yields by heating the dihalohydrazones (7, X = Cl) which were synthesised from 4-chloro-5-hydrazinopyridazinones (8). Heating the furylhydrazone (11) under similar conditions produced 3,6-dihydrofuro[3,4-*c*]pyridazine (12) (T. Yamasaki *et al.*, Heterocycles, 1996, 43, 1863).

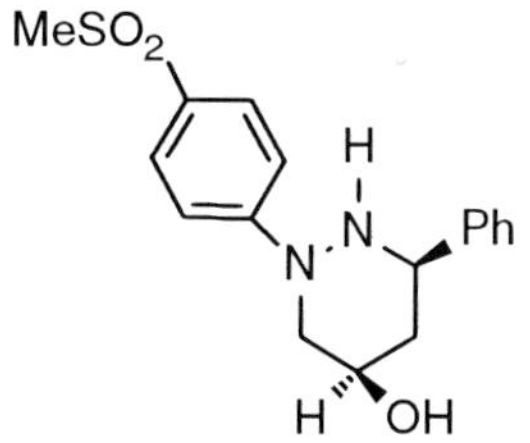

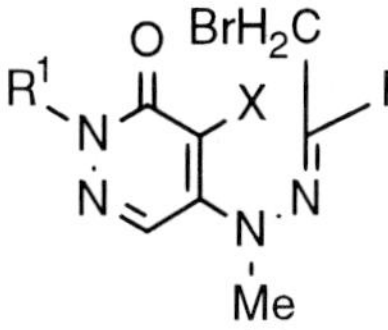

(7) 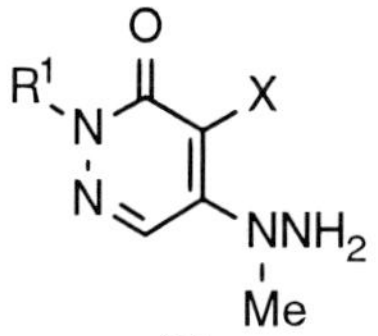 (8) 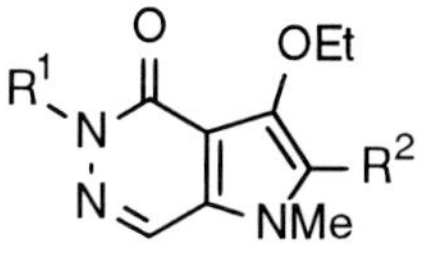 (9)

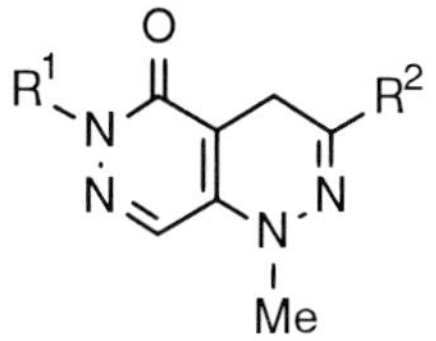

(10) 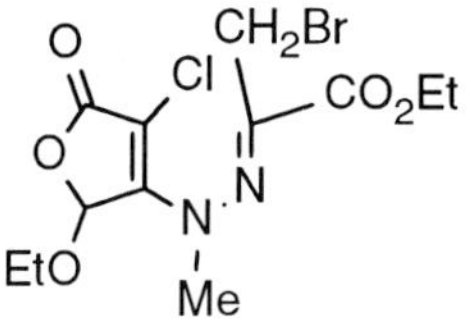 (11) 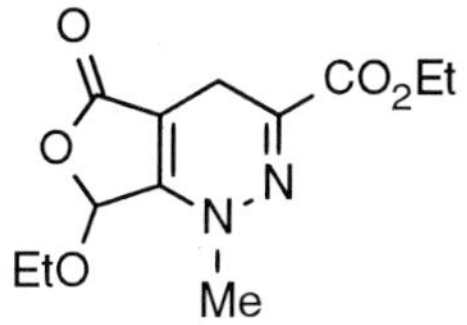 (12)

Several 5-trifluoromethyl-4-pyridazinones (13) (R = aryl, ethyl, heptyl, Me_2CH, R^1 = Me, Me_3C) were conveniently synthesised by the silica gel-mediated cyclisation of 1,1,1,2,2-pentafluoro-3,4-alkanedione 4-dialkylhydrazones, prepared from aldehyde dialkylhydrazones and $(C_2F_5CO)_2O$ (Y. Kamitori *et al.,* Tetrahedron Letters, 1993, 34, 5135).

Pyridazinones and tetrahydrocinnolines (e.g. 14, Ar = Ph, 4-MeC_6H_4, 4-$NO_2C_6H_4$) have been prepared via reaction of phosphonium ylides Ph_3P=CHR (R = CO_2Et, CO_2Me) with hydrazones (e.g. 15) (A.A. Nada *et al.,* J. chem. Research (S), 1997, 236).

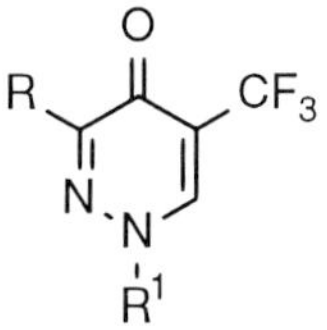

(13) 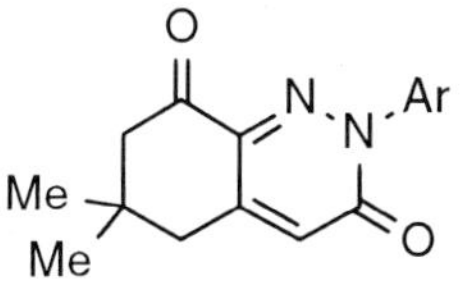 (14) 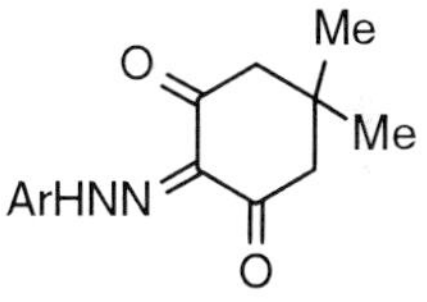 (15)

Pyridazines are obtained by the reaction of benzil monohydrazone with active keto methylene reagents, followed by base treatment of the chloro-substituted tetrahydropyridazines thus formed (M.G. Assy and E. Abd El-Ghani, Pol. J. Chem., 1995, 69, 685).

(ii) Cyclisations using hydrazines and hydrazones

The reaction of hydrazine hydrate with 1,5-diarylpent-1-yne-3,5-diones HOCR=CHCOC≡CPh (R = Ph, 4-MeC_6H_4, 4-$MeOC_6H_4$,

4-BrC_6H_4, 4-ClC_6H_4) afforded 4-pyridazinols (16) (M.G. Maru and M.M. Mishrikey, Rev. Roum. Chim., 1993, 38, 989).

The reaction of 2-(alkylamino)-3-(trifluoroacetyl)butene dioates $F_3COC(CO_2Me)=C(CO_2Me)NHR^1$ (R^1 = $CHMe_2$, CH_2Ph) with alkyl and aryl hydrazines in ether provides 1,6-dihydro-5-hydroxy-6-oxo-3-(trifluoromethyl)-4-pyridazine carboxylates as their alkylamine salts (17, R = H, Ph, 4-$MeOC_6H_4$, 2-pyridyl, CH_2Ph, CMe_3) and not the expected 5-alkylamino compounds. Acidification released the free 5-hydroxy compounds (S.G. Hegde and C.R. Jones, J. heterocyclic Chem., 1993, 30, 1501).

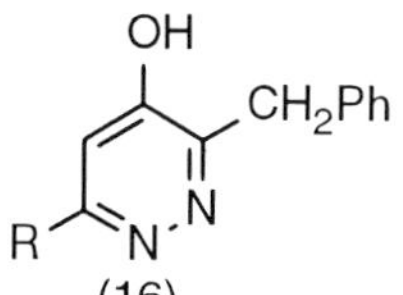

OH CH2Ph R N N (16)

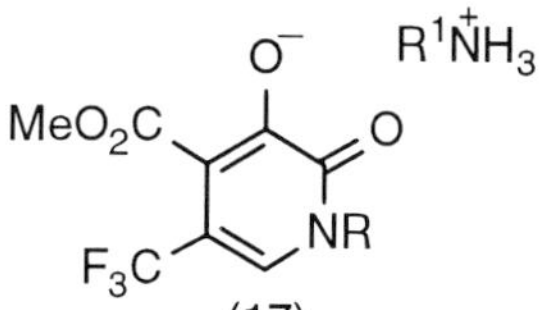

O⁻ R1NH3+ MeO2C O NR F3C (17)

A variety of 6-substituted and 5,6-disubstituted 3(2*H*)-pyridazinones (19, R^1 = Bu^t, aryl, heteroaryl, R^2 = H, Me) have been prepared by an efficient one-pot process starting from ketones. Cyclisation of the intermediate hydroxy 4-ketoacids (18) with hydrazine and subsequent elimination of water gives the pyridazinone directly (W.J. Coates and A. McKillop, Synthesis, 1993, 334).

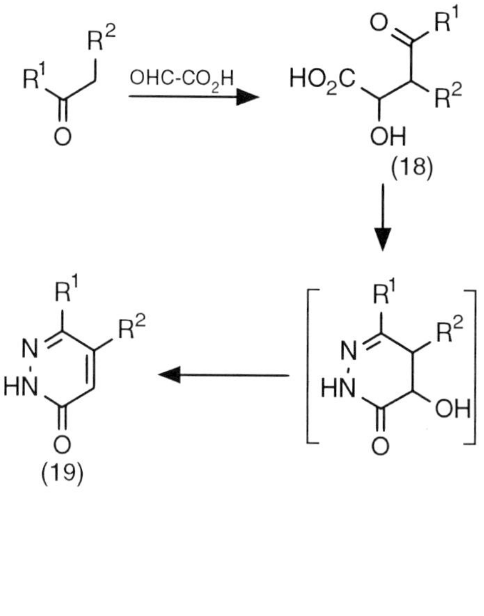

R2 R1 O OHC-CO2H O R1 HO2C R2 OH (18)

R1 R2 N HN O (19)

R1 R2 N HN OH O

Pyridazine-3-carboxylic acid derivatives e.g. (20) (R = CO_2CH_2Ph, R^1, R^2 = H or bond, R, R^2 = H, R^1 = CO_2CH_2Ph) were prepared via ring closure of α-hydrazino- and δ-hydrazino-pentanoates. Either optically active glutamic acid or an enantioselective catalytic hydrogenation was used to generate the chiral centre (U. Schmidt, C. Braun and H. Sutoris, Synthesis-Stuttgart, 1996, 223).

N-Substituted 1,4,5,6-tetrahydropyridazines (21, R = Ph, substituted Ph, CH_2Ph) can be prepared in a one-pot procedure from a hydrazine $RNHNH_2$ and 2,5-dimethoxy-tetrahydrofuran using sodium borohydride in aqueous THF acidic medium. Variable amounts of two side products are also formed : hexahydropyridazines (22) and 1-amino-pyrrolidines (23) (G. Verardo, N. Toniulti and A. G. Giumanini, Tetrahedron, 1997, 53, 3707).

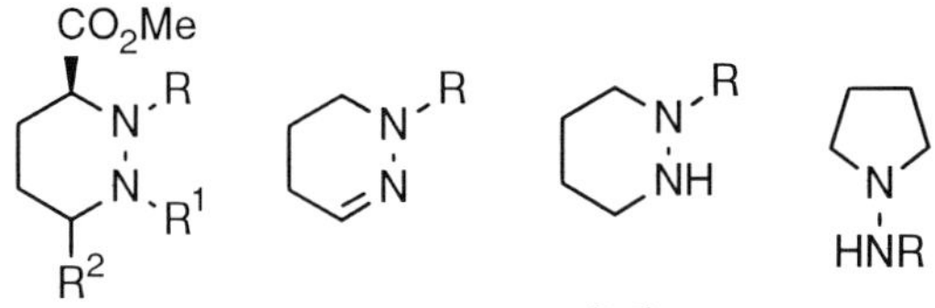

CO_2Me R N N R^1 R^2 (20) | R N N (21) | R N NH (22) | N HNR (23)

Pyridazinones with 2- and 3-thienyl substituents have been made from precursor 4-oxobutanoic acids via their 4,5-dihydro derivatives. The oxobutanoic acids $Ar^2COCH_2CHAr^1CO_2H$ (Ar^1 = Ph, Ar^2 = 2-thienyl and Ar^1 = 2-thienyl, Ar^2 = Ph) were obtained by addition of HCN to the chalcones $Ar^2COCH{=}CHAr^1$ followed by acid hydrolysis of the resulting nitrile. Nitriles $ArCOCHMeCH_2CN$ were prepared by the Michael-Stetter reaction and converted via the acids to the 4,5-dihydropyridazin-3(2*H*)-ones. SeO_2-mediated oxidation of the latter (or base-catalysed condensation of methanal or aryl aldehydes at the 4-position of the 4,5-di-hydrocompound) led to the pyridazin-3(2*H*)-ones (P. Powell and M.H. Sosabowski, J. chem. Research (S), 1995, 306). Similar compounds have been prepared by Pd(0)-catalysed coupling of organotin reagents with aryl, acyl and heteroaryl halides (M.H. Sosabowski and P. Powell, J. chem. Research (S), 1995, 402).

Amination of 3-(4-phenylbenzoyl)acrylic acid gives 2-(aryl-amino)-3-(4-phenylbenzyl)propionic acid (24, Ar = $PhCH_2$, 4-MeC_6H_4, 2-MeC_6H_4, 4-$NO_2C_6H_4$). Hydrazinolysis of (24) yields the corresponding 6-(4-biphenylyl)-4-(arylamino)-2,3,4,5-tetrahydropyridazin-3-one (25). Compounds (25) have also been prepared from hydrazinolysis of $\Delta^{\beta,\gamma}$-butenolides (26) which are obtained from dehydration of (24) (A.Y. Soliman and I.A. Attia, Indian J. Chem., 1997, 36B, 75).

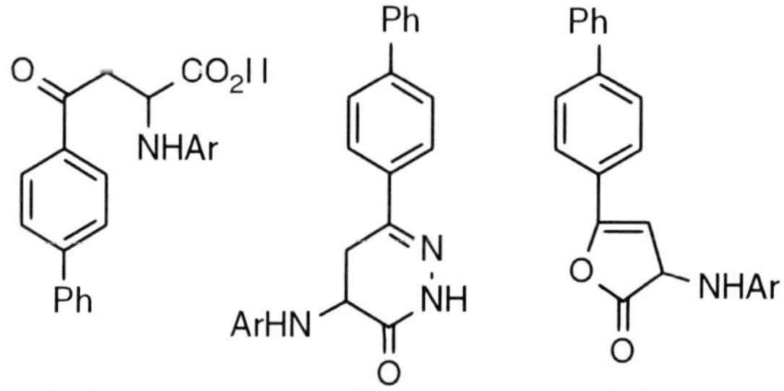

(24) (25) (26)

Cyclisation of hydrazine with Meldrum's acid derivatives (27) gives 73-91% of 4,6-disubstituted-4,5-dihydropyridazin-3(2*H*)-ones (28, R = H, Ph, 3-, 4-ClC_6H_4, 3,4-$(MeO)_2C_6H_3CH_2$, 2-thienylmethyl, R^1 = Ph, 4-ClC_6H_4) at room temperature (G. Toth *et al.*, Synth. Commun., 1997, 27, 3513).

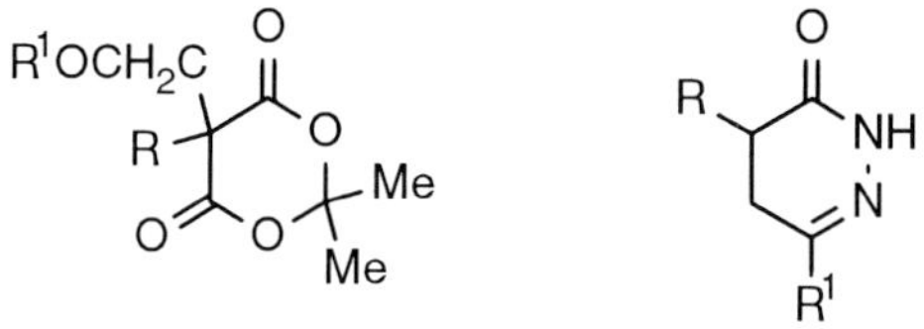

(27) (28)

Aryl- and alkyl-hydrazines $RNHNH_2$ (R = Ph, 4-MeC_6H_4, 4-$NO_2C_6H_4$, 2,4-$Me(Cl)C_6H_3$, Me_3C) react with the readily available cyclopropane derivative of Meldrum's acid (29) in refluxing acetonitrile to afford 30-68% of the corresponding 1-substituted-1,4,5,6-tetrahydro-pyridazin-3(2*H*)-ones (30) directly via N-C bond formation involving addition by the substituted hydrazine nitrogen atom followed by cyclisation and decarboxylation of the cyclised intermediates (K.-J. Hwang and K.-H. Park, Heterocycles, 1993, 36, 219).

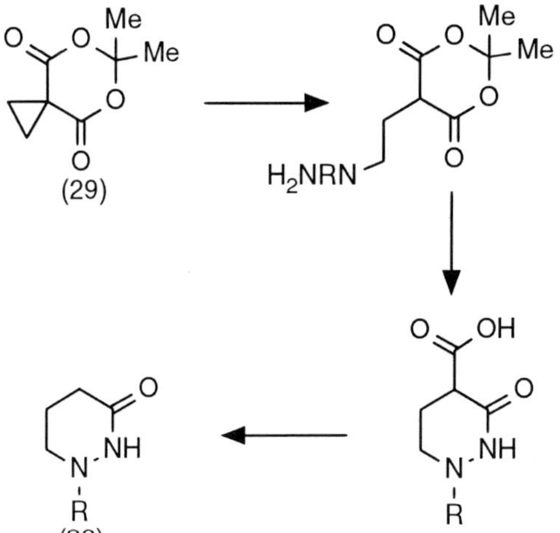

4-Substituted tetrahydro-3,6-pyridazinedione-3-hydrazones (31, R = H, Me, Et, Ph R^1 = H, Me) or their 3-hydrazino tautomers were prepared by the reaction of the corresponding 3-cyanopropionic acid derivative $RR^1C(CN)CH_2COR^2$ (R^2 = OEt, NH_2, same R, R^1), with hydrazine hydrate (J. Lange *et al.,* Synth. Commun., 1993, 23, 1371).

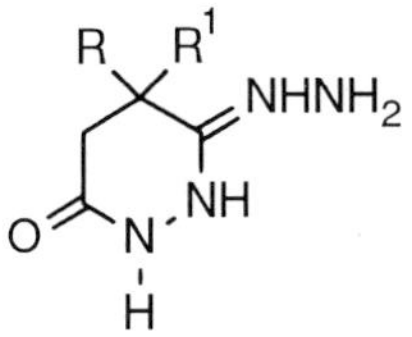

3-Chloropyridazine-6-carboxylic acid hydrazide was synthesised by employing hydrazine hydrate and methyl levulinate as starting materials (M. Morihsita *et al.,* Chem. pharm. Bull., 1994, 42, 371).

Condensation of β-(4-ethoxybenzoyl)-α-(4,5-dihydro-5-oxo-1,3-diphenylpyrazol-4-yl)propionic acid with hydrazine hydrate affords 4,5-dihydro-3(2*H*)-pyridazinone (32). Reaction of (32) with $POCl_3$ and P_2S_5 gives a dichloro derivative (33, X = Cl) and a dithione (33, X = SH), respectively. The behaviour of (33, X = Cl, SH) *in vitro* antibacterial screening reveals moderate activities against certain bacteria (E. Abd El-Ghani, M.G. Assy and H.Y. Moustafa, Monatsh. Chem., 1995, 126, 1265).

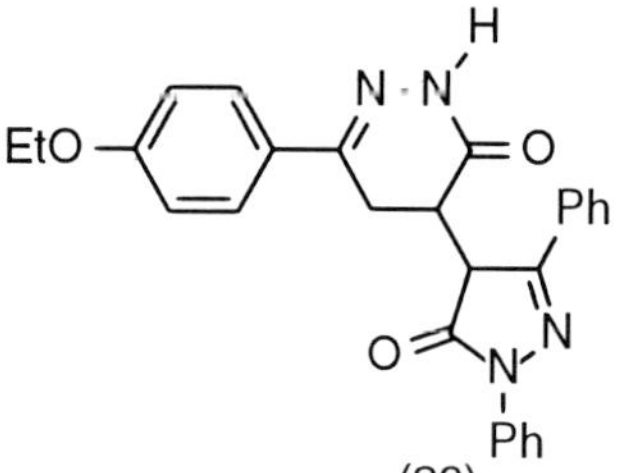

(32) 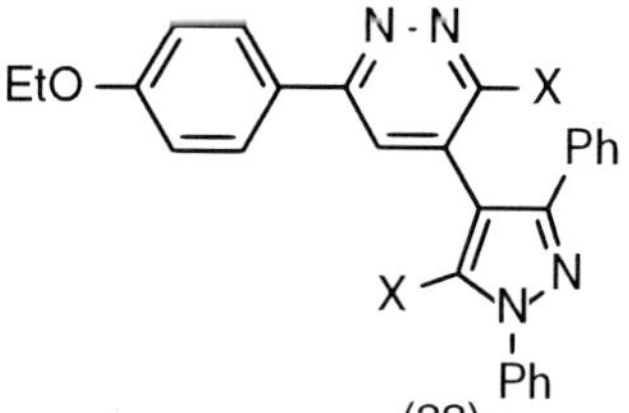 (33)

Benzopyrans (34, X = OTs, Cl) obtained from flavanone, were treated with KCN/Me_2SO_4 to give spiro compounds (R = CN) as three diastereoisomers, which were then hydrolysed to diastereoisomers of (35, R = COOH). Two of these acids (35, R = COOH) reacted with hydrazine to give stereoisomeric pyridazinones (36, R^1 = H). One stereoisomer of (36, R^1 = H) reacted with ethyl bromoacetate to give the corresponding isomer of (36, R^1 = CH_2CO_2Et) which was hydrolysed to the carboxylic acid (D. Litkey *et al.,* Khim. geterotsikl. Soedin., 1995, 616; Chem. Abstr., 1995, 124, 146036s).

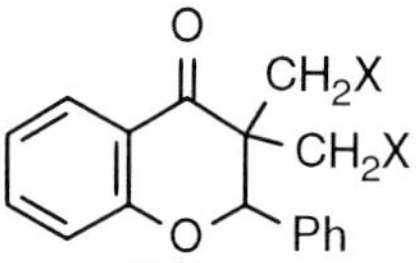

(34) 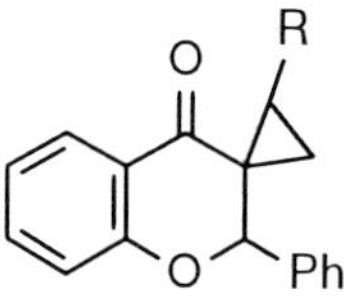 (35) 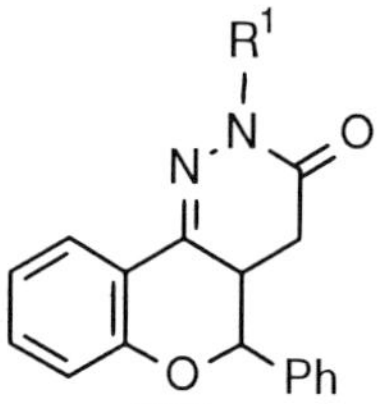 (36)

Dehydration of (37) with HCl/AcOH afforded the corresponding (arylethyl)furanones (38). Compounds of type (37) and (38) were suitable precursors for the formation of pyridazinones (39, R = H, Me, MeO) as they smoothly react with hydrazine (A. Amer *et al.,* J. prakt. Chem., 1997, 339, 20).

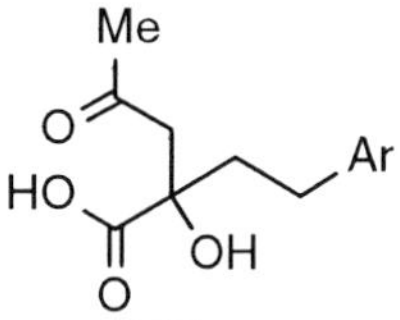

(37) 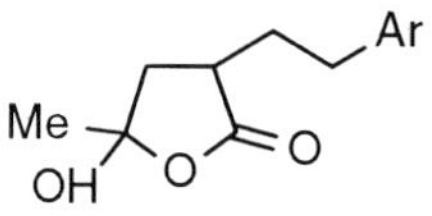 (38) 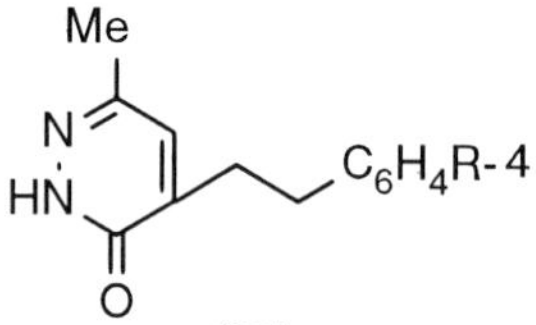 (39)

Hydrazine hydrate is utilised in an efficient method for the preparation of 6-substituted 3(2*H*)-pyridazinones and 6-substituted 4,5-dihydro-4-hydroxy-3(2*H*)-pyridazinones starting from 3,5-disubstituted 4,5-dihydroisoxazoles. N-O bond cleavage of the isoxazoline ring is promoted by molybdenum hexacarbonyl or by catalytic hydrogenation and affords the α-hydroxy γ-ketoesters $RCOCH_2CH(OH)CO_2Et$ (R = Me, Bu, 2-, 4-pyridyl, 4-HOC_6H_4) which are converted into 6-substituted 4,5-dihydro-4-hydroxy-3(2*H*)-pyridazinones on treatment with hydrazine hydrate at room temperature or reflux, in high yield (P.G. Baraldi *et al.*, Synthesis-Stuttgart, 1994, 1158).

Maleoyl chloride on cyclocondensation with $RCONHNH_2$ (R = Ph, substituted Ph, CH=CHPh), yields pyridazinones (40, R^1 = OH). Chlorination of (40, R^1 = OH) with $POCl_3$ followed by the action of hydrazine hydrate affords (40, R^1 = Cl, $NHNH_2$). Reaction of (40, R^1 = OH) with $ClCH_2CO_2H$ in aqueous NaOH furnishes (40, R^1 = OCH_2CO_2H). Compounds (40) have bactericidal and fungicidal activity (D.M. Purohit and V.H. Shah, Heterocyclic Commun., 1997, 3, 267).

3,5-Diaryl-2,5-diketopentanoates cyclise with hydrazine to give 4,6-diaryl-3-ethoxycarbonyl-1,4-dihydropyridazines which are oxidised to the corresponding pyridazines by manganese dioxide (A. Kino *et al.*, Chem. pharm. Bull., 1993, 41, 156).

Reaction of $RO_2CCH{=}CMeN{=}NCONH_2$ (R = Me, Et, Me_3C, $PhCH_2$) with β-tricarbonyl compounds $R^1COCH(COMe)_2$ (R^1 = Me, MeO) under mild conditions gave the corresponding 1,4-dihydropyridazines (41) in good to excellent yields. Treatment of (41) at room temperature with $Me_3PhN^+Br^-$ directly afforded the aromatic pyridazines in high yield. (O.A. Attanasi *et al.*, Synlett, 1997, 1361).

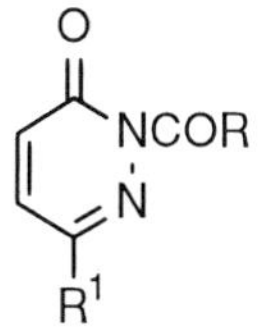

(40) 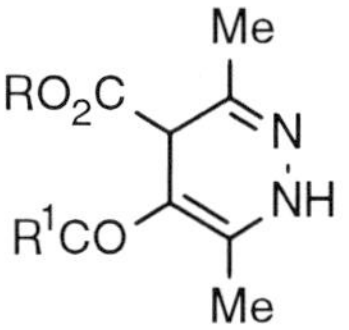 (41)

Ethyl (*Z*)-5-aryl-2-diazo-5-hydroxy-3-oxopent-4-enoates interact with triphenyl phosphine to give 6-aryl-3-ethoxycarbonyl-4-hydroxypyridazines (Ar = Ph, 4-MeC_6H_4, 4-ClC_6H_4) (Z.G. Aliev *et al.*, Russ. chem. Bull., 1997, 46, 2142).

Condensation of $ArNHN{=}C(CN)CO_2Et$ (Ar = Ph, 4-MeC_6H_4, 4-$MeOC_6H_4$) with $CH_2(CN)_2$ and $NCCH_2CO_2Et$ gave pyridazines (42) and (43) respectively (A.H.H. Elghandour *et al.*, Org. Prep. Proc. Int.,

1993, 25, 293).

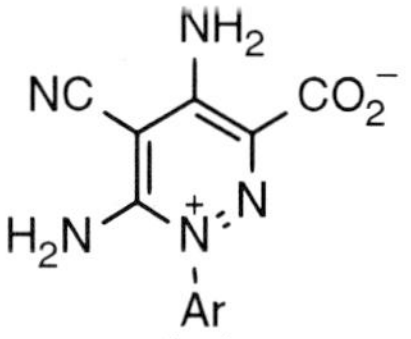

NC, NH_2, CO_2^-, H_2N, N, Ar (42)

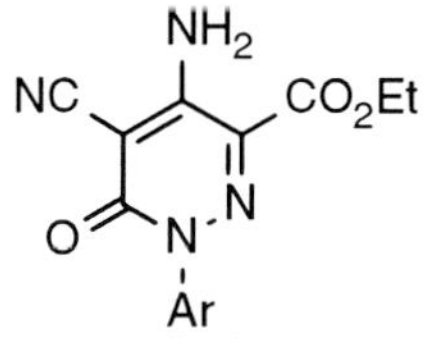

NC, NH_2, CO_2Et, O, N, Ar (43)

Appropriately protected di- and tripeptides constituting (*R,S*)-tetrahydropyridazine-3-carboxylic acids (44, R = Z-Val, R^1 = OEt, R = Z, R^1 = Leu-OMe, R = BOC-Ala-(*S,S*)-NHCH($CHMeNHCO_2CH_2CH=CH_2$)CO, R^1 = OMe, Z = $PhCH_2O_2C$, BOC = Me_3CCO_2C) were synthesised via protected γ-formyl-α-hydrazinobutyric acids (45, R^1 = Me, Et). The latter (45) were obtained from acetals of either Et γ-formylbutyrate or Et γ-formyl-α-oxobutyrate (U. Schmidt and B. Riedl, Synthesis-Stuttgart, 1993, 809).

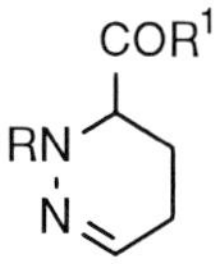

COR^1, RN, N (44)

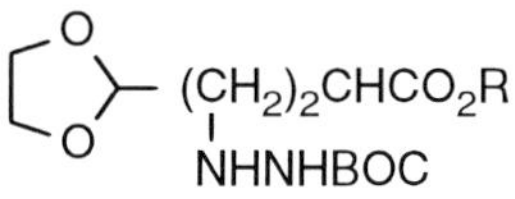

$(CH_2)_2CHCO_2R$, NHNHBOC (45)

(iii) Cycloadditions reactions

Aryl diazonium salts couple with the enones (*E*)-$RCOCH=CHNMe_2$ (R = Ph, 2-thienyl, 2-furyl) to yield, the corresponding $RCOC(CHO)=NNHR^1$ (R^1 = Ph, 4-$NO_2C_6H_4$), suitable precursors for the synthesis of pyridazinones. The coupling reaction of (*E*)-$(NC)_2C=CRCH=CHNMe_2$ (R = Ph, 2-thienyl) with benzene diazonium chloride in EtOH/NaOH gives the corresponding pyridazinaldehydes (46) (F. Al Omran *et al.*, Synthesis-Stuttgart, 1997, 91).

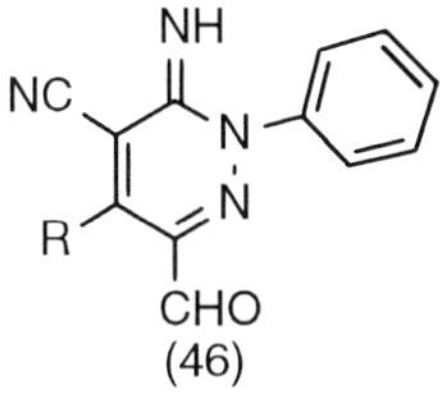

NH, NC, N, N, R, CHO (46)

The reaction of dichlorohydrazones (47) with Hunig's base (*N,N*-diisopropylethylamine) gives 4-chloroazodienes, which combine with a variety of electron-rich olefins (X = OR, NR_2) to yield chloro-substituted tetrahydropyridazines. These chloroazodiene cyclisations are best characterised as inverse electron demand [4+2]hetero Diels-Alder reactions that maintain a high degree of regio- and stereochemical control. The chloro tetrahydropyridazines that are formed give high yields of substituted pyridazines (48) upon treatment with base (M.S. South *et al.*, Tetrahedron Letters, 1995, 36, 5703; J. org. Chem., 1996, 61, 8921).

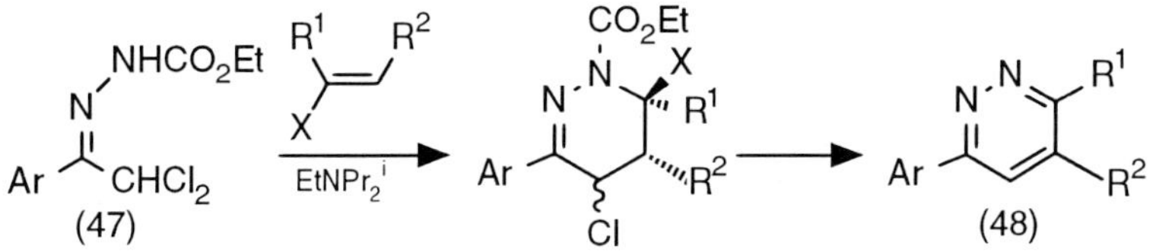

(47) (48)

A route for the synthesis of novel C-nucleosides (49, R = CF_3, R^1 = H; R = CO_2Me, R^1 = Pr, Bu, pentyl) involves [4+2]cycloaddition of the alkyne substituted precursors (50) with the 1,2,4,5-tetrazines (51), leading to the *O*-protected 4-(β-D-ribofuranosyl)pyridazines. Subsequent deprotection yields (49) in medium to good yields (M. Richter and G. Seitz, Arch. Pharm., 1994, 327, 365).

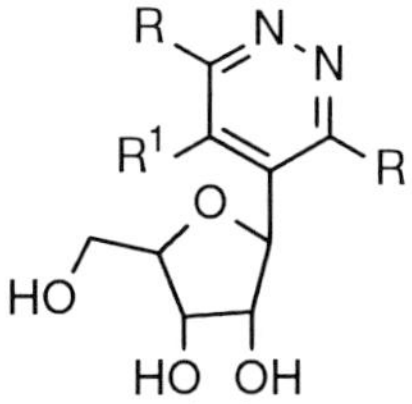

(49) 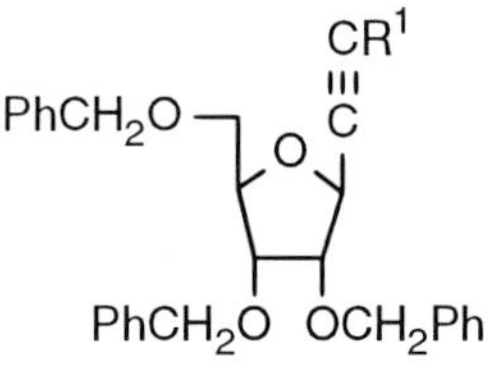 (50) 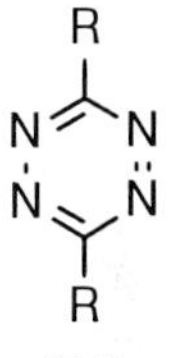 (51)

1,2-Diformylhexahydropyridazine has been prepared in 91% yield by refluxing $(NHCHO)_2$ with 1,4-dibromobutane in the presence of potassium carbonate in acetonitrile. Deformylation with HCl gas in methanol gave 88% of hexahydropyridazine (C. Ota, Jpn. Kokai Tokkyo Koho, JP 09124610; Chem. Abstr., 1997, 127, 34241k).

Alkynes $RC{\equiv}CR^1$ (R = Me_3C, Me_3Si, Me_3Ge, Me_3Sn, Bu_3Sn; R^1 = H, same as R) show high reactivity in inverse-type Diels-Alder reactions with the electron deficient 1,2,4,5-tetrazine to produce metallated pyridazines. Kinetic data prove the huge accelerating effect of the trialkyltin substituent, offering a simple access to new heteroaromatic organotin chemistry (D.K. Heldmann and J. Sauer, Tetrahedron Letters, 1997, 38, 5791).

A number of novel pyridazinomorphinans have been synthesised by the inverse electron demand Diels-Alder reaction of 3,6-disubstituted 1,2,4,5-tetrazines with enamines derived from dihydrocodeinone and with codeinone. Reduction of some of the pyridazinomorphinans did not furnish the expected pyrroloepoxymorphinans. In all cases investigated reductive cleavage of the epoxy bridge was observed to yield dihydropyridazino or pyrrolomorphinans (T. Klinder *et al.*, Arch. Pharm., 1997, 330, 163).

The asymmetric synthesis of (3*S*)-2,3,4,5-tetrahydropyridazine-3-carboxylic acid (54) has been achieved from the cycloadduct (53, R = 2,3,4,6-tetra-*O*-acetyl-D-glucopyranosyl) derived from di-tertiary butyl-azodicarboxylate and the diene (52, same R) by a hetero Diels-Alder reaction (I.H. Aspinall *et al.*, J. chem. Soc., chem. Commun., 1993, 1179).

(52) (53) (54)

(iv) From other heterocyclic compounds

Other heterocyclic compounds used in pyridazine syntheses and already mentioned are furans, pyrazoles, isoxazoles and tetrazines.

The nitrosation reaction of 3,5-dimethyl-2-phenyl-2*H*-1,2-thiazines with isoamyl nitrite and hydrogen chloride gave 3-cyano-thiazines. Mesoionic pyridazinium salts were formed as by-products. These salts (55, R = H, Me, MeO, etc.) became main products by the nitrosation of 3,5-dimethoxy-2-phenyl-2*H*-1,2-thiazines with sodium nitrite and hydrochloric acid (E. Fanghanel *et al.*, J. prakt. Chem., 1995, 337, 104).

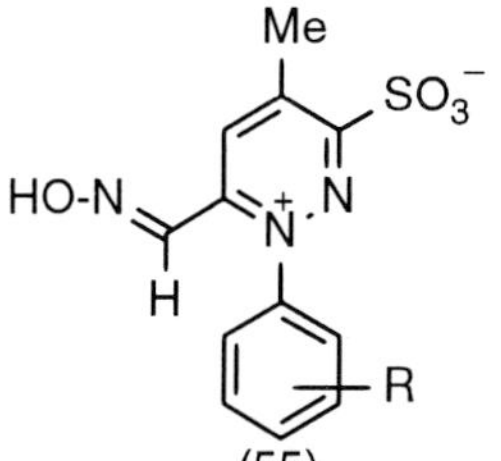

(55)

Strained polycyclic spiro[1-pyrazoline-3,1′-cyclopropanes] react with AcCl or BzCl at 0-15°C regioselectively to give 1-acyl-3-(2-chloroethyl)-2-pyrazolines in high yield. Thus reaction of spiro compound (56) with AcCl gives (chloroethyl)pyrazoline (57). Treatment of the (chloroethyl)pyrazolines with HCl gives 1,4,5,6-tetrahydropyridazines in high yields, e.g. (57) gives (58) (Y.V. Tomilov *et al.,* Russ. chem. Bull., 1995, 44, 2114).

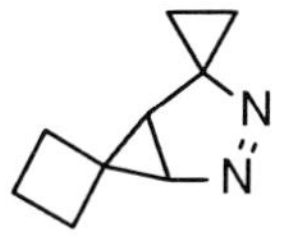

(56) 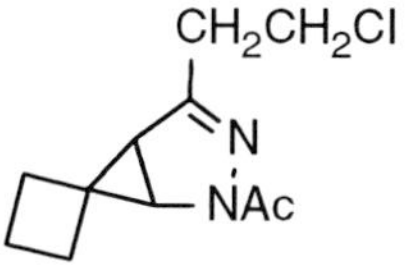 (57) 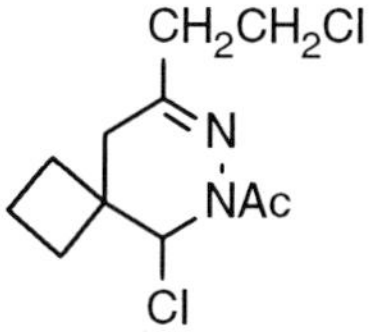 (58)

Rearrangement of 3-acetyl-4-aryl-2-pyrazolines at 0°C gave 3-methyl-5-aryl-4(1*H*)-pyridazinones in moderate yields (G.V. Subbaraju *et al.,* Indian J. Chem., 1995, 34B, 342).

A novel route to functionalised 3-aminopyridazines (59, R = Ph, CO_2Et, CN) by reaction of 3-chloro-6-phenyl-1,2,4-triazine with carbon nucleophiles RCH_2CN bearing a cyano substituent at a carbanion centre has been developed (A. Rykowski and E. Wolinska, Tetrahedron Letters, 1996, 37, 5795).

2-Bromoacetophenone azine reacts with carbanions from malononitrile and ethyl cyanoacetate under mild conditions to give, instead of the expected double substituted derivatives of acetophenone azine, cyclic 5,5-disubstituted-3,7-diphenyl-5,6-dihydro-4*H*-1,2-diazepines (60, R = CN, COOEt) which underwent ring contraction in boiling xylene to give 3,6-diphenylpyridazine (M. Potacek *et al.,* Molecules, 1996, 1, 152).

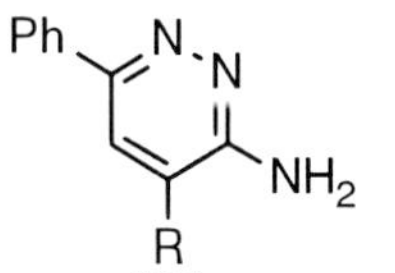

Ph, N, N, NH_2, R (59) N-N, Ph, Ph, NC, R (60)

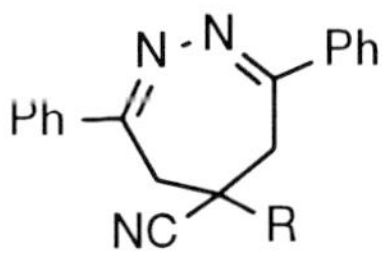

Reduction of diaryloxadiazolo[3,4-*d*]pyridazines (61) with sodium borohydride leads to symmetrically substituted 3,6-diaryl-4,5-diaminopyridazines (62, Ar = 4-alkylC_6H_4) (S. Mataka *et al.,* J. heterocyclic Chem., 1992, 29, 87).

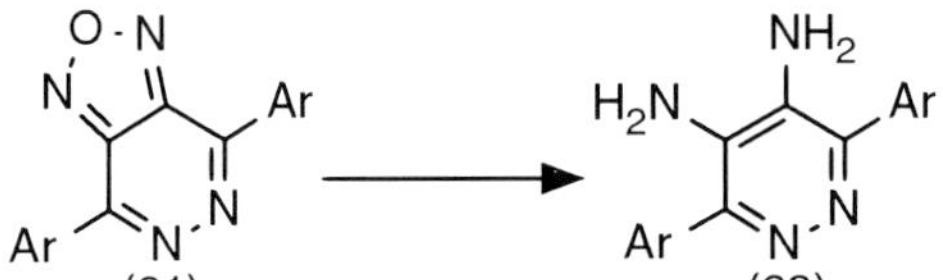

(61) → (62)

(v) Microwave techniques

Several syntheses have been carried out under microwave irradiation generally with solvent free conditions allowing short reaction times and providing high yields. For example, several polyfunctional pyridazine derivatives (63, R = H, Me, Cl, NO_2) have been prepared (E. Sotelo, Synth. Commun., 1997, 27, 2419).

Pyridazine diones, such as (64), have been synthesised in good to excellent yields under microwave irradiation by reaction of anhydrides, such as (65), with hydrazines, e.g. $PhNHNH_2$ (L. Bourel, A. Tartar and P. Melnyk, Tetrahedron Letters, 1996, 37, 4145).

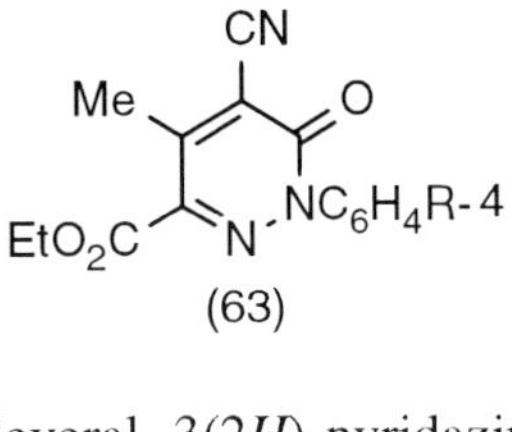

(63) 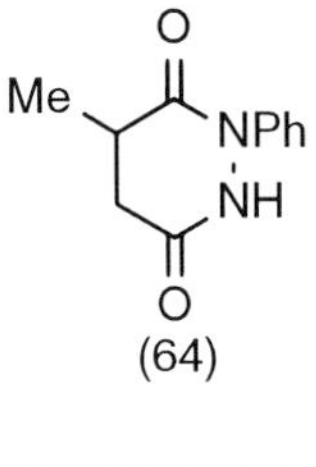 (64) 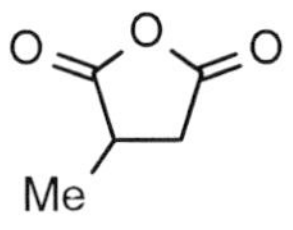 (65)

Several 3(2*H*)-pyridazinones were prepared from monophenylhydrazones of 1,2-dicarbonyl compounds and various active methylene groups within 1-15 minutes, without solvent, under microwave irradiation, in the presence of carefully adjusted amounts of piperidine or solid potassium *tert*-butoxide. The intermediate alkenes were successfully prepared in several cases (S. Jolivet, F. Texierboullet and J. Hamelin, Heteroatom. Chem., 1995, 6, 469).

Reactions induced by ultrasonic irradiation of methyl 9,12-dioxostearate with hydrazines in water in the presence of acidic alumina at 0-5°C provide high yields of pyridazine fatty esters (L.K. Jie, S.T. Marcel and P. Kalluri, J. chem. Soc., Perkin Trans. 1, 1997, 3485).

b. Reactions

3,6-Dimethoxypyridazine has been shown to react with hydrazine via 4-amination of 6-methoxy- and 6-hydrazino-3(2*H*)-pyridazinones to give 4-amino-6-methoxy- and 4-amino-6-hydrazino-3(2*H*)-pyridazinones, and not the corresponding 5-amino isomers, as previously reported. The published synthesis of the 5-amino isomers, used to confirm the earlier findings, is incorrect as regards 5-amino-6-hydrazino-3(2*H*)-pyridazinone, which has been prepared by an alternative route. The 4-amination reaction has also been extended to 6-chloro-3(2*H*)-pyridazinone (W.J. Coates and A. McKillop, Heterocycles, 1993, 35, 1313).

Reaction of *N*-(2-aminophenyl)-3,6-dichloro-4-pyridazinecarboxamide (66) with sodium hydride in DMF does not afford the expected pyridazino[3,4-*d*][1,5]benzodiazepine derivative (67) but instead results in a ring transformation to give 3-(5-chloro-3-pyrazolyl)-2(1*H*)-quinazolinone (68) via reaction at C-4 rather than C-3 (G. Heinisch, B. Matuszczak and K. Mereiter, Heterocycles, 1994, 38, 2081).

(66) (67) (68)

Reagents: (i) NaH, DMF, 100°C, 87%

Highly functionalized azidopyridazines have been reduced to the corresponding aminopyridazines in excellent yields by stirring with an excess of iron powder and ammonium chloride in ethyl acetate-water two-phase solution at room temperature (S.D. Cho *et al.,* Tetrahedron Letters, 1996, 37, 7059).

Treatment of 1,2,3,6-tetrahydropyridazine or 2,2-dimethylhexahydro-1,3-dioxo[4,5-*d*]pyridazine (69, R = H; R,R = $OCMe_2O$) with 2-fluoro-5-nitrophenyl isocyanate gave pyridazino[1,2-*a*][1,2,4]benzotriazin-6-ones (70, R = H; R,R = $OCMe_2O$). These latter (70) were converted into *N*-Me and nitro group reduction products (P.S. Ray *et al.*, J. heterocyclic Chem., 1993, 30, 45).

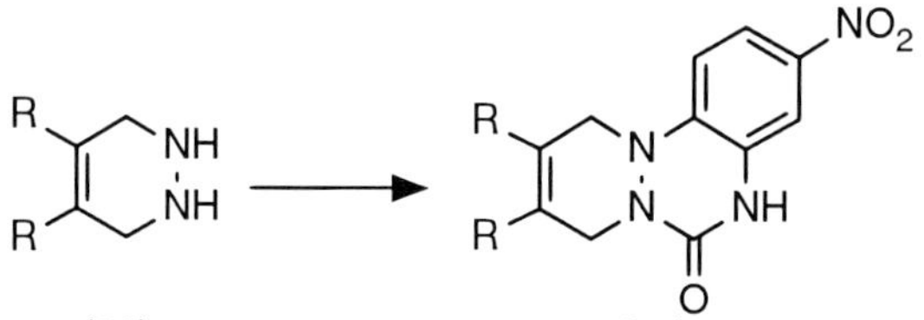

(69) (70)

Ethyl 3-hydroxy-5,6-diphenylpyridazine-4-carboxylate undergoes hydrazinolysis by hydrazine hydrate to give the corresponding carbohydrazide.The reaction of the latter compound with HNO_2, CS_2/NaOH and PhNCS followed by treatment with NaOH gives azide (71), oxadiazole (72, X = O) and triazole (72, X = NPh) (A.M.K. El Dean, S.M. Radwan and M.I.A. Moneam, Afinidad, 1996, 53, 387).

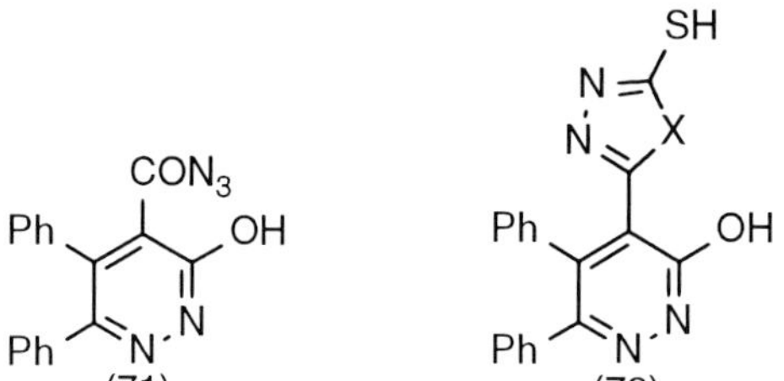

(71) (72)

Triazolo- and tetrazolopyridazines (73, X = CH, CMe, N) obtained in the reactions of 3-chloro-6-hydrazinopyridazine (74, R^1, R^2 = H, H) with formic acid, acetic acid and nitrous acid, were treated with hydrazine or morpholine to yield the corresponding derivatives substituted at C-6. Attempts to add a second azole ring to (73) proved unsuccessful (A.A. Katrusiak, S. Baloniak and A.S. Katrusiak, Pol. J. Chem., 1996, 70, 1279).

Hydrazino substituents in pyridazines and phthalazines (74, R^1, R^2 = H, H; H, Me; $(CH{=}CH_2)_2$) undergo oxidative cleavage under mild conditions with thallium trinitrate in methanol to give the corresponding methoxy derivatives in moderate yields (65-68%) (M. Kocevar, P. Mihorko and S. Polanc, J. org. Chem., 1995, 60, 1466).

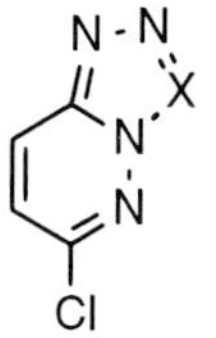

(73) 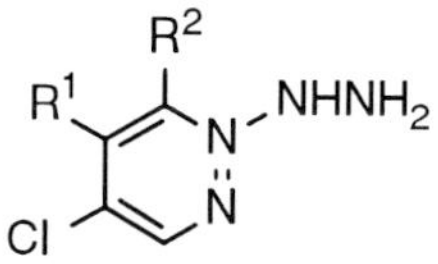 (74)

The reactions of 6-chloro-4-pyridazinones and 5-hydrazino-2-phenyl-3(2*H*)-pyridazinones with the Vilsmeier reagent (DMF/$POCl_3$) afforded pyrazolo[3,4-*d*]pyridazinones. Concurrently the formation of acyclic 6-chloro-4- and 5-(*N,N*-dimethylaminomethylidenehydrazino)-2-phenyl-3(2*H*)-pyridazinones was observed (A.A. Katrusiak, A.S. Katrusiak and S. Baloniak, Tetrahedron, 1994, 50, 12933).

Oxidative decyanation of α-(6-chloropyridazin-3-yl)phenyl-acetonitrile (77) can be achieved by passing a fresh stream of oxygen into an aqueous NaOH/DMSO solution of the substrate containing triethyl-benzylammonium chloride (TEBA), yielding the ketone (78) with concomitant hydrolysis or methoxylation of the chloro substituent. The direct one-pot preparation of 3,6-dibenzoylpyridazine (75) from 3,6-dichloro-pyridazine (76) is achieved in 48% yield (G. Heinisch and T. Langer, J. heterocyclic Chem., 1993, 30, 1685).

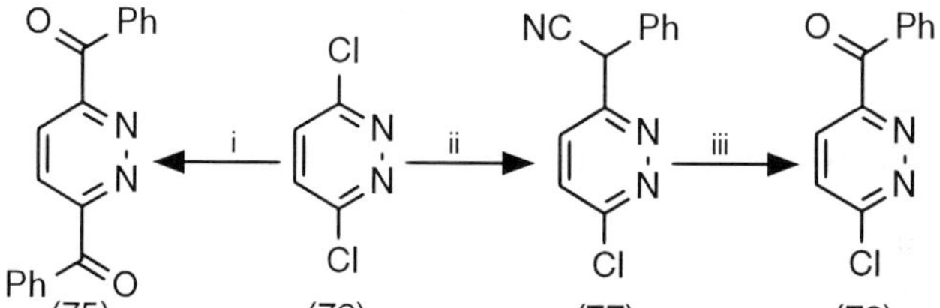

(75) (76) (77) (78)

Reagents: (i) $NCCH_2Ph$, 50% aq. NaOH, TEBA, O_2, (ii) $NCCH_2Ph$, aq. NaOH, DMSO, TEBA; (iii) aq. NaOH, DMSO, TEBA, O_2

5-Hydrazinopyridazin-3(2*H*)-ones (79) react with diesters of α-ketodicarboxylic acids in acetic acid to give dihydropyridazino[4,5-*c*]-pyridazin-5(1*H*)-ones (80). Pyrrolo[2,3-*d*]pyridazin-4(5*H*)-ones (81) are also formed as by products (2-8% yield) when R = H in (79) (T. Yamasaki *et al.*, J. heterocyclic Chem., 1992, 29, 1313).

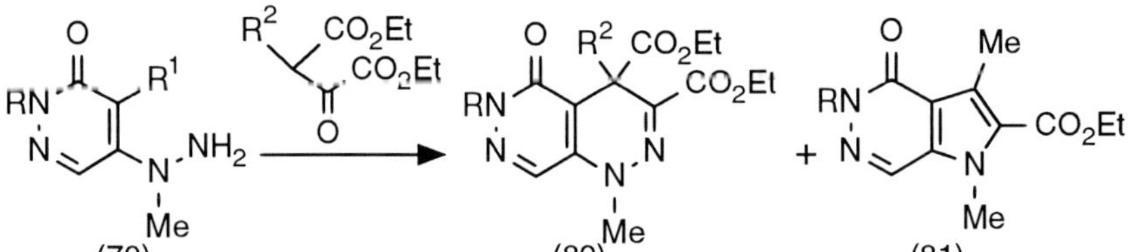

(79) (80) (81)

Alkylations of 4,5-dichloropyridazin-6-one (82) and of 4,5-dichloro-1-hydroxymethylpyridazin-6-one (83) with α,ω-dibromoalkanes $Br_2(CH_2)_nBr$ (n = 1-4, 6) or 4,5-dichloro-1-(ω-bromoalkyl)pyridazin-6-ones (84, same n) in the presence of potassium carbonate and acetonitrile or Bu_4NBr/KOH gave mainly *N*-alkylation products such as (84) or (85) depending on the conditions (S.K. Kim *et al.,* J. heterocyclic Chem., 1997, 34, 209; 1997, 34, 1135).

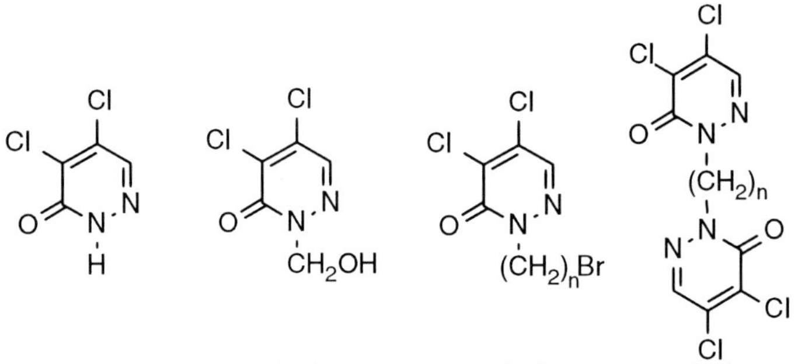

(82) (83) (84) (85)

The hydroxypyridazinone (86) on the other hand with a variety of alkyl halides in acetone/potassium carbonate gave high yields of the *O*-alkyl derivatives (87) as the only products. Steric factors and the electron withdrawing property of the trifluoromethyl group appear to hinder *N*-alkylation (S.G. Hegde and C. R. Jones, J. heterocyclic Chem., 1993, 30, 1501).

(86) (87)

Reagents: (i) alkyl halide, K_2CO_3, acetone, 72-83%

3-Chloropyridazines are relatively unreactive in palladium coupling reactions and the bromo- and iodo-analogues are not readily available so the triflate esters of pyridazine (88) have been developed. These are easily prepared in high yield by the reaction of triflic anhydride either with 3(2*H*)-pyridazinone in pyridine at room temperature or with the lithium salts in THF at -78°C. The triflates undergo Pd(0) cross-coupling reactions with terminal alkynes to give high yields of 3-alkynylpyridazines (89) (D. Toussaint, J. Suffert and C.G. Wermuth, Heterocycles, 1994, 38, 1273).

(88) (89)

Reagents: (i) Tf_2O, pyridine, 20°C or (a) BuLi, -78°C, THF, (b) Tf_2O, -78°C, THF, (ii) $HC{\equiv}CR^3$ (2 equiv.), $PdCl_2(PPh_3)_2$, 4% mol, CuI, 15% mol, 3:1 THF:$Pr^i{}_2NH$, 0.5-6 h, 0°C, 71-96%

Another approach for the synthesis of unsymmetrical 3,6-disubstituted pyridazines is to use the Pd(0)-catalysed coupling reactions of 3-iodopyridazines. Thus starting from inexpensive, commercially available 3,6-dichloropyridazine, several 6-substituted-3-iodopyridazines were prepared and these were subjected to a variety of palladium cross-coupling reactions to give unsymmetrical 3,6-disubstituted pyridazines in 50-84% overall yield (T.L. Draper and T.R. Bailey, J. org. Chem., 1995, 60, 748.

Treatment of 6-(3-aryl-2-hydroxypropyl)-4,5-dihydro-3(2*H*)-pyridazinones (90) (R,R^1 = alkoxy, R^2 = H, Me) with trimethylsilyl triflate in refluxing dichloromethane gave the unexpected 3-(1-naphthyl)-propionic acid ethyl ester derivatives (91, same R, R^1, R^2). The isolation of a spirotetrahydropyridazinone intermediate suggests an intramolecular aminoalkylation key step, followed by ring opening and elimination of hydrazine and water promoted by the triflate (P.G. Baraldi *et al.*, Tetrahedron Letters, 1993, 34, 3777).

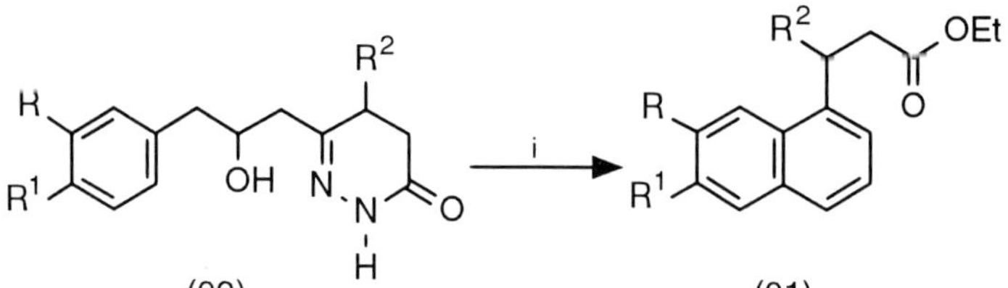

(90) (91)

Reagents: (i) trimethylsilyl triflate, $ClCH_2CH_2Cl$, EtOH

The methoxycarbonylation of several pyridazin-3-yl triflates (92) with CO and MeOH is catalysed by $Pd(OAc)_2$ and 1,1′-bis(diphenylphosphino)ferrocene (M. Rohr *et al.*, Heterocycles, 1996, 43, 1459).

The homolytic substitution of protonated 3-chloro-6-methylpyridazine followed by hydrolysis with acetic acid affords 4- or 4,5-functionalised pyridazinones (93). Less reactivity is found for 3-chloro-6-phenylpyridazine (V. Dal Piaz, M.P. Giovannoni and G. Ciciani, Tetrahedron Letters, 1993, 34, 3903).

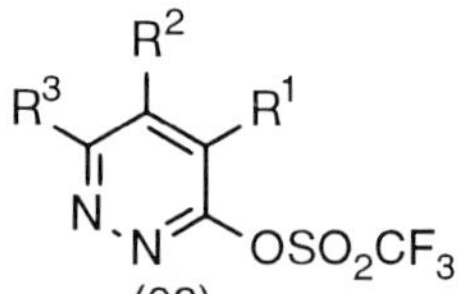

(92) 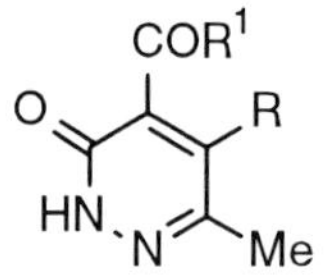 (93)

Reaction of 4,5-dichloro-3-nitropyridazin-6-one with dimethylchloromethyleneammonium chloride gave 3,4,5-trichloropyridazin-6-one, 3,4,5,6-tetrachloropyridazine and 4,5-dichloro-3-(*N,N*-dimethylamino)-pyridazin-6-one (D.H. Kweon *et al.*, J. heterocyclic Chem., 1996, 33, 1915).

4-Amino-5-chloro-2-phenylpyridazin-3(2*H*)one on reaction with AcCl or Ac_2O gives the 4-acetylamino derivative or the diacetylamino derivative, respectively. These acetyl derivatives react with dialkylamines and cyclic amines to yield the appropriate 4-acetylamino-5-(disubstituted amino)-2-phenylpyridazin-3(2*H*)-ones, which on subsequent alkaline hydrolysis give 4-amino-5-(disubstituted amino)-2-phenylpyridazin-3-(2*H*)-ones (94) (V. Konecny *et al.*, Coll. Czech. chem. Commun., 1997, 62, 800).

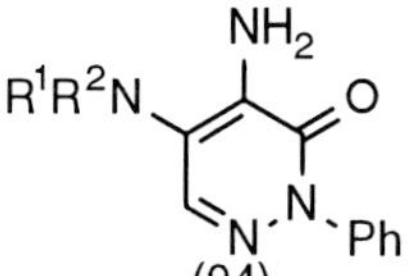

(94)

3,6-Dichloro-4-pyridazinyl phenyl ketone (95, $R^1 = R^2 = Cl$) can be converted to the corresponding 3-amino-6-chloro (95, $R^1 = NH_2$, $R^2 = Cl$) and 3-amino compounds (95, $R^1 = NH_2$, $R^2 = H$). These derivatives are valuable precursors for the construction of pyrido[2,3-*c*]pyridazines (96, R = OEt, NR^3R^4) (N. Haider, G. Heinisch and J. Moshuber, Pharmazie, 1992, 47, 679).

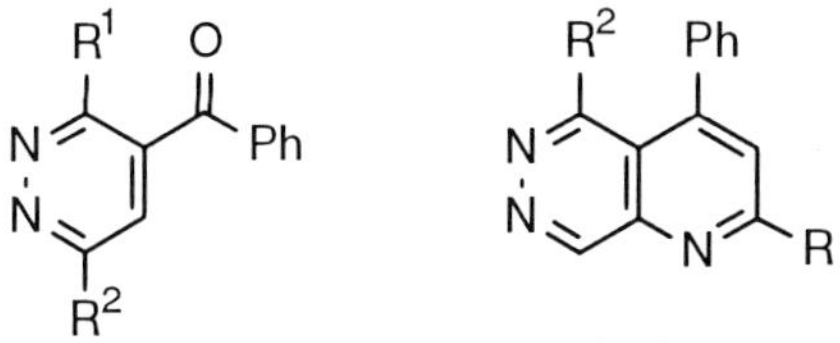

(95) (96)

Metalation of 3-acylaminopyridazine (97, R = H) using lithium 2,2,6,6-tetramethylpiperidide (LTMP) followed by reaction with an aldehyde R^1CHO, introduced a carbon side chain exclusively *ortho* to the substituted amino group to give the 4-functionalised pyridazine derivatives (98). Similar reaction of 3-acylamino-6-chloropyridazine (97, R = Cl) using lithium diisopropylamide (LDA) followed by R^1CHO permitted exclusive functionalization *ortho* to the chlorine atom to yield the 5-substituted pyridazine derivatives (99) (A. Turck *et al.*, Tetrahedron, 1993, 49, 599).

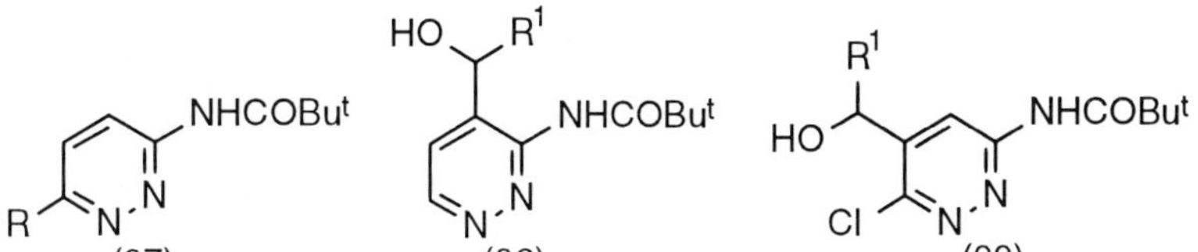

(97) (98) (99)

With the *ortho*-directing group at C-4 (100, R = $NHCOBu^t$, $CONHBu^t$), metalation with LTMP, followed by reaction of the *ortho*-lithiated species thus obtained with aldehydes R^1CHO as electrophiles produced the 4,5-disubstituted pyridazines (101) (*idem,* J. heterocyclic chem., 1995, 32, 841).

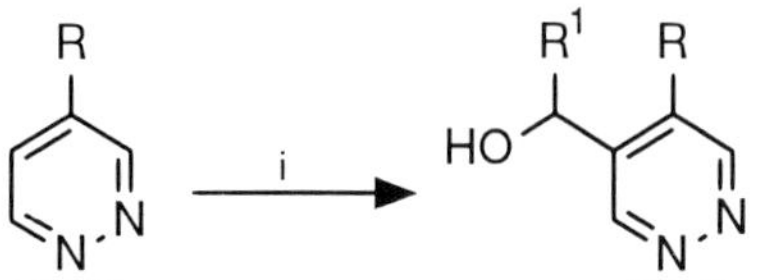

(100) (101)

Reagents: (i) LTMP, then RCHO

Side-chain metalation of pyridazines (102, X = Cl, OMe, NHCOBu^t) with LDA or LTMP and subsequent alkylation with various electrophiles has been investigated (J.M. Sitamze, A. Mannand and C.G. Wermuth, Heterocycles, 1994, 39, 271).

Regioselective metalation has also been used in the synthesis of 5,6-diarylpyridazin-3(2*H*)-ones (F. Trecourt *et al.,* J. heterocyclic Chem., 1995, 32, 1057).

Very hindered bases give improved regioselectivity for the metalation of 3-chloro-6-methoxypyridazine (L. Mojovic *et al.*, Tetrahedron, 1996, 52, 10417).

3-Chloro-6-phenylpyridazine can be converted into the corresponding 3-fluoro derivative in 89% yield upon treatment with an excess of neat tetrabutylphosphonium hydrogen difluoride at 100°C (Y. Uchibori, M. Umeno and H. Yoshioka, Heterocycles, 1992, 34, 1507).

2-Fluorophenyl 3-pyridazinyl ketone (103) is a valuable precursor for the synthesis of 3-pyridazinyl-substituted 1,4-benzodiazepines, quinolines, 2,1-benzisoxazolines and two novel diazaacridones (104 and 105) (N. Haider, G. Heinisch and J. Moshuber, Heterocycles, 1994, 28, 125).

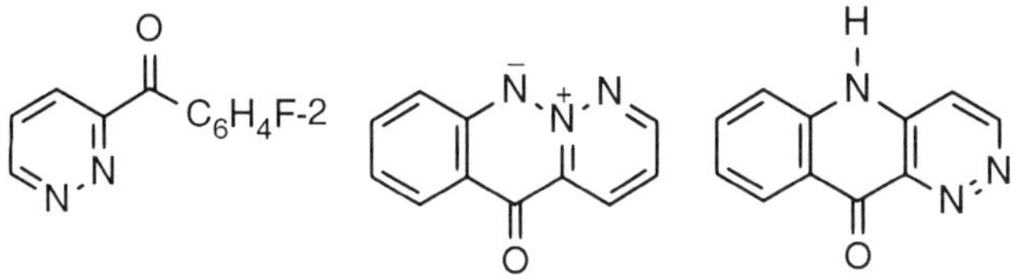

(103) (104) (105)

Oxidation of 3-chloro-6-methylpyridazine (106) with acid dichromate gives 35% of 3-chloro-6-pyridazine carboxylic acid together with 31% of the corresponding pyridazinone formed by hydrolysis of the chloro group, and this becomes the sole product if the reaction is allowed to proceed for a longer period.

The acid chloride of (107) is unstable, but direct reaction of the unpurified material in Friedel-Crafts acylation of aromatics gives moderate to excellent yields of aroylpyridazines (108) (A.E. Mourad, D.S. Wise and L.B. Townsend, J. heterocyclic Chem., 1992, 29, 1583).

Cl Cl Cl Cl
N N N N
i ii iii
N N N N
Me CO_2H COCl COAr
(106) (107) (108)

Reagents: (i) K_2CrO_7, H_2SO_4, (ii) $SOCl_2$, (iii) ArH, $AlCl_3$

The 6-[*N*-(3-ethoxycarbonylpropyl)amido]tetrahydropyridazin-3(2*H*)-one (109) and the corresponding 1,6-dihydro compound (110) are formed by the amidation of the corresponding carboxylic acids with γ-aminobutyric ethyl ester hydrochloride (F.R. Xu *et al.,* Chem. J. Chin. Univ., 1995, 16, 1254).

$CONH(CH_2)_3CO_2Et$ $CONH(CH_2)_3CO_2Et$
N N
NR NR
O O
(109) (110)

Treatment of ethyl pyridazine-4-carboxylate and phenoxyacetic acid, with silver nitrate/sulfuric acid/toluene two-phase system gave ethyl 5-(phenoxymethyl)pyridazine-4-carboxylate (72%). Similarly, ethyl 5-[(phenylthio)methyl]pyridazine-4-carboxylate was prepared from ethyl pyridazine-4-carboxylate and (phenylthio)acetic acid. In the absence of an organic layer unexpected dimeric reaction products were formed (G. Heinisch *et al.,* Heterocycles, 1995, 41, 1461).

Diels-Alder reactions of 5-acetyl-2-methyl-4-nitro-6-phenyl-pyridazin-3(2*H*)-one (111) with 2,3-dimethylbuta-1,3-diene (DMB) and cyclohexa-1,3-diene (CHD) have been shown to yield the phthalazinones (112 and 113), and the pyridazinone (114) (V. Dal Piaz *et al.,* Tetrahedron Letters, 1993, 34, 161; Synthesis-Stuttgart, 1994, 669).

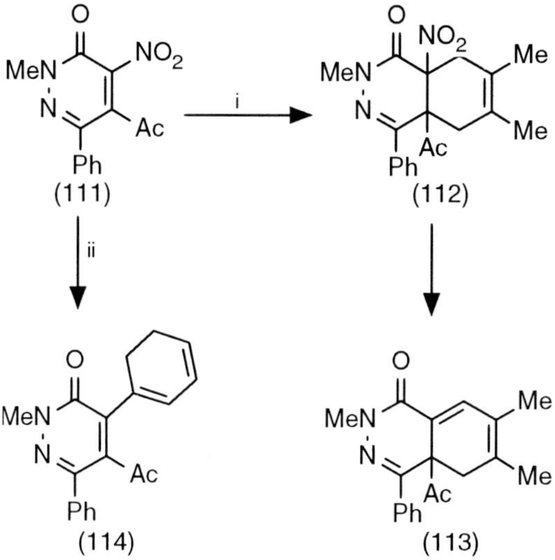

Reagents: (i) DMB, toluene, 125-130°C, 25%, (ii) CHD, toluene, 125-130°C, 75%

The Diels-Alder reaction of 5-substituted 2-methylpyridazin-3(2*H*)-ones (115, R = I, $EtSO_2$) with DMB afforded the phthalazinone (116) and its 5,8-dihydro derivative (P. Matyus, K. Fuji and K. Tanaka, Heterocycles, 1993, 36, 1975).

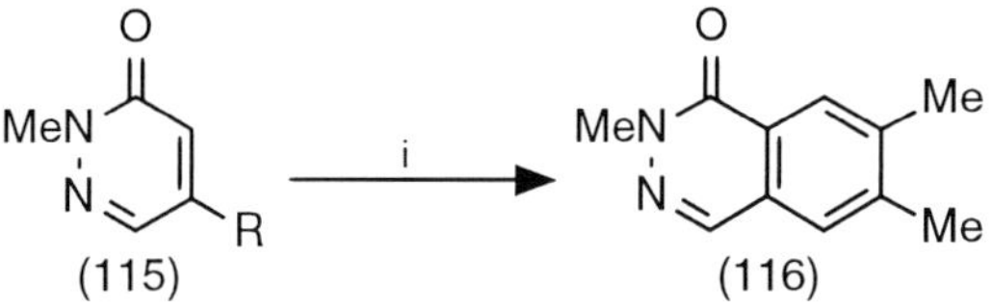

Reagent: (i) DMB

Diels-Alder reactions of 2-methylpyridazin-3(2*H*)-ones substituted with ethoxysulfonyl, cyano and methoxycarboxyl groups at the 4- and 5-positions with 1,3-dienes gave high yields of the expected adducts only after prolonged reaction times. 5-Substituted pyridazinones are more reactive than the 4-substituted isomers (F. Farina, M.V. Martin and M. Romanach, Tetrahedron, 1994, 50, 5169).

Due to the exceptional facility of its [4+2]- and [2+4]-cycloaddition reactions with unactivated dienophiles, 4,5-dicyano-pyridazine has been labelled a "superheterodiene". Thus reaction under relatively mild conditions with DMB yielded (117) and (118) (R. Nesi *et al.*, Tetrahedron, 1994, 50, 9189; J. chem. Soc., chem. Commun., 1995, 2201).

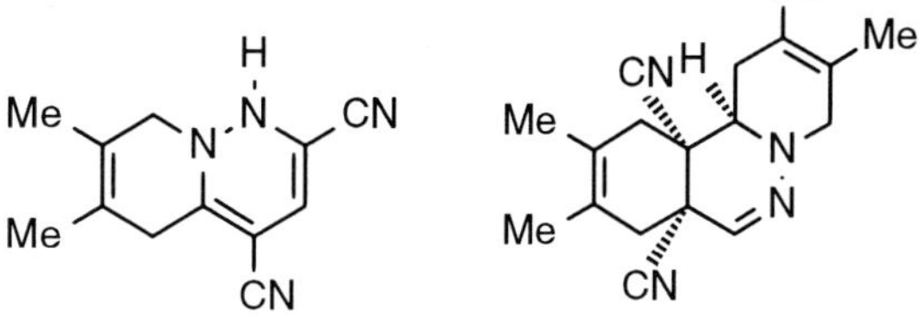

(117) (118)

3-Substituted-4-pyridazinecarbonitriles (119, X = O, NH) undergo thermally induced intramolecular Diels-Alder reactions with inverse electron demand affording fused benzonitriles (120). Incorporation of a 1,2-phenylene unit into the side chain as in (121, X = O, NAc), resulted in a more favourable conformation of the dienophilic sub-structure and thus in a pronounced acceleration of the [4+2]cycloaddition reaction (G. Fulep and N. Haider, Molecules, 1998, 3, 10).

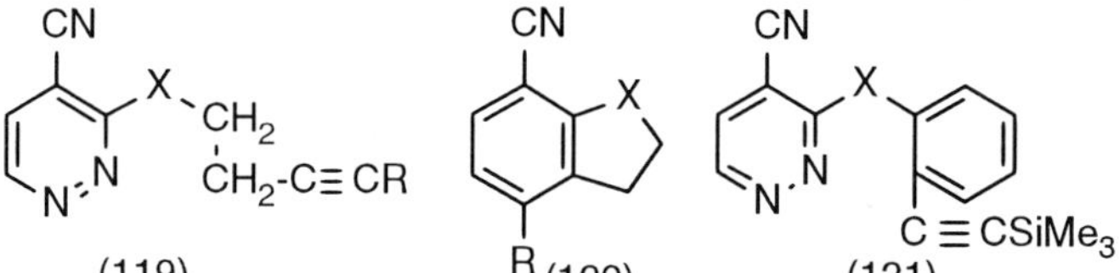

(119) (120) (121)

The reactions between 3-(4-halophenyl)pyridazinium ylides (122) and an aryl isocyanate do not give dipolar cycloaddition products, but instead yield disubstituted carbanion ylides by *C*-acylation (I. Mangalagiu, Acta chem. Scand., 1995, 49, 778).

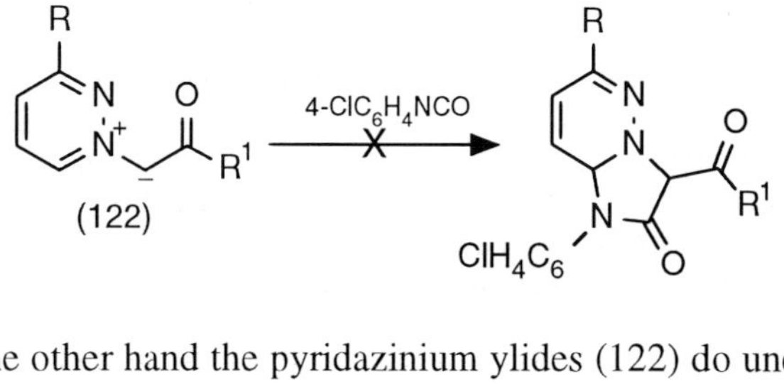

(122)

On the other hand the pyridazinium ylides (122) do undergo [3+2]-

dipolar cycloaddition with ethyl acrylate, ethyl propiolate and acrylonitrile. The reactions are regio-selective and lead to several new pyrrolopyridazines (I. Mangalagiu *et al.,* Tetrahedron, 1996, 52, 8853; Acta chem. Scand., 1997, 51, 927).

The phenyl (or *p*-tolyl)sulfonylmethyl group is introduced with complete regiospecificity to the C-4 position of 3-substituted pyridazines using vicarious nucleophilic substitution of pyridazinium dicyanomethylides (T. Itoh *et al.,* J. chem. Soc., chem. Commun., 1995, 2067).

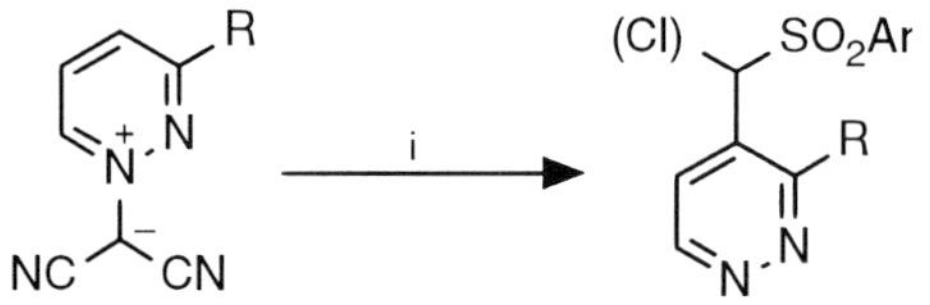

Reagent: (i) $ArSO_2CH_2Cl$, $KOBu^t$, then $(NH_4)_2S_2O_8$, MeOH

Cycloaddition of benzonitrile oxide to pyridazine affords a mono-cycloadduct (123) (A. Corsaro *et al.,* Tetrahedron, 1996, 52, 6421).

A new procedure has been described for the preparation of pyridazinones (124, R = H, Me, F, Cl, Br, R^1 = H, $(CH_2)_4$) from the corresponding 4,5-dihydropyridazinones under mild conditions with $Cu(II)Cl_2$ in acetonitrile via halogenation and spontaneous HCl elimination (F. Csende *et al.,* Synthesis-Stuttgart, 1995, 1240).

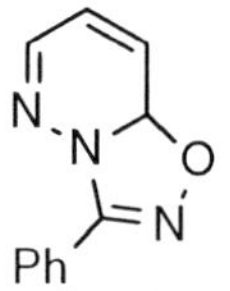

(123) 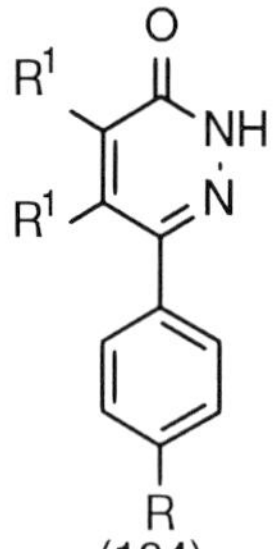 (124)

2-Methyl-1-phenyl-1,2-dihydro-3,6-pyridazinedione reacts with tertiary phosphite esters $P(OR)_3$ (R = Me, Et) in the absence of solvent to give (125), a tricyclic product, whereas 1-phenyl-1,2-dihydro-3,6-pyridazindione reacts with the phosphite esters in the absence of solvent to give alkylation products (126) (A.A. Nada *et al.,* Phosphorus, Sulfur and Silicon and the Related Elements, 1996, 119, 27).

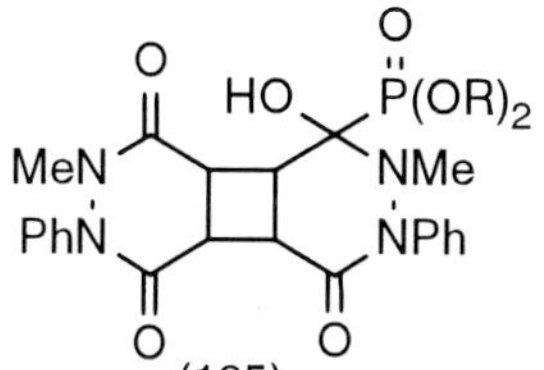

(125) 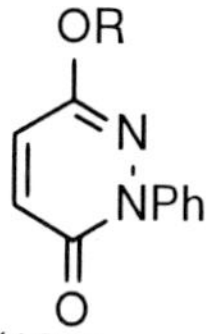 (126)

Oxidation of 3-amino-6-phenylthiopyridazine (128) with one equivalent of *m*-CPBA gives the sulfoxide (127) in 82% yield with the sulfone (129) as a by-product. With 2.5 equivalents of *m*-CPBA the sulfone (129) is formed in 44% yield (A.E. Mourad, D.S. Wise and L.B. Townsend, J. heterocyclic Chem., 1993, 30, 1365).

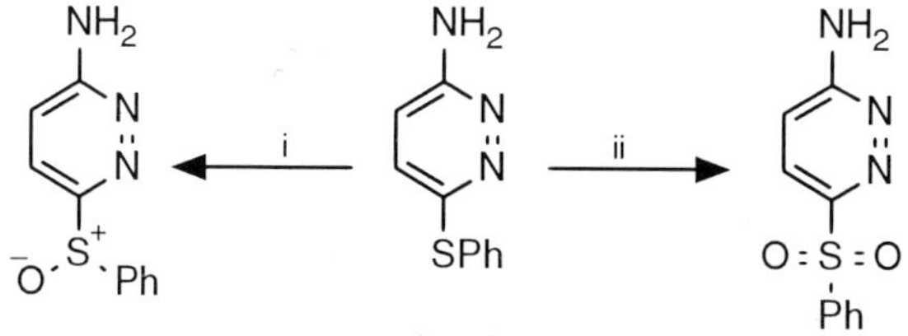

(127) (128) (129)

Reagents: (1) *m*-CPBA (1 equiv.), DCM, (ii) *m*-CPBA (2.5 equiv.), DCM

Protodesilylation of a series of 2-*tert*-butyl-4-chloro-5-(silylmethylthio)pyridazin-3(2*H*)-ones (130) proceeds smoothly with cleavage of the sp^3 hybridised C-Si bond in the presence of a catalytic amount of sodium hydroxide at room temperature. The presence of the aryl group on the Si atom and the 3(2*H*)-pyridazinon-5-yl group on the S atom play critical roles in the occurrence of this reaction (Z.J. He and Z.M. Li, Chinese chem. Letters, 1996, 7, 603; Phosphorus, Sulfur and Silicon and the Related Element, 1996, 117, 1).

1,2-Diethoxycarbonyl-1,2,3,6-tetrahydropyridazine (131) and the correspondingly substituted 1,2,3,4-tetrahydropyridazine are hydroformylated and hydrocarbethoxylated (hydrocarboxymethylated) in the presence of some known Co, Pd and Pt catalysts. The hydroformylation reaction can be tuned by a suitable choice of catalyst and reaction conditions, thus allowing the preparation with high selectivity of one of the two possible isomeric aldehydes. The carbonylation reaction is less synthetically useful, since it shows low activity and unsatisfactory chemoselectivity and regioselectivity. However, the ester 1,2,4-tricarbethoxyhexahydropyridazine can be prepared in good yield from (131) by using [$Pd(Cl)_2(PPh_3)_2$] as catalyst (G. Menchi *et al.*, J. organomet.

Chem., 1993, 450, 229).

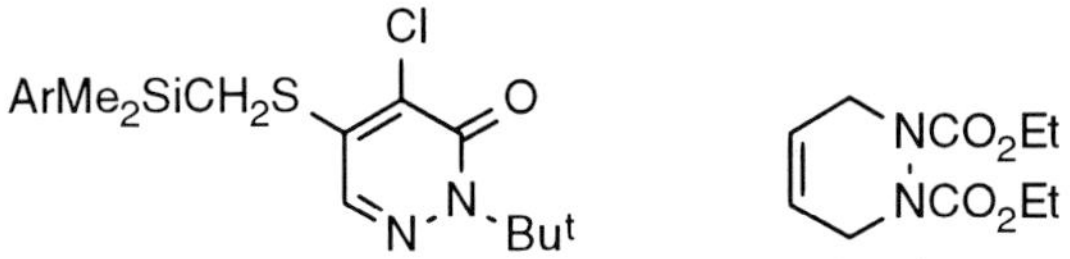

(130) (131)

Ethynylation of pyridazine can be achieved using bis(tributylstannyl)acetylene via *N*-alkoxycarbonyl quaternary salts of the substrate, followed by treatment with trifluoroacetic acid (T. Itoh *et al.,* Synlett, 1994, 557).

Reaction of pyridazines with chloroformate and allyltributyltin gives good yields of the allyl substituted dihydropyridazines (132) (T. Itoh *et al.*, Heterocycles, 1994, 37, 709).

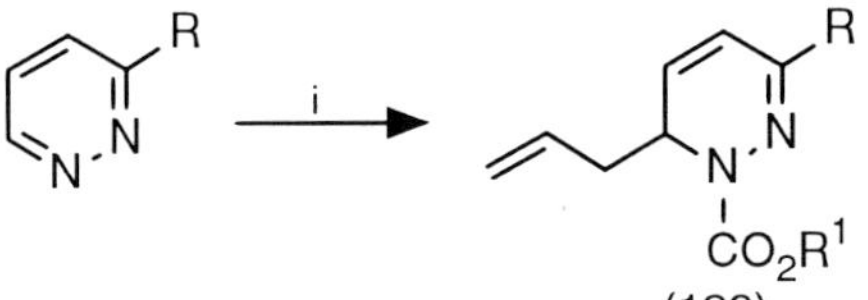

(132)

Reagents: (i) $ClCO_2R^1$, DCM, 0°C, then $Bu_3SnCH_2CH{=}CH_2$

c. Complexes with metal ions

A variety of metal ion-pyridazine complexes have been prepared and their structures, properties and reactions studied.

The reactions of copper(II) salts with pyridazine give a range of complexes (T. Otieno *et al.,* Inorg. Chem., 1995, 34, 1718) including polymeric species (L. Carlucci *et al.,* J. chem. Soc., Dalton Trans., 1994, 2397). Electron spin echo envelope modulation spectroscopy (ESEEM) was used to study the nuclear quadrupole interactions of the remote nitrogen atom of pyridazine in some complexes (F. Jiang and J. Peisach, Inorg. Chem., 1994, 33, 1348). Complexes of Cu(I) have also been studied (M. Maekawa *et al.,* J. chem. Soc., Dalton Trans., 1994, 603; M.B. Ferrari *et al.,* J. chem Crystallography, 1997, 27, 257).

Dinuclear Cu(II) and Ni(II) complexes containing Schiff base ligands and pyridazine or phthalazine bridges have been studied (C. Li *et al.,* Bull. chem. Soc. Jpn., 1997, 70, 2429) as have Cr(III) complexes with thiocyanate and pyridazinone ligands (C. Guran, I. Jitaru and I. Ciocoiu, Rev. Roum. Chim., 1994, 39, 291).

Complexes of Pt and Re containing pyridazine or methylpyridazine ligands have been studied extensively (P.A. Kozmin *et al.,* Zhur. neorganicheskoi Khim., 1993, 38, 859; E.W. Abel *et al.,* J. chem. Soc., Dalton Trans., 1994, 445; E.W. Abel *et al.,* Polyhedron, 1994, 13, 2501, 2907; E.W. Abel, J. chem. Soc., Dalton Trans., 1995, 3165; E.W. Abel, P.J. Heard and K.G. Orrell, Inorg. chim. Acta, 1997, 255, 65) as have complexes of tungsten and rhodium (E.W. Abel *et al.,* J. organomet. Chem., 1994, 464, 163).

Site-specific metal-ligand interactions lead to the self-assembly of coordination polymeric chains in the crystal structures of Ag(I) complexes with 3,6-bis(diphenylphosphino)pyridazine (S.M. Kuang *et al.,* J. chem. Soc., chem. Commun., 1998, 581).

Condensation of 3,6-diformylpyridazine with 1,3-diaminopropane in the presence of Pb(II) template ions gave complexes of lead containing either the macrocycles (133) or (134) depending on ratio of the three reactants. The abililty of Pb(II) perchlorate to template the formation of the two macrocyclic ring sizes simply by alteration of the reagent ratio is unprecedented. The first dicopper(II) complex, containing bridging macrocyclic pyridazine units was prepared via transmetalation of the (134)-lead complex with $Cu(ClO_4)_2$ where the ring had contracted to a [2+2]macrocycle (133) in the reaction (S. Brooker *et al.,* J. chem. Soc., chem. Commun., 1994, 487; J. chem. Soc., Dalton Trans., 1996, 2117; J.chem. Soc., chem. Commun., 1996, 2579).

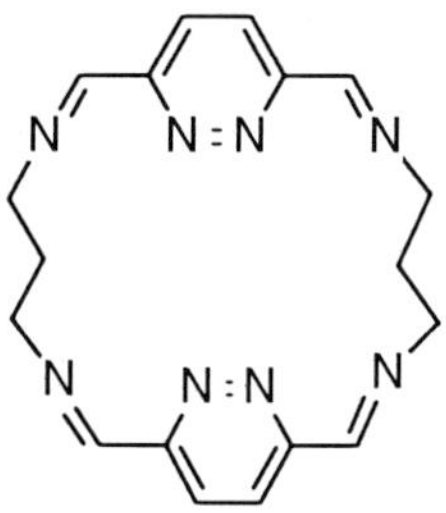

(133)

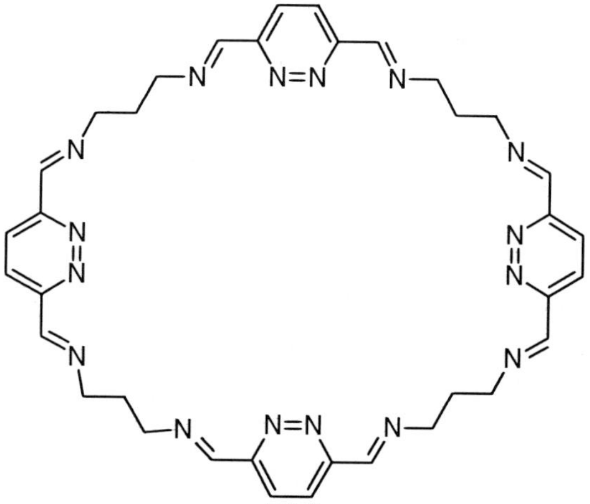

(134)

d. Physical Properties and Spectra

Studies on the protonation of 4,5-dichloro-2-methyl-3(2*H*)-pyridazinone by density functional theory and by *ab initio* quantum chemical methods as well as ^{13}C-NMR relaxation measurements revealed that *O*-protonation took place (P. Matyus *et al.*, Heterocycles, 1994, 38, 1957).

The Moeller-Plessett method and charge density functional theory calculations were applied to pyridazine for the calculation of molecular frequencies using different basic sets (F. Billes and H. Mikosch, J. mol. Struct., 1995, 349, 409).

Principal component analysis has been used to demonstrate the correlation between aromaticity of pyridazines and IR parameters obtained by *ab initio* MO calculations (S.E. Galembeck *et al.*, J. mol. Struct., 1993, 282, 97).

A comparative study has been made of the vibrational spectroscopy of pyridazine, pyrimidine and pyrazine (F. Billes, H. Mikosch and S. Holly, J. mol. Struct., 1998, 423, 225).

Infrared spectroscopic studies have been made on the polarity and lactam-lactim tautomerism of three 3(2*H*)-pyridazinone derivatives (135, R, R^1, R^2; H, H, H; H, Cl, Cl; Me, Cl, Cl). The molecules exist in *oxo-enol* tautomeric equilibrium in dioxane solution. The intermolecular cyclic hydrogen bonded dimer of the molecules was supported by *ab initio* charge distribution calculations (E. Matrai, J. mol. Struct., 1997, 408, 467).

On the basis of ^1H NMR NOE measurements and an X-ray crystal structure determination, it has been shown that a number of pharmacologically active pyridazines should be represented as aromatic pyridazine tautomers rather than the previously reported arylidene-4,5-dihydropyridazines (J.A.M. Guard and P.J. Steel, Aust. J. Chem., 1995, 48, 1601).

Intercalation compounds of α-Zr phosphate with pyridazine have been studied by solid state NMR, to gain information about the proton conduction mechanism. ^{13}C Shifts, obtained under CP MAS conditions, show that the heterocycle guest molecules are in the protonated state (A.L. Blumenfeld *et al.*, Solid state Ionics, 1994, 68, 105).

The ^{13}C NMR spectra of pyridazine and its methyl derivatives have been simulated using parametric techniques (M. Jalali Heravi and M. Moosavi, Aust. J. Chem., 1995, 48, 1267).

^1H and ^{13}C NMR spectral assignment of the *N*-methylpyridazines have been made by long-range coupling to the methyl group (R.A. Newmark, A. Tucker and L.C. Hardy, Magnetic Resonance in Chemistry, 1996, 34, 728).

The EI-MS of ten cycloalkane-condensed 4,5-dihydro-3(2*H*)-

pyridazinones have been studied. The molecular ions are very stable and are generally also the base peaks (E. Maki *et al.,* Rapid commun. Mass Spectrometry, 1994, 8, 1021).

The EI-MS of some *O*-(3-oxo-2-phenyl-3(2*H*)-pyridazinone-6-yl)-thiophosphoro organic esters (136) have been interpreted (R. Popescu *et al.*, Phosphorus, Sulfur and Silicon and the Related Elements, 1994, 91, 9).

The influence of the substitution pattern on the position of the maxima in the electronic spectra of pyridazinium ylides (137) has been demonstrated (M. Petrovanu *et al.*, Rev. Roum. Chim., 1993, 38, 759).

The electronic and IR spectra of some 6-(4-bromophenyl)-3(2*H*)-pyridazinones (138, R = H, CH_2CO_2Et) were determined, and the behaviour on reduction studied by cyclic voltammetry. The experimental data indicated the presence in neutral media of the lactam form, the dissociation in strongly alkaline media of the lactim form and the formation on reduction of an unstable anion-radical, rapidly undergoing a follow-up reaction (V.E. Sahini *et al.*, Rev. Roum. Chim., 1994, 39, 991).

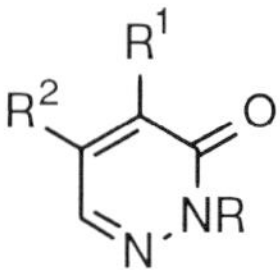

(135) 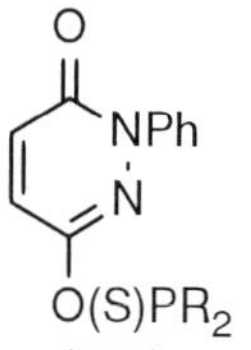 (136) 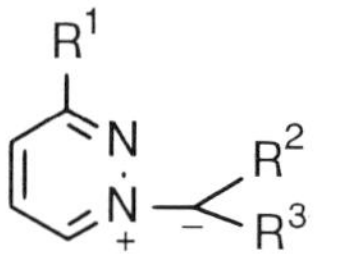 (137) 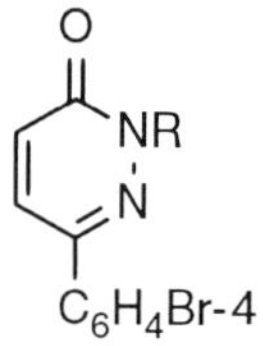 (138)

The electronic absorption and diffusion spectra of pyridazinium-ylides have been analysed. These substances undergo orientation-induction interactions in polar solvents and specific interaction of the proton donor-acceptor type, in protic solvents (D. Dorohoi, J. chim. Phys. et phys.-chim. Biol., 1994, 91, 419).

The laser-excited Raman spectrum of 3,6-dichloropyridazine was recorded using an Ar laser source and frequency-assignments made (S. Mohan and R. Murugan, Ind. J. pure appl. Physics, 1993, 31, 496).

Results from degenerate four wave mixing (DFWM) and laser induced fluorescence (LIF) of pyridazine have been presented. The low fluorescence yield of pyridazine prohibited direct LIF measurements, but the DFWM spectrum is in good agreement with previous absorption spectra. DFWM is an absorption based technique, so it is ideal for high resolution absorption spectroscopy studies when the upper state dynamics are unfavourable for LIF detection (H. Li, P. Dupre and W. Kong, Chem. phys. Letters, 1997, 273, 272).

A study has been carried out of the photoelectric conversion of a novel molecule with strong non-linear optical properties (second harmonic generation), (*E*)-*N*-methyl-4-{2-[4-(dihexadecylamino)phenyl]-ethenyl}pyridazinium iodide, in LB films fabricated on a transparent SnO_2 glass electrode (T.R. Cheng, C.H. Huang and L.B. Gan, J. materials Chem., 1997, 7, 631). This work has recently been extended to a series of (*E*)-*N*-alkyl-4-{2-[4-dialkylamino)-phenyl]ethenyl}pyridazinium iodides, their Langmuir-Blodgett film forming properties and second harmonic generation were studied. The iodide having alkyl = methyl, dialkyl = dihexadecyl appeared to be the best among the four congeners (T.R. Cheng *et al.*, J. materials Chem., 1998, 8, 931).

Liquid chromatography has been used to analyse pyridazines with easy detection by UV. Reverse-phase HPLC is particularly useful for the more polar derivatives for which GC is unsuitable (C. Yamagami and N. Takao, Chem. pharm. Bull., 1992, 40, 925).

Two crystalline forms of 6-chloro-4-(*N,N*-dimethylamino-methylidenehydrazine)-2-phenyl-3(2*H*)-pyridazinone, the unsolvated crystal and an inclusion compound with water, were studied by X-ray diffraction. In both structures, the molecules are linked into dimers by a pair of hydrogen bonds between the hydrazine and carbonyl groups (A.A. Katrusiak and A.S. Katrusiak, J. mol. Structure, 1996, 374, 251).

The structures of four simple pyridazine-3(2*H*)-ones (4-methyl, 5-methyl monohydrate, 6-methyl monohydrate and 2-(4-tolyl)) were determined (A.J. Blake and H. McNab, Acta Cryst., C, 1996, 52, 2622).

The hemisulfate monohydrate of DL-6-(2-[3-(*tert*-butylamino)-2-hydroxypropoxy]phenyl)-3-pyridazinyl-hydrazine, Prizidilol, is orthorhombic (K. Prout, K. Burns and A.M. Roe, Acta Cryst., B, 1994, 50, 68).

5-(2-Chlorobenzyl)-6-methyl-3(2*H*)-pyridazinone is monoclinic (S. Moreau *et al.*, Acta Cryst., C, 1995, 51, 1834).

3-Chloro-5-tosylmethylpyridazine is orthorhombic (M. Ratajczak-Sitarz and A.A. Katrusiak, Pol. J. Chem., 1994, 68, 255).

In 6-(hydroxyiminomethyl)-4-methyl-1-phenylpyridazinium-3-sulfate, the pyridazinium ring is almost planar and is twisted about 50° out of the benzene ring plane. The oxime hydroxyl group forms an intermolecular hydrogen bond to one of the sulfonate oxygen atoms (O. Simonsen *et al.*, Acta Cryst., C, 1994, 50, 1150).

The crystal and molecular structure of eleven 6-substituted-3(2*H*)-pyridazinones have been determined (K. Prout *et al.*, Acta Cryst., B, 1994, 50, 71).

e. Biological Activity

A review describes the use of the pyridazine system as a potential substitute for aromatic moieties in antithrombotic, antiviral and antitumour agents (G. Heinisch, Acta pharm. Hung., 1996, 66, 59).

An extensive review on aminopyridazines as muscarinic agonists with no cholinergic syndrome has been published (C.G. Wermuth, Farmaco, 1993, 48, 253).

In a review several derivatives of 3-hydrazinopyridazine are reported to possess properties as chemotherapeutics or antiinflammatory agents, CNS depressants and stimulants, antihypertensives, vasodilators and β-adrenoceptor blockers (M. Pinza and G. Pifferi, Farmaco, 1994, 49, 683).

Biological structure-activity relationship studies have been made on a large number of pyridazine derivatives. A new series of 5-(4-piperazinyl)-3(2*H*)-pyridazinones (139, R = H, Me, Ph, R^1 = Cl; R = Me, R^1 = H) and, (140, R = R^1 = H, R^2 = OMe; R = Me, R^1 = H, R^2 = H, OMe) has been synthesised. The α-blocking activity of these compounds was determined (S. Corsano *et al.*, Eur. J. med. Chem., 1993, 28, 647).

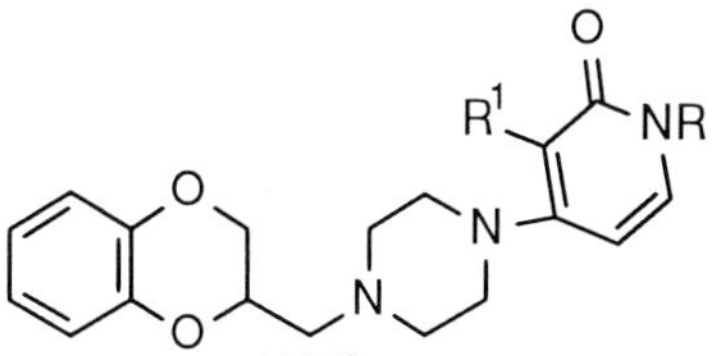

(139) 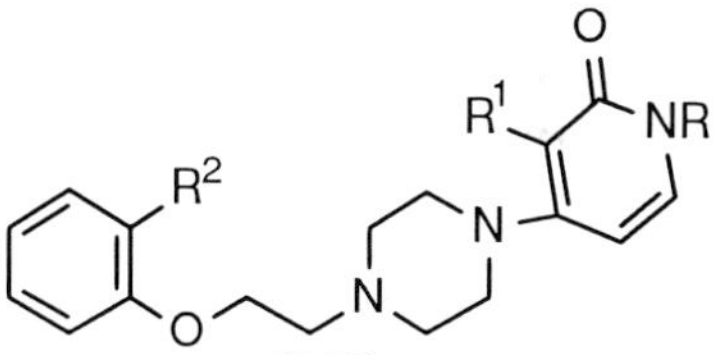 (140)

This work was later extended to the 4-[4-(phenoxyethyl)-1-piperazinyl]-3(2*H*)-pyridazinones and their alkane-bridged dimers (S. Corsano *et al.*, Eur. J. med. Chem., 1995, 30, 71). The same workers have determined the bronchodilating activity of some 6-piperazinyl-3(2*H*)-pyridazinones; (141, R = NH_2, R^1 = Cl; R = Cl, R^1 = H showed good bronchospasmolytic activity) (S. Corsano *et al.*, Bioorg. Med. chem. Lett., 1993, 3, 2713).

A series of *N*-acetic acid derivatives of 4-carboxy-6-aryl-pyridazin-3(2*H*)-one (142, R = H, $CHMe_2$, Cl, etc.) have been made. All the prepared compounds showed a significant *in vitro* aldose reductase-inhibitory effect (most active: 142, R = $CHMe_2$). The antioxidant action of several derivatives was also studied *in vitro*. Compounds (142, R = Cl, H) were most active (P. Coudert *et al.*, Eur. J. med. Chem., 1994, 29, 471).

Tricyclic pyridazinones have also been tested as aldose reductase inhibitors (L. Costantino *et al.*, J. med. Chem., 1996, 39, 4396).

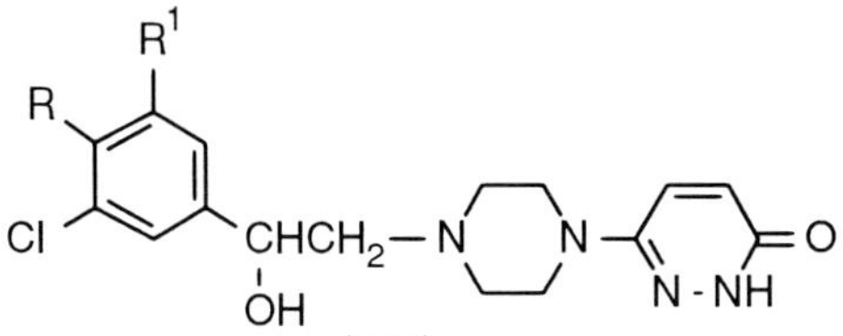

(141)

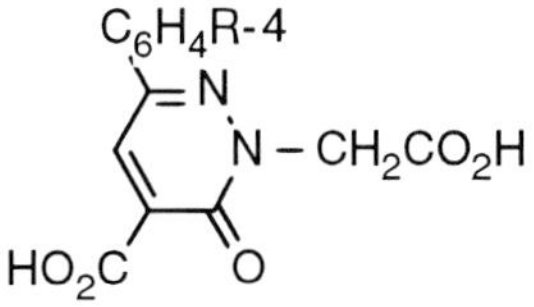

(142)

3-Amino-6-phenylpyridazines in which the amino substituent is a linear butyrophenone moiety, acyclic butyrophenone moiety or a phenylpiperazine fragment have been synthesised and tested for antipsychotic-antidepressant activity (M.E. Castro *et al.*, Eur. J. med. Chem., 1994, 29, 831).

A series of 4,6-diarylpyridazin-3(2*H*)-ones substituted in the 2-position by 4-[(4-aryl)piperazin-1-yl]but-2-ynyl moieties was synthesised and evaluated for antidepressant activity (L. Castanier *et al.*, Arzneimittel-Forschung/Drug Research, 1995, 45-2, 947).

The potential antidepressant effects of two pyridazine derivatives, 5-benzyl-6-methyl-2-[4-(3-trifluoromethylphenyl)piperazin-1-yl]methylpyridazin-3(2*H*)-one and 5-benzyl-6-methyl-2-[4-(3-chlorophenyl)piperazin-1-yl]methylpyridazin-3(2*H*)-one were evaluated (C. Rubat *et al.*, J. Pharmacy and Pharmacology, 1995, 47, 162).

A new synthetic route to the pyridazine antidepressant minaprine, has been developed using *ortho*-metalation and cross-coupling with transition metals (A. Turck *et al.*, Bull. Soc. chim. France, 1993, 130, 488).

A series of thienocycloheptapyridazines, structurally related to minaprine, has been synthesised and the compounds tested for their affinity towards muscarinic receptors (D. Barlocco *et al.*, Drug. des. Discovery, 1997, 14, 273).

The ureidopyridazine (143) and the triazolopyridazine (144, R = Ph, 2,6-$Cl_2C_6H_3$; R^1 = H, NH_2) derivatives were synthesised and shown to have anticonvulsant properties (S. Moreau *et al.*, J. med. Chem., 1994, 37, 2153).

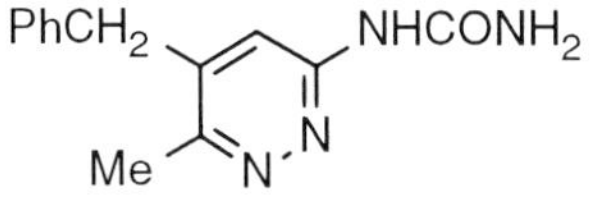

(143)

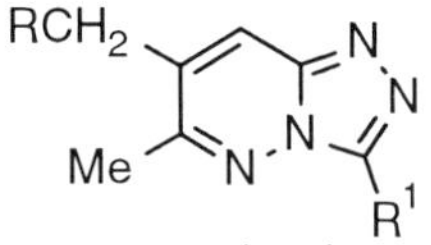

(144)

Several new aryloxypyridazinyl-*N*-(4-arylsulfonyl)acetamides have been made and found to be highly active endothelin receptor antagonists. Compound (145) with a simple dimethylpyridazinone moiety was one of the most potent compounds *in vitro* (D. Dorsch *et al.,* Bioorg. med. Chem. Letters, 1997, 7, 275).

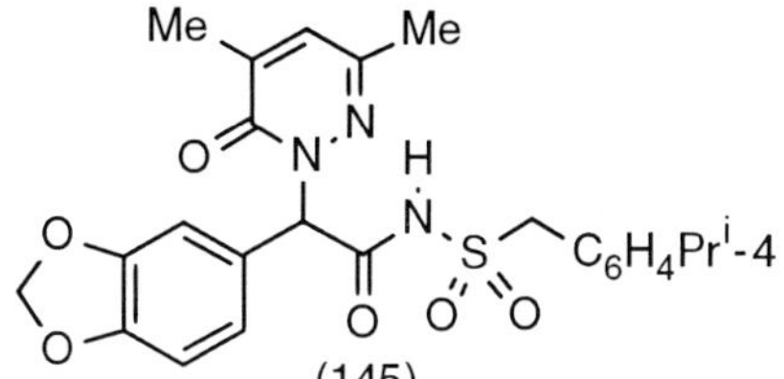

(145)

6-Substituted-3(2*H*)-pyridazinones (146, R^1 = alkyl; R^2 = H, alkyl; X = CO; alk = bond, alkylidene; dotted line indicates optional double bond) were prepared and found to be effective in the treatment of endotoxic shock and kidney diseases (A. Ishida *et al.,* Jpn. Kokai Tokkyo Koho JP, 09711534; Chem. Abstr., 1997, 127, 5099q).

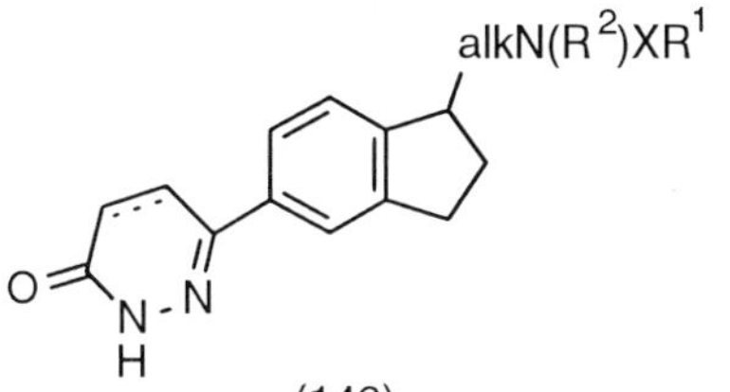

(146)

4,5-Disubstituted 6-phenyl-3(2*H*)-pyridazinones (147, X = amino, morpholino, nitro, Cl, Br, SMe, SOMe,; R = H, alkyl) have been synthesised and evaluated as hypotensive agents (V. Dal Piaz *et al.,* Eur. J. med. Chem., 1994, 29, 249).

A series of 4,5-functionalised 3(2*H*)-pyridazinones (148, R = NO_2, NH_2, Br, SOMe) have been evaluated as PGE_2 and interleukin-1 release inhibitors (V. Dal Piaz *et al.,* Pharmacol. Res., 1994, 29, 367).

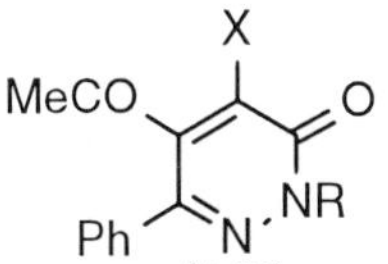

(147) 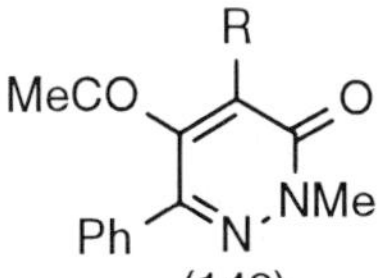 (148)

6-Aryl-3-chloropyridazine carboxamides such as (149) are a non-peptide class of interleukin-1β converting enzyme inhibitors (R.E. Dolle *et al.*, Bioorg. Med. chem. Letters, 1997, 7, 1003).

3-(3,4-Dichlorophenyl)-1-(4-nitrobenzyl)-1,4,5,6-tetrahydropyridazine and similar tetrahydropyridazines (150, R^1 = H, F, Cl, Br; R^2 = Cl, Br, NO_2) represent a new class of selective nonsteroidal progesterone-receptor ligands (D.W. Combs, K. Reese and A. Phillips, J. med. Chem., 1995, 38, 4878).

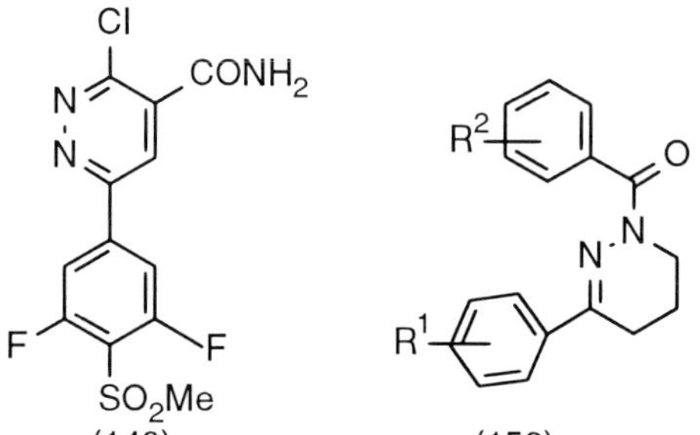

Cl CONH2 N N F F SO2Me (149)

R2 O N N R1 (150)

Pyridazinyl oestradiol derivatives (152, R = Ph, 4-BrC_6H_4, 4-MeC_6H_4, 4-$MeOC_6H_4$) were found to be devoid of oestrogenic and anti-oestrogenic activity unlike the parent steroidal 1,4-diketones (151, same R) (K.A. Ismail *et al.*, Arch. Pharm., 1996, 329, 433).

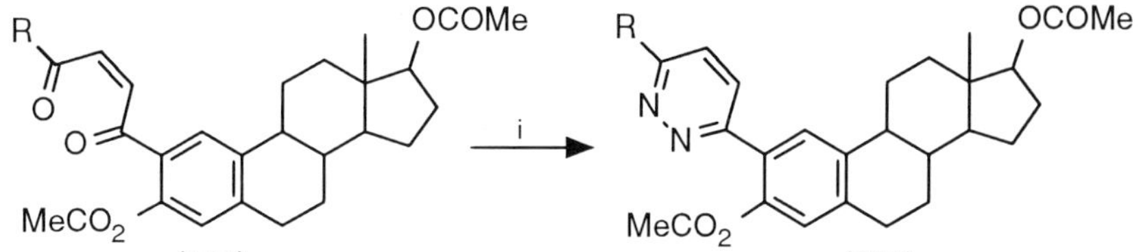

R O O OCOMe $MeCO_2$ (151) —i→ R N N OCOMe $MeCO_2$ (152)

Reagent: (i) hydrazine hydrate

O-Nitrophenylsulfinyl- and *o*-nitrophenyl-sulphonyl substituted 2-phenyl-3-(2*H*)-pyridazinones were shown to have antibacterial and anti-fungal activity (M. Takaya, J. pharm. Soc. Japan, 1993, 113, 676).

3-Chloro-4-carbamoyl-5-aryl-6-methylpyridazine *N*-oxides (153, R = H, NO_2) showed antimicrobial activity (E. Gavini *et al.*, Farmaco, 1997, 52, 67).

2-Aroyl-6-substituted 3(2*H*)-pyridazinones (154, R = Ph, substituted Ph, CH=CHPh; R^1 = OH, Cl, $NHNH_2$, OCH_2CO_2H) have antimicrobial activity (D.M. Purohit and V.H. Shah, Heterocyclic Commun., 1997, 3, 267).

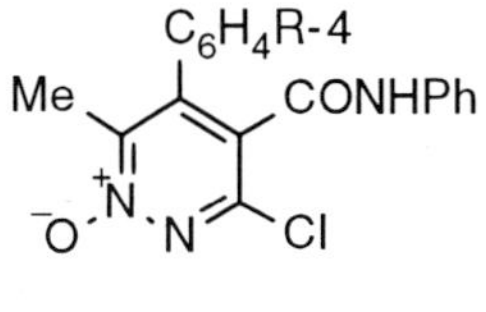

(153)

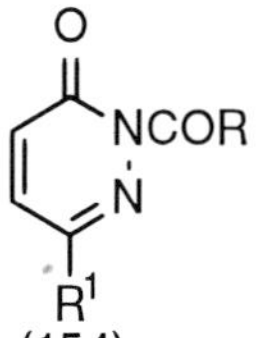

(154)

Fused thiazolo[4,5-d]pyridazines have been evaluated as antibacterials (M. Makki and H.M. Faidallah, J. Chin. chem. Soc., 1996, 43, 433).

A series of 3-arylpiperazinyl-5-benzylpyridazines have been evaluated for analgesic activity (S. Moreau *et al.,* Arzneimittel-Forschung/Drug Research, 1996, 46, 800).

Analgesic activity has been reported for a series of novel 5-arylidenetetrahydropyridazin-3(2*H*)-ones substituted by an arylpiperazinyl moiety (155, R^1 = H, F, Me; R^2 = H, 3-Cl, 3-CF_3, 4-F, n = 1, 2) (C. Rubat *et al.*, J. pharm. Sci., 1992, 81, 1084).

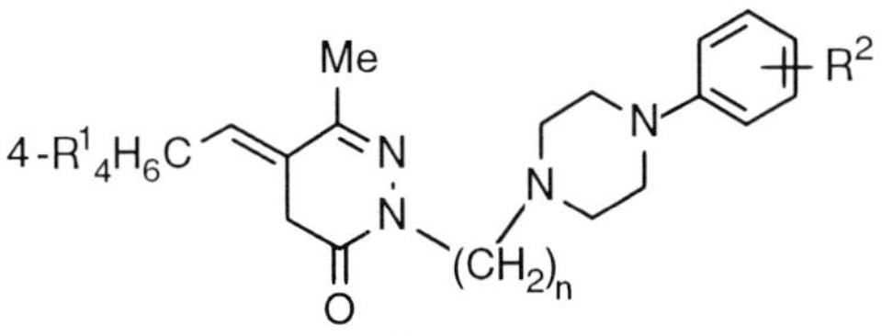

(155)

Imidazo[1,2-*b*]pyridazine-2-acetic acids have been evaluated as analgesics (E. Luraschi *et al.*, Farmaco, 1995, 50, 349).

In order to examine analgesic and antiinflammatory activities of the position isomers of emorfazone (4-ethoxy-2-methyl-5-morpholino-3(2*H*)-pyridazinone) various 5- and 6-alkoxy-, 6-allyloxy-2-alkyl- or 2-cyclohexyl- or 2-phenyl-4- or 5-morpholino-3(2*H*)-pyridazinones were prepared and examined for their activities. As a result, 6-ethoxy-2-methyl-5-morpholino-3(2*H*)-pyridazinone (156) and 2-methyl-5-morpholino-6-*n*-propoxy- and 6-*n*-butoxy-3(2*H*)-pyridazinone were revealed to be more potent than emorfazone itself (M. Takaya and M. Sato, J. pharm. Soc. Japan, 1994, 114, 94).

Of a number of 2-substituted 4,5-functionalized 6-phenyl-3(2*H*)-pyridazinones synthesised and evaluated for their antinociceptive activity, compounds (157 and 158) were more active than the reference drug emorfazone (V. Dal Piaz *et al.*, Eur. J. med. Chem., 1996, 31, 65).

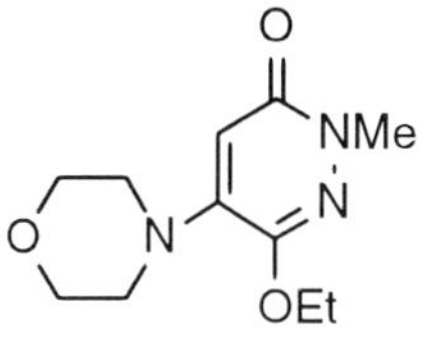

(156) 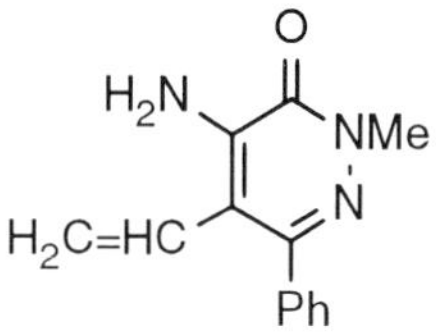 (157) 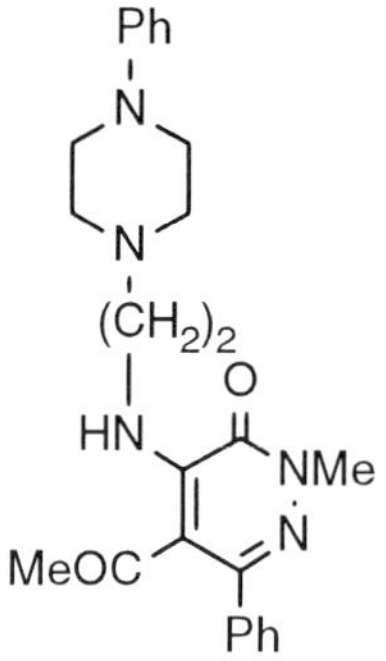 (158)

A new series of 4,6-diarylpyridazines substituted in the 3-position by arylpiperazinyl moieties was synthesised and evaluated for analgesic activity. The most active derivatives were (159, T = H, Cl, F) (F. Rohet *et al.*, Bioorg. med. Chem., 1997, 5, 655).

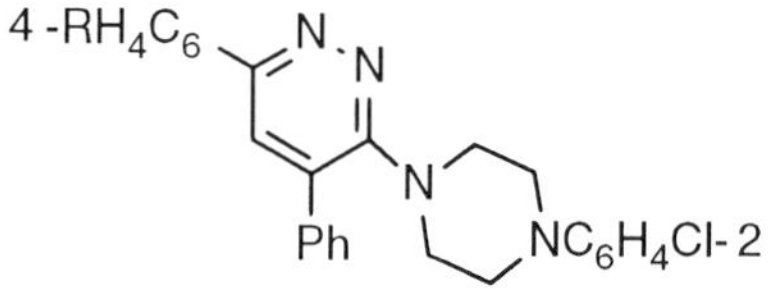

(159)

(3,4-*Trans*-4,5-*trans*)-4,5-dihydroxy-3-(hydroxymethyl)hexahydropyridazine, 1-azafagomine (160, R = OH) was synthesised and found to be a potent inhibitor of α- and β-glucosidase (M. Bols, R. G. Hazell and I.B. Thomsen, Chemistry, 1997, 3, 940). Fluorinated analogues of (160) including the 5-fluoro derivative (160, R = F) were found to be considerably weaker glucosidase inhibitors than 1-azafagomine itself (I.B. Thomsen *et al.*, Tetrahedron, 1997, 53, 9357).

2,9-Dihydro-6-(5-methyl-3-oxo-2,3,4,5-tetrahydropyridazin-6-yl)-pyrazolo[4,3-*b*][1,4]benzoxazine (161) shows cardiotonic activity (D.W. Combs, Bioorg. Med. chem. Letters, 1993, 3, 1663).
6-[4-(Benzylamino)-7-quinazolinyl]-4,5-dihydro-5-methyl-3(2*H*)-pyridazinones have also been tested for cardiotonic activity (Y. Nomoto *et al.*, J. med. Chem., 1996, 39, 297).

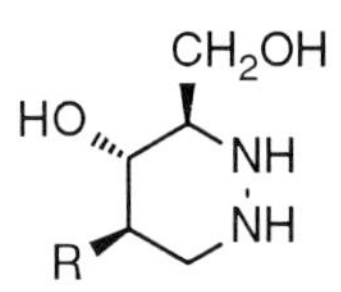

(160) 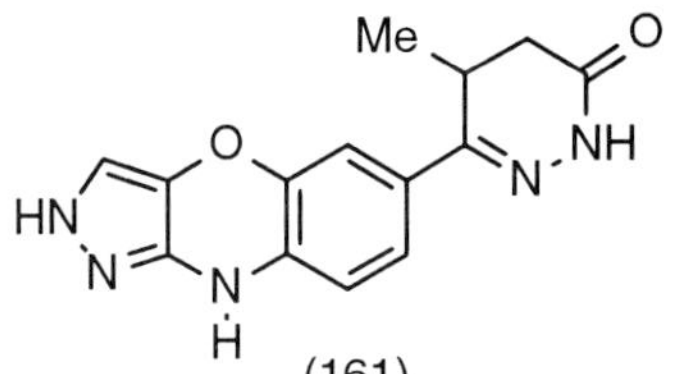 (161)

ω-Sulfamoylalkoxyimidazo[1,2-*b*]pyridazines are antiasthmatics (M. Kuwahara *et al.*, Chem. pharm. Bull., 1996, 44, 122).

Several 6-aryl-5-substituted-3(2*H*)-pyridazinones have been synthesised and evaluated as platelet aggregation inhibitors. The group present at the 5-position seems to be particularly important in determining the activity (R. Laguna *et al.*, Acta pharm. Hung., 1996, 66, S43; Chem. pharm. Bull., 1997, 45, 1151).

6-Aryl-4,5-functionalised-2-methyl-3(2*H*)-pyridazinones have also been synthesised and evaluated as platelet aggregation inhibitors (V. Dal Piaz, G. Ciciani and M.P. Giovannoni, Drug des. Discovery, 1996, 14, 53; Farmaco, 1997, 52, 173).

Fourteen 6-(4-substituted phenyl)-4,5-dihydro-3(2*H*)-pyridazinones were synthesised. All inhibited platelet aggregation induced by ADP in rabbits. The compound containing the benzimidazole group showed the strongest inhibitory activity (W. Li *et al.*, Khongguo Yaowu Huaxue Zazhi, 1997, 7, 12; Chem. Abstr., 1997, 127, 2148596).

Other biological studies include imidazo[1,2-*b*]pyridazines as CNS agents (G. B. Barlin *et al.*, Aust. J. Chem., 1995, 48, 1031; 1996, 49, 443); tetra- and hexa-hydropyrido[3,4-*d*]pyridazines as CNS agents (H. Sladowska *et al.*, Farmaco, 1996, 51, 431); oxazolo[3′2′:1,2]pyrrolo[3,4-*d*]pyridazines and imidazolo[1′,2′:1,2]pyrrolo[3,4-*d*]pyridazines as potential non-nucleoside HIV-1 reverse transcriptase inhibitors (B. Barth *et al.*, Arch. Pharm., 1996, 329, 403).

2′,3′-Dideoxy and 3′-azido-2′,3′-dideoxypyridazine nucleosides (162, R = NH_2, OMe, $R^1 = R^2 = H$; R = OH, $R^1 = Cl$, $R^2 = N_3$) have been synthesised as potential antiviral agents but were inactive (B. Kasnar *et al.*, Nucleosides and Nucleotides, 1994, 13, 459).

The stereocontrolled synthesis of pyridazine 2′-deoxy-β-nucleoside analogues (163) by intramolecular glycosylation of pyridazine has been described (H. Sugimura, M. Motegi and K. Sujino, Nucleosides and Nucleotides, 1995, 14, 413).

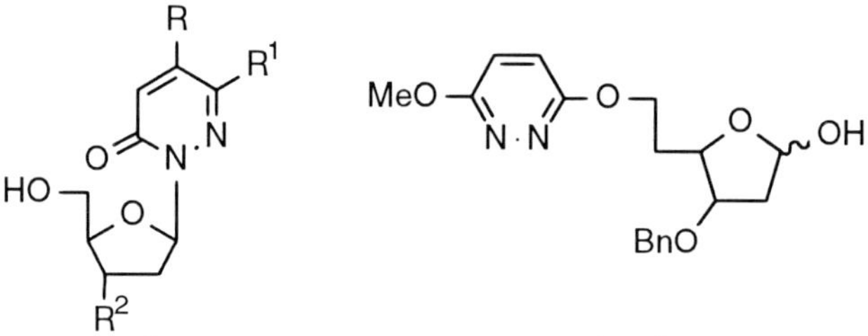

(162) (163)

Podophyllotoxin A-ring pyridazine analogue (164) has been constructed from (-)-podophyllotoxin (E. Bertounesque, T. Imbert and C. Monneret, Tetrahedron, 1996, 52, 14235).

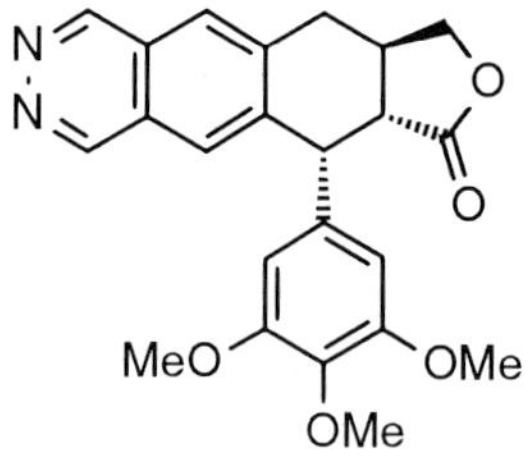

(164)

f. Herbicides

The synthesis and biological activity of the heterocycle fused tetrahydropyridazine herbicides such as (165, X = O, S) has been reviewed (J. Satow *et al.*, ACS Symposium Series, 1995, 584, 100).

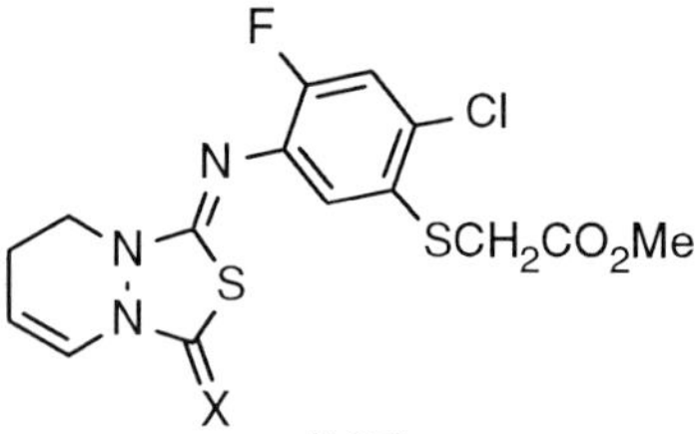

(165)

The *N*-demethylation of the pyridazinone pro-herbicide metflurazon into norfluazon implies a toxification in photosynthetic organisms (F. Thies *et al.*, Plant Physiology, 1996, 112, 361).

Substituted 2,6-diphenylpyridazin-3(2*H*)-ones have been synthesised as bleaching herbicides using molecular graphics at the design stage (K. Bowden, A. Brownhill and F. Malik, J. Serb. chem. Soc., 1997, 62, 951).

A new microbial phytotoxin pyridazocidin (166) represents the first reported natural product that appears to act via reversible oxidation-reduction linked to photosynthetic electron transport (B.C. Gerwick *et al.*, Weed Science, 1997, 45, 654).

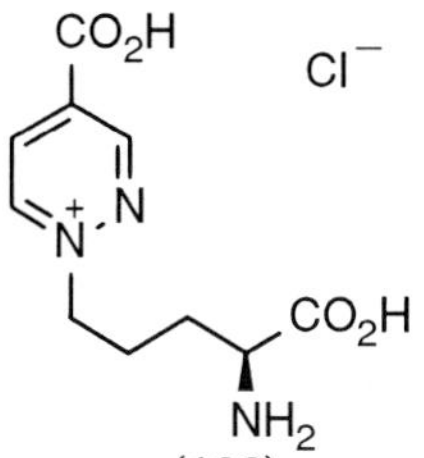

CO_2H Cl^- N $\overset{+}{N}$ CO_2H NH_2

(166)

g. Miscellaneous Applications

Pyridazine derivatives such as (167, R = H, Me, OMe, Cl) 3,6-dianilinopyridazine, 3,6-di-β-naphthylaminopyridazine and 3-β-naphthylamino-6-isopropylaminopyridazine have been prepared and found to be good antioxidants and antifatigue agents in NR and SBR vulcanizates, their efficiency being comparable with that of the industrially used 1,4-diphenylphenylene diamine (M.N. Ismail *et al.*, J. Elastomers & Plastics, 1993, 25, 266; Plast. Rubber Compos. Process. Appli., 1993, 19, 241).

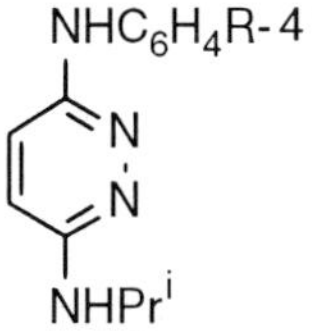

$NHC_6H_4R\text{-}4$ N N $NHPr^i$

(167)

New fluorinated poly(imide-pyridazineamide)s have been synthesised and their properties compared with those of similar polymers not containing pyridazine rings, as well as those of other related previously reported heterocyclc polymers (M. Bruma, B. Schulz and F.W. Mercer, Polymer, 1994, 35, 4209).

2. Cinnoline and its Derivatives

a. Methods of Synthesis

(i) From diazonium salts

Diazotization of suitable *o*-substituted anilines is a well established procedure leading to cinnoline derivatives and when *o*-aminophenylalkynes (2) are used and the reaction mixture is heated (the Richter reaction) the products are 4-cinnolinones (1). It is now recognised that the cinnolinones are formed by hydrolysis of the first formed 4-chlorocinnolines (S.F. Vasilevsky, E.V. Tretyakov and H.D. Verkruijsse, Synth. Commun., 1994, 24, 1733). If the diazotization and cyclisation step is carried out at -15°C, the respective 4-halogenocinnolines (3) can be isolated in moderate to good yields when hydrochloric or hydrobromic acid is used (S.F. Vasilevsky and E.V. Tretyakov, Liebigs Ann., 1995, 775).

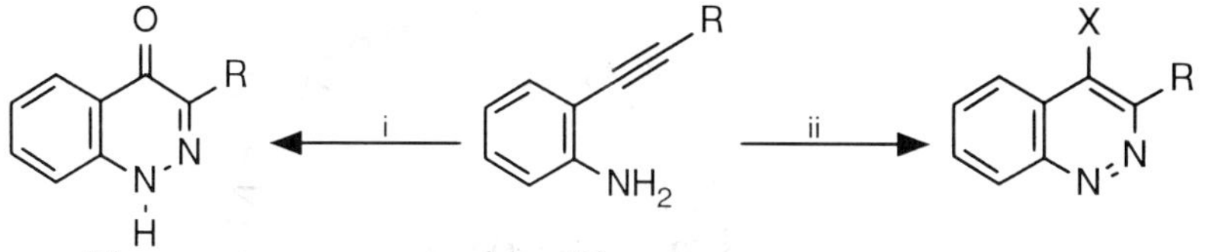

(1) (2) (3, X = Cl or Br)

Reagents: (i) aq. HCl or HBr, $NaNO_2$, 0°C then 90°C, (ii) aq. HCl or HBr, $NaNO_2$, -15°C

Certain *o*-arylanilines carrying bicyclic aryl substituents yield tetracyclic cinnolines (G. Cirrincione *et al.*, Farmaco, 1995, 50, 849) and 6-amino-3-methylpyridazin-4-ones carrying a substituted phenyl group as a 5-substituent undergo diazotization and cyclisation to give tricyclic cinnoline (1,2,9,10-tetraazaphenanthrene) derivatives (Y.M. Volovenko, Khim. geterosikl. Soedin., 1997, 854).

(ii) From phenylhydrazones

The phenylhydrazones (4) undergo aluminium chloride catalysed cyclisation through reaction at the nitrile function to give the 4-aminocinnoline 3-carboxamides (5) in high yield (A. Stanczak *et al.*, Pharmazie, 1994, 49, 884; L.V.G. Nargund, V. Virupax and S.M. Yarnal, Indian J. Chem., 1994, 33B, 764; L.V.G. Nargund, Y.N. Manohara and V. Hariprasad, Indian J. Chem., 1994, 33B, 1001; R.G. Menon and E. Purushothaman, Indian J. Chem., 1996, 35, 1185).

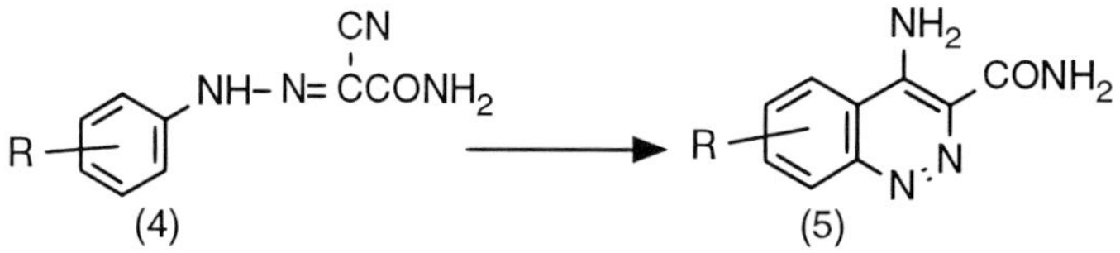

(4) (5)

4-Amino-3-arylcinnolines (7) are available in good yield from phenylhydrazones of aryl trifluoromethyl ketones (6) by cyclisation at low temperature with potassium bis(trimethylsilyl)amide (A.S. Alexander, Tetrahedron Letters, 1995, 36,1383).

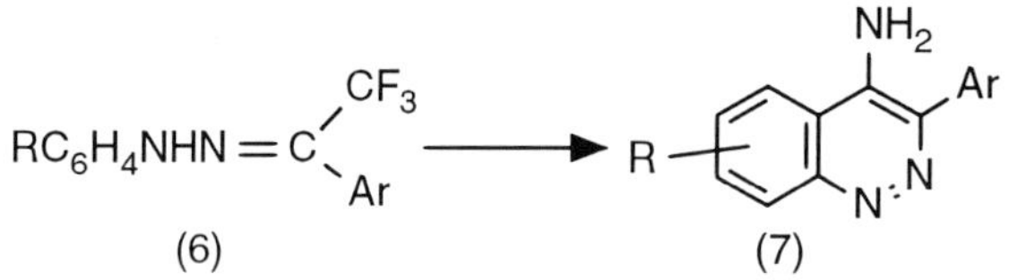

(6) (7)

The 3,4-disubstituted cinnoline (9) is obtained in 40% yield from 4-*p*-tolylhydrazino-2-phenyl-2-oxazolin-5-one (8) under basic conditions (A.M. Radwan *et al.*, J. chem. Soc. Pakistan, 1995, 17, 113).

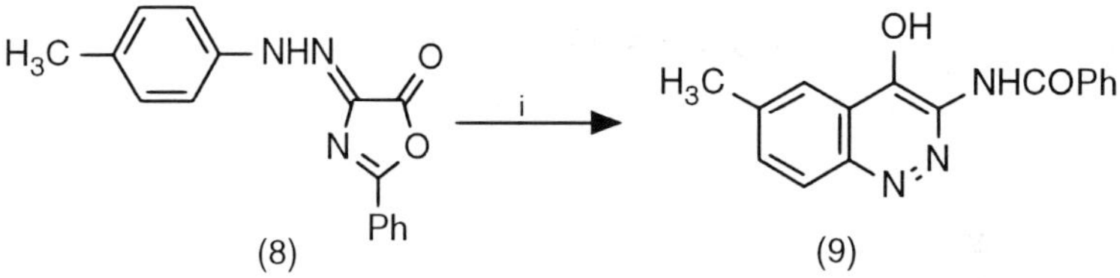

(8) (9)

Reagents: (i) $PhCOCH_3$, Na, dioxane, reflux

(iii) Replacement of fluorine

Formation of partially reduced cinnoline derivatives by intramolecular cyclisation processes at an activated fluorine substituent on an aromatic nucleus have been used. The *N*-aminoquinolone derivative (10) gave the tricyclic (11) on treatment with cesium carbonate in DMSO (D. Barrett *et al.*, J. org. Chem., 1995, 60, 3928).

A novel class of betaines, the 1,1-disubstituted cinnolinium 3-oxides (13) have been obtained by cyclisation of the hydrazide (12) in aqueous potassium carbonate (V.J. Arán *et al.*, J. chem. Soc., Perkin 1, 1997, 2229).

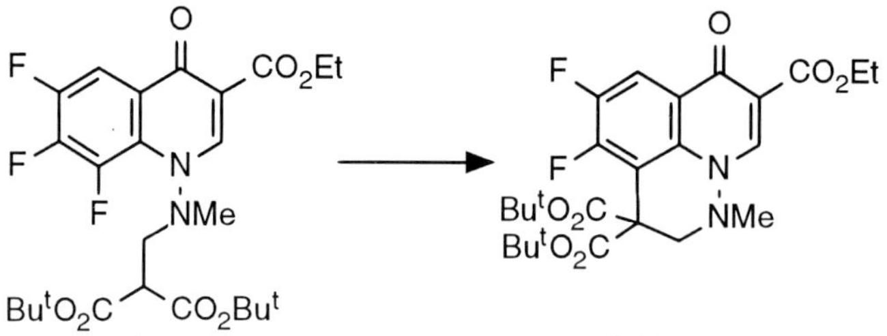

(10) (11)

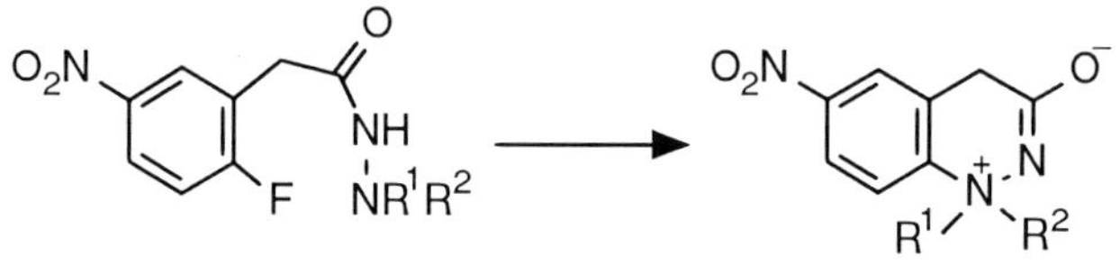

(12) (13)

(iv) Addition reactions

The addition of trishydrotris(triphenylphosphine)cobalt or (dinitrogen)hydrotris(triphenylphosphine)cobalt to a melt of azobenzene and diphenylacetylene (tolane) causes regioselective double addition of tolane to yield 2,6-distilbenylazobenzene and the 2,3-dihydrocinnoline (14) (G. Halbritter *et al.,* Angew. Chem., 1994, 106, 1676; G. Halbritter, F. Knoch and H. Kisch, J. organomet. Chem., 1995, 492, 87).

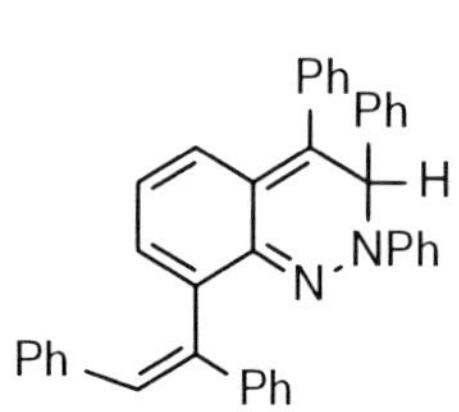

(14) 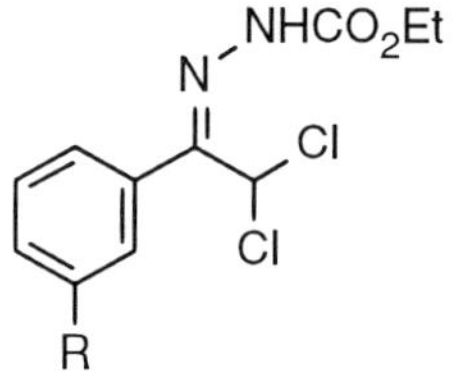 (15) 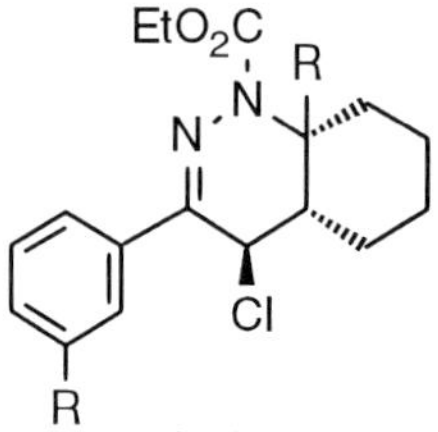 (16)

A [2+4]cycloaddition occurs in moderate yield when a diene is generated by addition of Hunig's base to (15) in the presence of the enamine, 1-morpholinocyclohexene, to give (16, R = 1-morpholino) and treatment of this with aqueous base gives the 3-substituted 5,6,7,8-tetrahydrocinnolines in high yield (M.S. South and T.L. Jakuboski, Tetrahedron Letters, 1995, 36, 5703).

(v) By formation of the N-N *bond*

Photolysis of 2-(*o*-azidophenyl)ethylamine gives cinnoline in poor yield through nitrene formation and insertion followed by oxidation in air of the 1,2,3,4-tetrahydrocinnoline (S. Murata, M. Miwa and H. Tomioka, J. chem. Soc., chem. Commun., 1995, 1255).

A novel strategy for the synthesis of derivatives of cinnoline 4-carboxylic acids uses the base induced cyclisation of *o*-(aminohetero-aryl)nitrobenzenes. Thus, readily accessible 3-amino-5-hydroxy-4-(2-nitroaryl)pyrazoles (17) undergo base induced cyclisation to the tricyclic (18) in moderate to good yield. The five-membered ring is oxidatively cleaved in good yield to give cinnoline-4-carboxylic acid 1-oxides (19) (M. Scobie and G. Tennant, J. chem. Soc., chem. Commun., 1993, 1756).

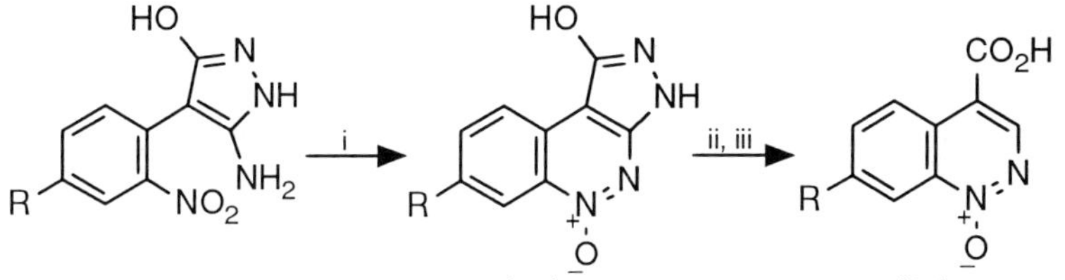

(17) (18) (19)

Reagents: (i) aq. NaOH, (ii) aq NaOCl, NaOH, (iii) $Na_2S_2O_4$, aq. DMF

In a similar way, 3-amino-4-(2-nitrophenyl)-2*H*-isoxazolin-5-ones (20) are cyclised to the tricyclic (21) but these undergo an unusual reductive hydrazinolysis to give 3-aminocinnoline-4-carboxylic acid *N*-oxides (22) (M. Scobie and G. Tennant, J. chem. Soc., chem. Commun., 1994, 2451).

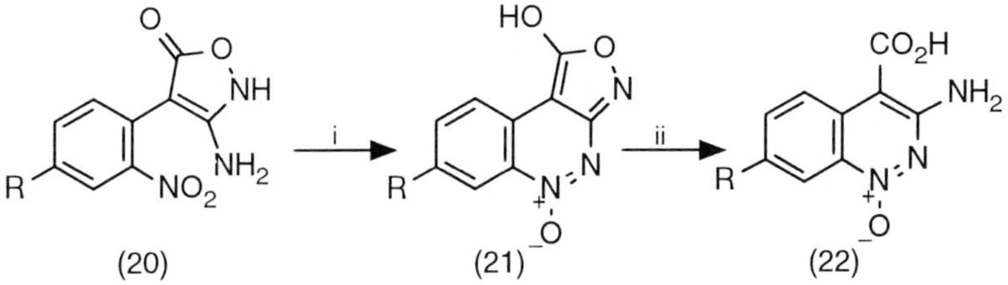

(20) (21) (22)

Reagents: (i) NaH, DMF, (ii) N_2H_4, H_2O, EtOH

(vi) From γ-ketoesters or cyclohexan-1,3-diones

Routes to partially reduced cinnolines derivatives have been mentioned in (*iii*) and (*iv*). In addition, δ-ketoester derivatives of 4,5,6,7-

tetrahydrothiophene (23) have been condensed with hydrazine to give (24, $R^2 = H$) (G. Cignarella *et al.*, Acta chim. Slov., 1994, 41, 173; G.A. Pinna *et al.*, Eur. J. med. Chem., 1994, 29, 447; G.A. Pinna *et al.*, Farmaco, 1997, 52, 29) and aryl hydrazines to give (24, R^2 = aryl) (H. Tanaka *et al.*, Eur. J. med. Chem., 1997, 32, 607).

(23) (24)

Benzil monohydrazone condenses with cyclohexan-1,3-diones to give derivatives of 3,4-diphenyl-5,6,7,8-tetrahydrocinnolin-5-one (25) (M.G. Assy and E. Abd El-Ghani, J. Pol. Chem., 1995, 69, 685; M.G. Assy and F.M. Abd El-Motti, Indian J. Chem., 1996, 35B, 608; M.G. Assy and F.M. Abd El-Motti, Egypt. J. Chem., 1996, 39, 581).

(b) Physical properties

Molecular orbital calculations with specific parameterization (MOSP) (R. Bacaloglu, I. Bacaloglu and Z. Simon, Rev. Roum. Chim., 1992, 37, 819) and by the SCF method (H.J.M. Soscum *et al.*,Theochem., 1994, 113, 1) show good correlation between the experimental and theoretical basicity of cinnoline.

Mass spectrometry has been used to explore rearrangement processes of negative ions in the gas phase and the resonance stabilization of the ions (M.V. Muftakhov *et al.*, J. mass Spectrom., 1995, 30, 275).

Several crystal structure studies of cinnolines have been reported. The 4-cinnolones (26, $R^1 = Cl$, $R^2 = H$) (M.L. Glowka, I. Iwanicka and A. Olczak, Acta Crystallogr., 1994, C50, 1770) (26, $R^1 = H$, $R^2 = CH_2CO_2Et$) (M.L. Glowka, I. Iwanicka and A. Stanczak, J. chem. Crystallogr., 1994, 24, 397) and (26, $R^1 = H$, $R^2 = CH_2CONHNH_2$) (M.L. Glowka and A. Stanczak, Polish J. Chem., 1996, 70, 1512) show a *para*-quinonoid form of the diazine ring. The 4-iminocinnoline (27) shows bond lengths and angles similar to those in the cinnolinones (M.L. Glowka *et al.*, Polish J. Chem., 1997, 71, 170). The X-ray crystal structure for 2-methylcinnolin-3-one shows the short and long bond pattern expected of the *ortho*-quinonoid classical structure (28). Studies of the ^{14}N nuclear quadrupole coupling constants found for (29) showed that these are little different from those of pyrazole and it is concluded that there is little bicyclic aromatic character in (29). The carbocyclic ring appears to be a butadiene-like substituent on the heterocycle (M.H. Palmer *et al.*, Chem. Phys., 1992, 168, 41).

Crystal structures for the cinnolines (14) (G. Halbritter, F. Knoch and H. Kisch, J. organomet. chem., 1995, 492, 87) and the tetrahydrate of (29) are available (M.L. Glowka, A. Alczak and A. Stanczak, Polish J. Chem., 1994, 68, 1631).

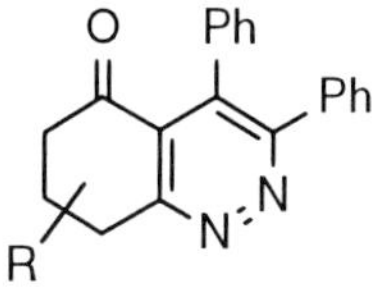

(25) 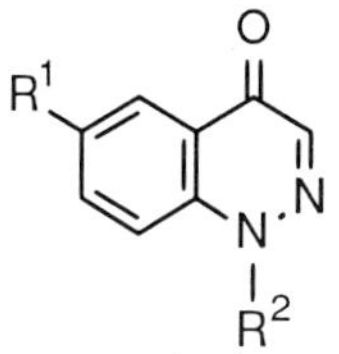 (26) 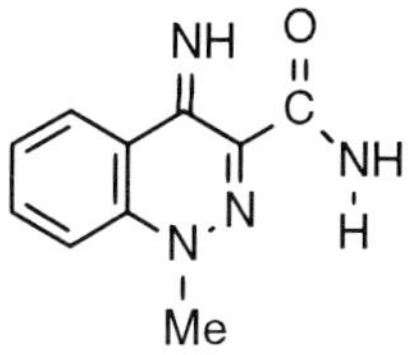 (27)

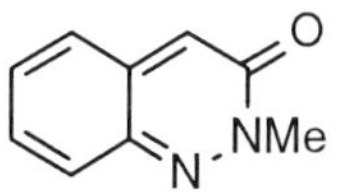

(28) 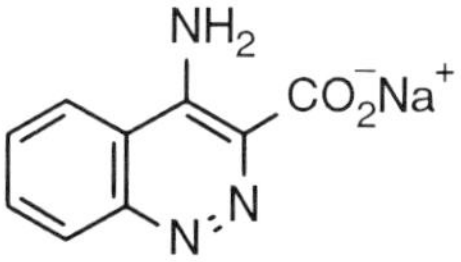 (29)

(c) Reactions

The first *o*-directed lithiation of cinnolines is reported. Treatment of 4-chlorocinnoline with either lithium diisopropylamide or lithium 2,2,6,6-tetramethylpiperidine at low temperature followed by an electrophile gives 3-substituted cinnoline in moderate to good yields. In this way, 4-chlorocinnoline and *o*-methoxybenzaldehyde provides a secondary alcohol which can be oxidised to the ketone (30); a useful source of tetracyclic cinnoline derivatives, eg. (31).

The reaction of 3-chlorocinnoline with lithium amide and then aldehyde is more complex because the expected product (32) is formed (in good yield) but is accompanied by the dimer (33) (A. Turck *et al.*, Tetrahedron, 1995, 51, 13045).

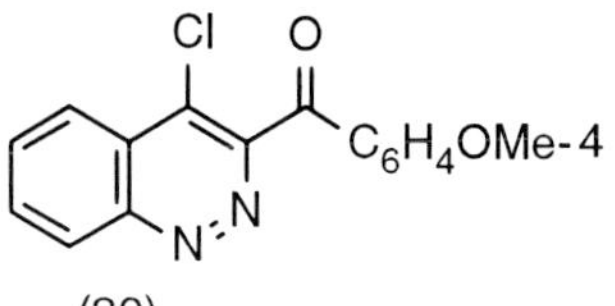

(30) 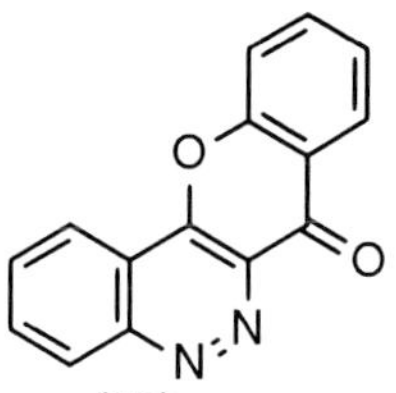 (31)

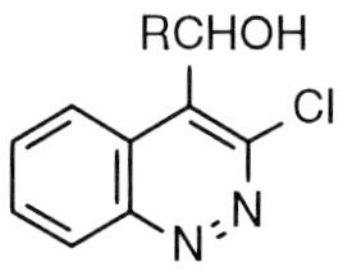

(32) 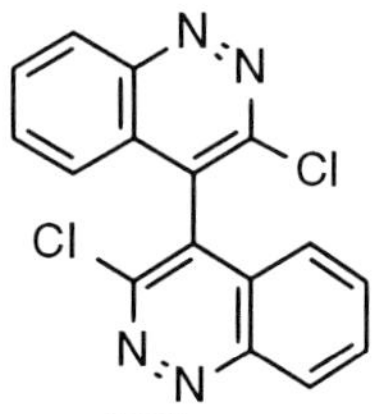 (33)

The hydrolysis of ethyl 6-ethyl-4-hydroxycinnolin-3-carboxylate (34, R^1 =OH, R^2 = OEt, R^3 = H) has been studied in detail and shown to probably not involve intramolecular general base catalysis (R.G. Button and P.J. Taylor, J. chem. Research (S), 1996, 218). The hydrolysis of (34, R^1 = NH_2, R^2 = OH, R^3 = H) in aqueous potassium hydroxide gives (34, R^1 = OH, R^2 = OH, R^3 = H) in good yields (A. Stanczak *et al.*, Pharmazie, 1997, 52, 91). The amide (34, R^1 = R^2 = NH_2, R^3 = H) cyclises readily with potassium *O*-ethyldithiocarbonate to give (35) in high yield (G.R. Menon and E. Purushothaman, J. Indian chem. Soc., 1995, 72, 731). 3-Acetylcinnolinones (34, R^1 = OH, R^2 = Me) are readily converted into tricyclic compounds such as pyrazolo[4,3-*c*]cinnolines (36) (M.S. Abbady, S.M. Radwan and E.A. Bakhite, Indian J. Chem., 1993, 32B, 1281).

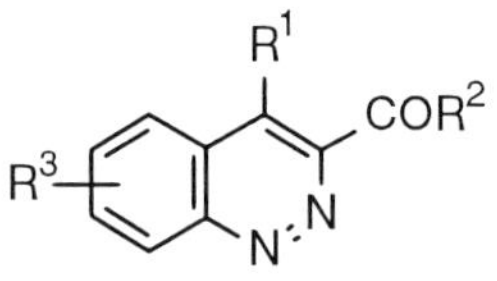

(34) 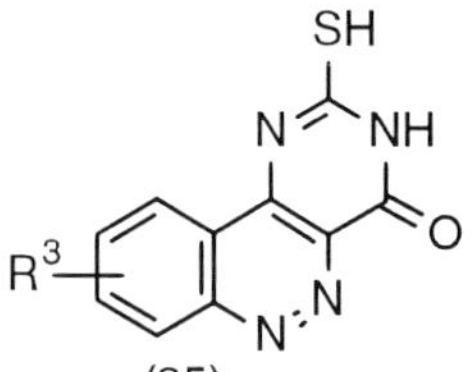 (35) 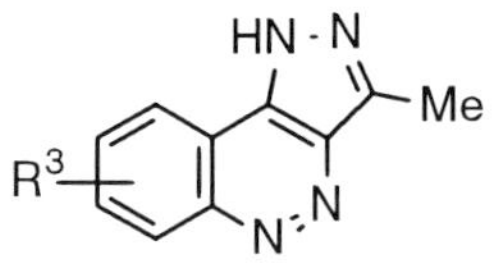 (36)

The [4+2]cycloaddition of ynamines (eg. $R^2C{\equiv}CNEt_2$) to cinnoline or 4-substituted cinnoline may occur by either formation of a 3,8a-adduct which collapses by loss of dinitrogen to give an α-aminonaphthalene (37) or through a 1,4-adduct which loses HCN to give the 2-aminoquinoline (38) (K. Iwamoto *et al.,* Heterocycles, 1996, 43, 2409).

Acylation of 1,2,3,4-tetrahydrocinnolin-4-one occurs at N2 first and, under more vigorous conditions, at both N1 and N2. Hence, use of 3-chloropropanoyl chloride under vigorous conditions gives (39) (W. Kwapiszewski, A. Stanczak and S. Groszkowski, Acta Pol. Pharm., 1992, 49, 47).

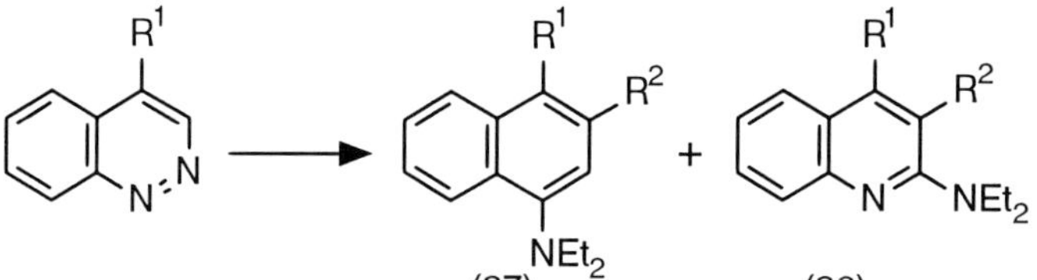

(37) (38)

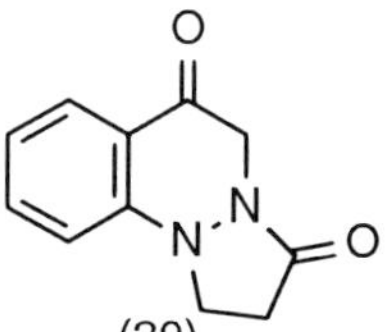

(39)

Alkylation of 4-hydroxycinnoline with ethyl bromoacetate gave the 2-substituted derivative which may be converted to the 4-oxo-1,2,3,4-tetrahydrocinnoline acetic acids (A. Stanczak *et al.*, Pharmazie, 1994, 49, 406).

3. The Benzocinnolines

a. Benzo[c]*cinnoline*

(i) Methods of preparation

The reductive cyclisation of 2,2′-dinitrobiphenyl to yield benzo[*c*]-cinnoline (1) is an attractive idea which continues to draw attention. Electrochemical reduction in ethanolic sulfuric acid in the presence of titanium oxysulfate gives a 14% yield of (1) (V. Danciu and G. Mousset, Can. J. Chem., 1993, 71, 1136). The benzocinnoline (1) is formed in less than 10% yield in experiments designed primarily to investigate the triruthenium dodecacarbonyl-catalysed reductive carbonylation of the dinitro compound (A. Bassoli, S. Canini and F. Farina, J. mol. Catal., 1994, 89, 121; E. Bolzacchini *et al.,* J. mol. Catal., 1996, 110, 227).

However, activated molybdenum (prepared readily from molybdenum pentachloride) catalyses the reductive cyclisation in the presence of hydrazine hydrate to give (1) in 88% yield (S.H. Pyo and B.H. Han, Bull. Korean chem. Soc., 1994, 15, 1012).

The reduction of 2,2′-dinitrobiphenyl to give benzo[*c*]cinnoline *N*-oxide (2) is accomplished using metallic zinc in the presence of bismuth(III) chloride (H.H. Borah *et al.*, Tetrahedron Letters, 1994, 35, 3167) or cadmium(II) chloride (B. Baruah *et al.*, Chem. Letters, 1996, 351). The dinitrobiphenyl and its derivatives can be reduced to (2) or the di-*N*-oxide (3) or to a mixture of both (2) and (3) by the action of potassium hydroxide in isopropanol. The product depends on the molar ratio of the hydroxide (larger amounts favouring formation of the mono-*N*-oxide) and the nature of substituents on the biphenyl (electron withdrawing substituents tending to promote the production of (2)) (C. Paradisi, G. Gonzalez-Trueloa and G. Scorrano, Tetrahedron Letters, 1993, 34, 877).

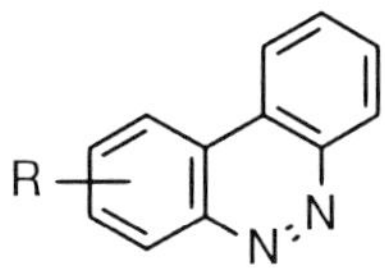

(1) 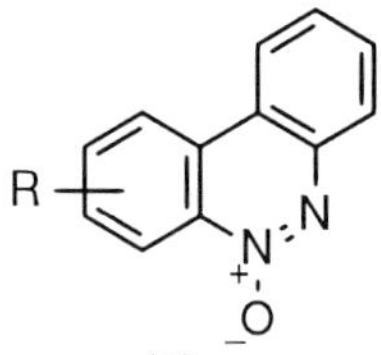 (2) 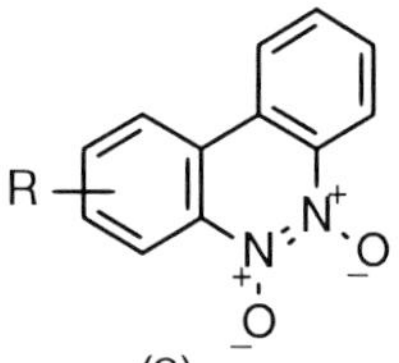 (3)

Benzo[*c*]cinnoline is formed in moderate yield by photolysis of 2-ethylamino-2′-azidobiphenyl in cyclohexane (S. Murata, H. Tsuji and H. Tomioka, Bull. chem. Soc. Japan, 1994, 67, 895) or from *trans*-azobenzene by photolysis in the presence of Nafion (a perfluorinated polymeric sulfonic acid) (C.H. Tung and J.Q. Guan, J. org. Chem., 1996, 61, 9417).

(ii) Physical properties

Quantitative resonance theory and the group additivity-resonance energy model have been applied to benzo[*c*]cinnoline and the latter gave a close approach to the experimental value for the heat of formation (P. Yang and W. Hao, J. chem. Research (S), 1996, 204).

The absorption and emission spectra of benzo[*c*]cinnolines have been studied as part of series including diazahelicenes and the powerful influence of the solvent on the phosphorescence:fluorescence ratio noted (G. Greiner, H. Rau and R. Bonneau, J. photochem. photobiol. A, 1996, 95, 115). The effect of solvent cage formation on the absorption and fluorescence spectra has been investigated and the effects are discussed in terms of the solvent cage perturbation, potential functions for photo-dissociation, excited state intramolecular torsion and excited state twisting-intramolecular-charge-transfer (M. Kasha, A. Sytnik and B. Dellinger, Pure applied Chem., 1993, 65, 1641).

Chemical ionization mass spectra of benzo[*c*]cinnoline using dimethyl ether and perdeuteriodimethyl ether are reported (E.P. Burrows, J. mass. Spectrom., 1995, 30, 312). Benzo[*c*]cinnoline ions are found in the negative ion mass spectrometry of halogenated azobenzenes (M.V. Muftakhov *et al.,* J. mass Spectrom., 1995, 30, 275).

(iii) Reactions

Benzo[*c*]cinnoline 5-oxide (2) can be deoxygenated in high yield by baker's yeast (W. Baik *et al.,* Tetrahedron Letters, 1997, 38, 845) or by irradiation in the presence of triethylamine, 2,3-dimethyl-2-butene and triethyl phosphite (E. Fasani, A.M. Amer and A. Albini, Heterocycles, 1994, 37, 985).

The dinitration of benzo[*c*]cinnoline with mixed nitric and sulfuric acids gives 1,7-dinitrobenzo[*c*]cinnoline (4) but a similar reaction with 3,8-di-*tert*-butylbenzo[*c*]cinnoline (5, $R^1 = R^2 = C(CH_3)_3$, $R^3 = H$) produces 3-*tert*-butyl-8-methyl-1,7-dinitrobenzo[*c*]cinnoline (5, $R^1 = C(CH_3)_3$, $R^2 = CH_3$, $R^3 = NO_2$) and 3,8-dimethyl-1,7-dinitrobenzo[*c*]-cinnoline (5, $R^1 = R^2 = CH_3$, $R^3 = NO_2$). A suggested mechanism for the degradation of the *tert*-butyl group is given (J.W. Barton and D.K.M. Sum, J. chem. Research (S), 1992, 386).

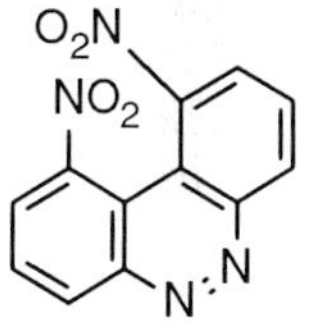

(4) 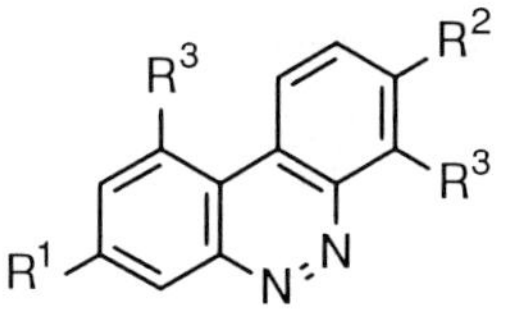 (5)

1-Nitro- and 4-nitro-benzo[*c*]cinnolines have been converted to the corresponding fluoro compounds by reduction to the amine and application of the Schiemann reaction, but the yields were low (E. Kilic and C. Tuzun, Synth. Commun., 1992, 22, 545).

5,6-Dihydrobenzo[*c*]cinnoline and its 5,6-disubstituted derivatives undergo electrochemical oxidation to their corresponding dications and, in some cases, to the trications (M. Dietrich *et al.,* J. Amer. chem. Soc., 1996, 118, 5020).

Studies of the coordination of benzo[*c*]cinnoline with transition and lanthanide metals, the structure of the products and sometimes reactions of the complex include: the formation of cobalt complexes from tris(trimethylphosphine)cobalt(I) complexes (H.-F. Klein *et al.,* Z. Naturforsch, Section B, 1993, 48, 778; H.-F. Klein *et al., ibid.*, 1993, 48, 785); the synthesis and NMR studies of the 1,2-metallotropic shifts in complexes of the type [W(CO)$_5$L] (where L = benzo[*c*]cinnoline) (E.W. Abel *et al.*, J. organomet. Chem., 1994, 464, 163); the reaction of platinacycles with benzo[*c*]cinnoline (A.D. Ryabov, Inorg. Chem., 1993, 32, 3166); the [2+2]cycloaddition of benzo[*c*]cinnoline and the carbyne salt [Cp(CO)$_2$Re:CTol]$^+$ (L.A. Mercando *et al.,* Organometallics, 1993, 12, 1559); and the formation and crystal structure of a monomeric complex formed from metallic ytterbium and benzo[*c*]cinnoline in THF (Y. Nakayama, A. Nakamura and K. Mashima, Chem. Letters, 1997, 803).

*b. Benzo[*f*]cinnolines*

The benzo[*f*]cinnolines (6, R = NH_2 or $NHCOCH_3$) have been synthesised and both showed lower hypotensive and antihypertensive properties than the isomeric benzo[*h*]cinnolines (G.A. Pinna, Farmaco, 1996, 51, 653).

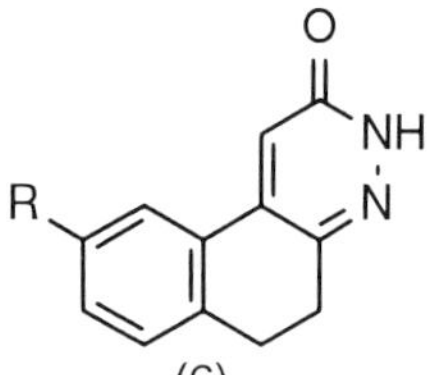

(6) 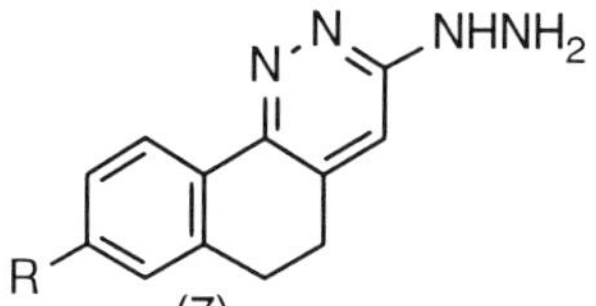 (7)

*c. Benzo[*h*]cinnolines*

Several 3-hydrazino-5,6-dihydrobenzo[*h*]cinnolines (7) have been synthesised and tested for hypotensive and diuretic activity (E. Ravina *et al.,* Pharmazie, 1992, 47, 574) while derivatives of 5,6-dihydrobenzo[*h*]-cinnolin-3-one (8) have been prepared and tested for anti-ulcer activity (G. Cignarella *et al.,* Eur. J. med. Chem., 1992, 27, 819) and for hypotensive activity and inhibition of collagen-induced platelet aggregation (G.A. Pinna *et al.*, Farmaco, 1997, 52, 25).

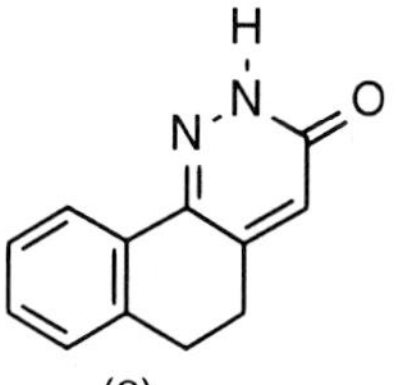

(8)

4. Phthalazine and its derivatives

a. Methods of synthesis

(i) From phthalic acid derivatives

The hydrazide (1) cyclises in the presence of acid to give (2) in good yields (T. Otto-Jaworska, Acta Pol. Pharm., 1992, 49, 81). *Ortho*-arylbenzoic acids (3, X = O or S) react with hydrazine to give the phthalazinones (4, X = O or S) which have been used in the preparation of all-aromatic phthalazine-containing polymers (S. Yoshida and A.S. Hay, Macromolecules, 1997, 30, 2254; R. Singh and A.S. Hay, *ibid.*, 1992, 25, 1025).

Similar reactions have been used to make 4-heteroaryl-phthalazinones which show bronchodilatory properties (M. Yamaguchi *et al.*, J. med. Chem., 1993, 36, 4061) and the highly substituted 4-aryl-phthalazinones (5, R^1 = Br) (A.A. El-Sawy *et al.*, Egypt. J. Chem., 1990, 33, 13) and (5, R^1 = Ph) (F.A. Yassin *et al.*, Egypt. J. Chem., 1990, 33, 199).

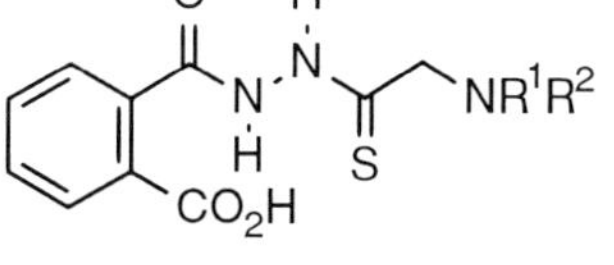

(1)

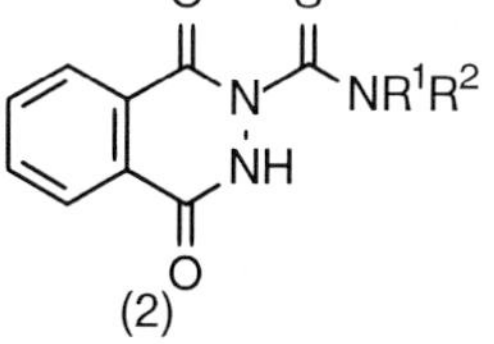

(2)

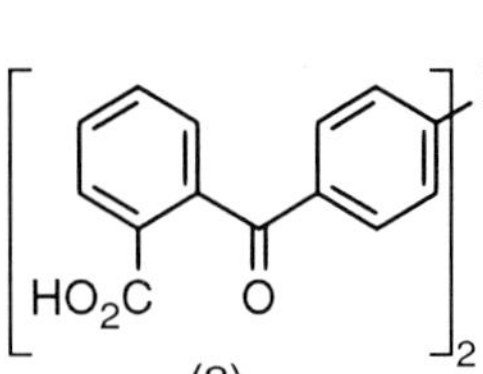

(3) 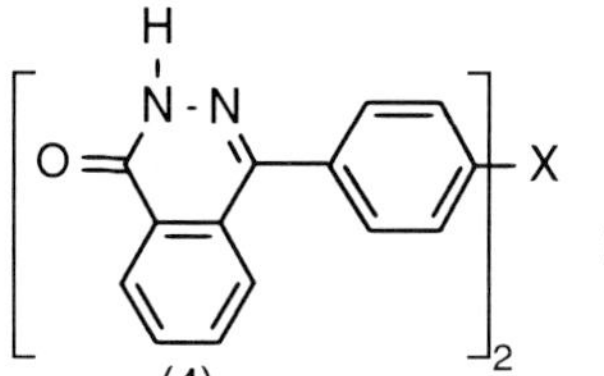 (4) 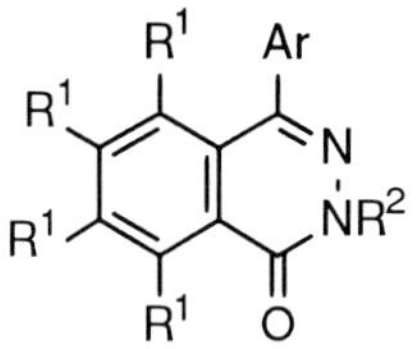 (5)

The furan (6) is unusually susceptible to attack by hydrazine because of activation by the quinonoid system and consequently the tricyclic phthalazine (7) is formed in high yield after aerial oxidation under alkaline conditions of the initial product (M.D. Coburn *et al.*, J. heterocyclic Chem., 1993, 30, 1593).

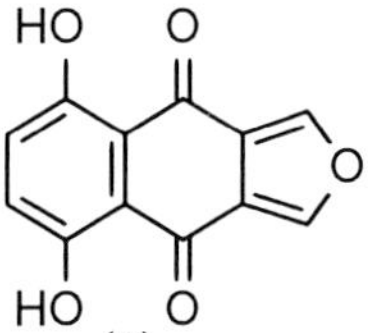

(6) 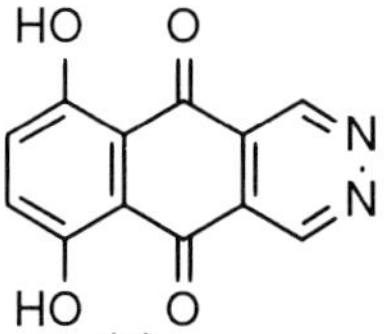 (7) 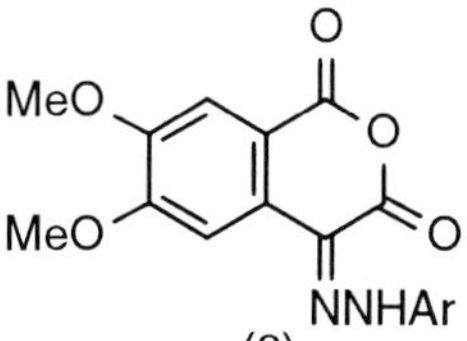 (8)

The homophthalic acid derivative (8) on treatment with a secondary amine (R^1R^2NH) or aqueous sodium hydroxide affords (9, R = NR^1R^2) or (9, R = OH), respectively (J.A. Patankar, R.S. Verma and S.S. Athalye, Indian J. heterocyclic Chem., 1996, 5, 239).

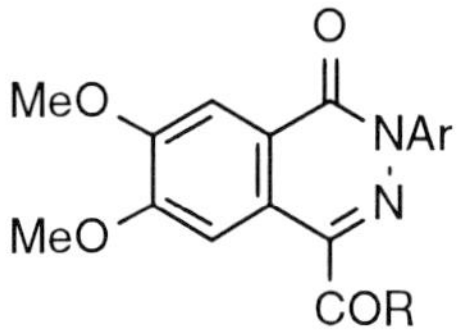

(9) 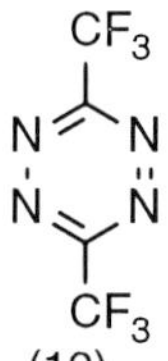 (10) 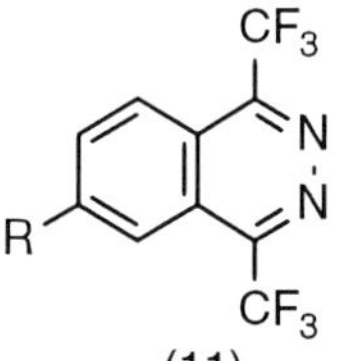 (11)

(ii) From cycloaddition reactions

Substituted benzenes carrying electron donating groups can take part in [4+2]cycloaddition reactions in which they act as dienophiles. In this way, 3,6-bis(trifluoromethyl)-1,2,4,5-tetrazine (10) reacts with anisole to give (11, R = OMe) in good yield after elimination of dinitrogen and oxidation in the work-up. When *N,N*-dimethylaniline is used, a lower yield of the corresponding phthalazine (11, R = NMe_2) is obtained but 1,2-bis(bis[3,6-trifluoromethyl]pyridazin-4-yl)ethylene is isolated as a byproduct (R. Hoferichter and G. Seitz, Liebigs Ann., 1992, 1153).

A similar reaction between (10) and cyclopropabenzene (12) yields 4a,8a-methanophthalazine (13), the first propellane with both an electron-rich and an electron-deficient 4π-diene system. Electron-rich dienophiles and cycloalkenes undergo [4+2]cycloaddition to the diaza-diene system (J. Laue, G. Seitz and H. Wassmuth, Z. Naturforsch, Section B, 1996, 51, 348).

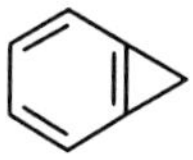

CF_3 N N CF_3

(12) 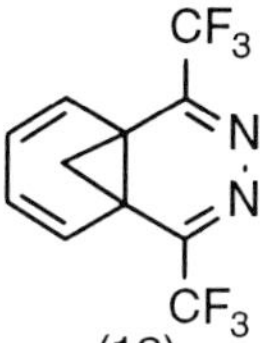 (13)

The synthesis of [*f*]-annelated phthalazines (15, X = O or S) can be achieved by thermolysis of the pyridazino[4,5-*d*]pyridazine (14, X = O or S) which undergoes intramolecular [4+2]cycloaddition and subsequent loss of dinitrogen (N. Haider and C. Loll, J. heterocyclic Chem., 1994, 31, 357).

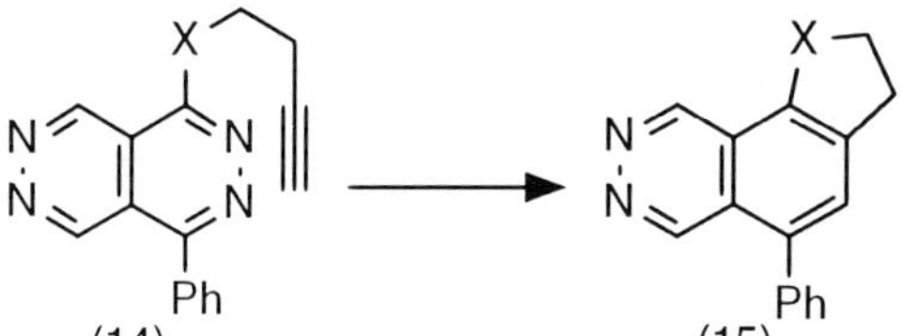

X N N N N Ph X N N Ph

(14) (15)

The synthesis of polyfunctionally substituted phthalazinones by use of [4+2]cycloaddition reactions has been extensively investigated.

4-Substituted 7-amino-2-arylthieno[3,4-*d*]pyridazin-1(2*H*)-ones (18) undergo [4+2]cycloaddition reactions with electron-poor olefins such as maleic anhydride (A.H.H. Elghandour *et al.,* J. prakt. Chem., 1992, 334, 723), *N*-phenylmaleimide (M.H. Elnagdi *et al.*, J. chem. Research (S), 1993, 130), olefins or Mannich bases (used as precursors of olefins) (A.M. Hussein *et al.*, Heteroatom Chem., 1997, 8, 29). In these cases, the initially formed adduct loses hydrogen sulfide to give the substituted phthalazinone. For example, *N*-phenylmaleimide gives the adduct (16) (not isolated) which then yields (17). In some cases, loss of hydrogen sulfide did not occur and a 5-mercaptophthalazin-1(2*H*)-one (19) or a tricyclic compound (20) containing sulfur is isolated (H. Al-Awadhi *et al.*, Tetrahedron, 1995, 51, 12745). If a tetrasubstituted olefin is used, e.g. tetracyanoethylene, the product is a thione, e.g. (21) (A.M. Hussein *et al.*, Pol. J. chem., 1996, 70, 589). These reactions proceed in good yields.

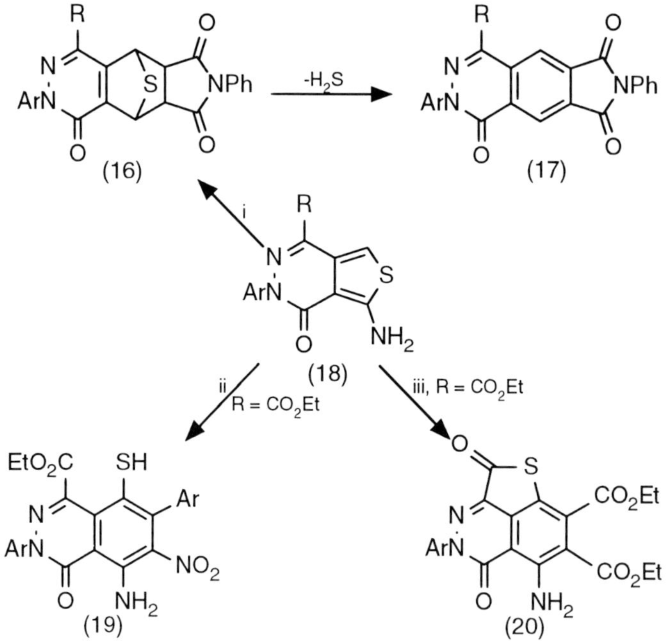

Reagents: (i) *N*-phenylmaleimide, (ii) ArCH:CHNO$_2$, (iii) (*E*)-EtO$_2$CCH:CHCO$_2$Et

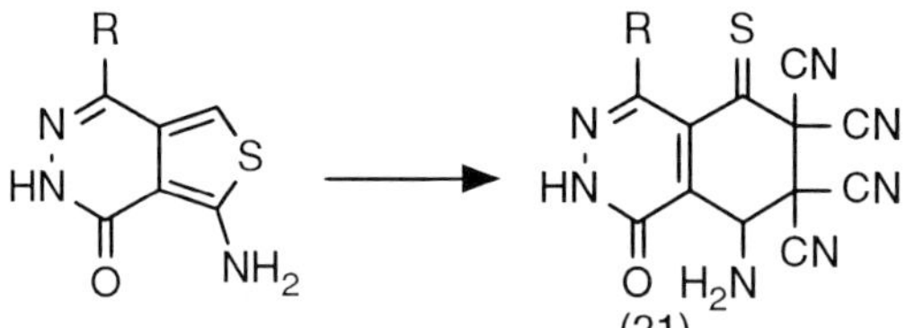

In similar processes, the meso-ionic thienopyridazinium (22) undergoes cycloaddition reactions with acrylonitrile and the adduct then loses hydrogen sulfide to give the phthalazinium ylide (23) (R.M.

Mohareb *et al.*, Gazz. chim. Ital., 1992, 122, 503).

CH_2:CHCN

(22) (23)

(iii) From amidines

A novel route to 4-(aryloxymethyl)phthalazin-1(2*H*)-ones (27) is by acid induced cyclisation of (26). The intermediates (26) are readily available from nitriles (24) and hydrazides (25) (A.M. Barnard *et al.*, Synthesis, 1998, 317).

$R^1C_6H_4OCH_2CN$ + $R^2C_6H_4CONHNH_2$

(24) (25)

$R^1C_6H_4OCH_2C(NH_2)=NNHCOC_6H_4R^2$ (26)

(27)

Reagents: (i) NaOEt, EtOH, (ii) *p*-TsOH, PrOH, reflux

b. Physical Properties

Theoretical calculations of a number of properties of phthalazines have been made and correlations with experimental results sought and found, in most cases. A good linear relationship was found between the HOMO energy calculated by *ab initio* SCF-MO methods and both the experimental pK values and the first π-electron ionization energy (H.J.S. Machado and A. Hinchliffe, Theochem., 1995, 339, 255) and there is a good agreement between the experimental ^{13}C-NMR chemical shifts and the values calculated by the CNDO method (D. Zhang *et al.*, Guenghu Shiyanshi, 1995, 12, 24; Chem. Abstr., 1996, 125, 167194c).

The electronic absorption spectrum of phthalazine has been investigated by calculation and experiment, and the vapour phase spectrum reported for the first time. *Ab initio* CI-singles calculations predict two neighbouring singlet (π*, n) states: $^1A_2$ followed by $^1B_1$. The first 6000 cm^{-1} of the absorption spectrum is assigned to the forbidden n→π* transition. The higher energy regions of this band show no clear evidence of a second n→π* transition, despite theoretical predictions and circular-dichroism evidence (G. Fischer and P. Wormell, Chem. Phys., 1995, 198, 183). The results of calculations using quadratic response theory for singlet and triplet operators have been compared with the measured total radiative lifetime of phthalazine in the vapour phase and with phosphorescence microwave double resonance data obtained for matrix isolated samples. The spin-sublevel activities are well reproduced by the calculations (S. Knuts, H. Agren and B.F. Minaev, Theochem., 1994, 117, 185).

The measured standard molar enthalpy of formation of phthalazine at 298.15 K in the gaseous state is 329.9±3.3 $kJmol^{-1}$ (M.A.V. Ribeiro da Silca, A.R.M. Matos and V.M.F. Morais, J. chem. Soc., Faraday Trans., 1995, 91, 1907).

Interest in surface-enhanced Raman spectra (SERS) of phthalazines has continued. Elevated pressure has been shown to increase the intensity of the absorptions (E. Janssen, L. Smeller and K. Heremans, Colloq. INSERM, 1992, 224 (High Pressure Biotechnol.), 431; Chem. Abstr., 1994, 120, 87524s). SERS have been recorded for phthalazine adsorbed on individual, selected clusters of colloidal silver particles and good quality spectra obtained with 5 pmol of the heterocycle (B. Vlckova *et al.,* J. phys. Chem., 1996, 100, 3169; B. Vlckova, X.J. Gu and M. Moskovits, *ibid*., 1997, 101, 1588). Evidence has been obtained for the anion-induced formation of a surface–Ag^+–phthalazine complex (M. Muniz-Miranda and G. Sbrana, J. raman Spectros., 1996, 27, 105). However, a warning has been given that explanations of unusual excitation wavelength and coverage dependence may be incorrect. The reasons for these uncertainties stem from the finding that phthalazine adsorbed on a colloidal silver surface undergoes a photochemical change in which the *N-N* bond is broken and the adsorbed species resembles an *o*-disubstituted benzene. The photochemical rate constant for this ring opening is large in the visible spectrum and increases towards shorter wavelength. The SERS spectra of phthalazine previously reported show evidence of the presence of the photoproduct (J.S. Suh *et al.,* J. phy. Chem., 1996, 100, 805). The photochemical reaction is a one-photon process in green light and the reaction rate is considerably increased in an aqueous medium (J.S. Suh, N.H. Jang and T.K. Kang, J. raman Spectros., 1997, 28, 567).

A review explains the mechanisms and experimental design for the oscillation chemiluminescence of luminol (5-aminophthalazin-1,4-dione) (H. Brandl, S. Albrecht and M. Haufe, Chem. unserer Zeit, 1993, 27, 303). Three tricyclic compounds (28) (29, R = C_3H_7) and (29, R = CH_2Ph) are said to produce 1.7 to 5.5 times greater chemiluminescence than luminol (J. Ishido *et al.,* Dyes Pigm., 1995, 27, 1).

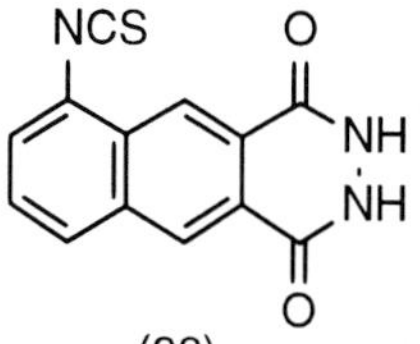

(28) 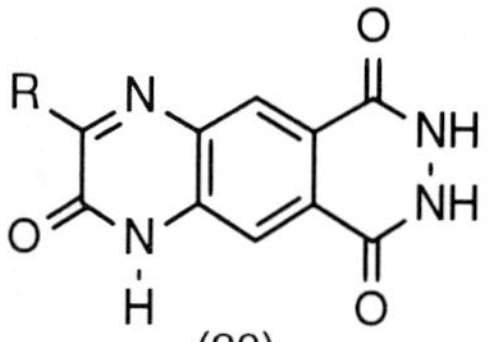 (29)

c. Reactions of the heterocycle

N-Acylated phthalazines are susceptible to nucleophilic attack at the adjacent carbon atom. Ethyl chloroformate has been used to cause acylation of phthalazine (30) and in the presence of 6-aryl-2,3-dihydro-imidazo[2,1-*b*]thiazoles (31) the 1-heteroarylphthalazine (32) is formed in good yield. This can be oxidised by chloroanil to (33) but in poor yield (D. Staiger and L. Webb, J. heterocyclic Chem., 1993, 30, 1663).

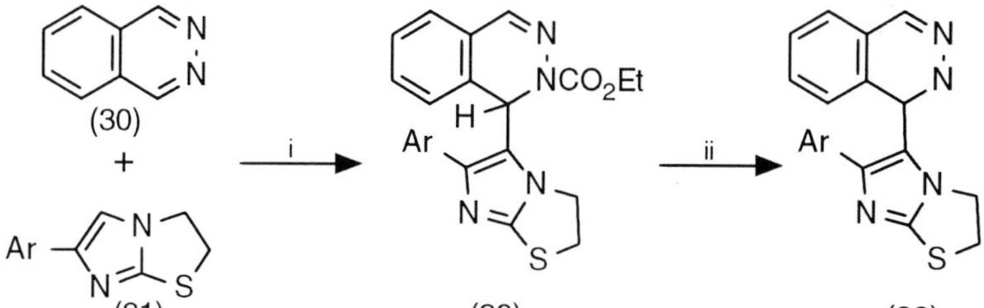

(30) (31) (32) (33)

Reagents: (i) $ClCO_2Et$, 0-5°C; (ii) *o*-chloroanil, toluene, reflux

In a similar way, good yields of 1-allyl-2-alkoxycarbonyl-phthalazine (34, $R^1 = CH_2CH{:}CH_2$) are obtained from (30) by the action of chloroformate and allyltributyltin (T. Itoh *et al.,* Chem. pharm. Bull., 1994, 42, 1768; T. Itoh *et al.,* Heterocycles, 1994, 37, 709) and 1-ethynyl-2-alkoxycarbonylphthalazine (34, $R^1 = C{:}CH$) when bis(tributyl-stannyl)acetylene is used (T. Itoh *et al.*, Tetrahedron, 1994, 50, 13089).

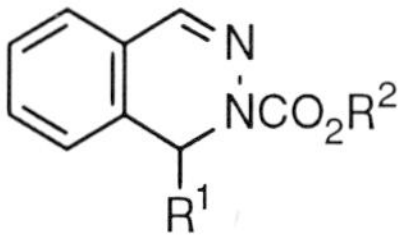

(34) 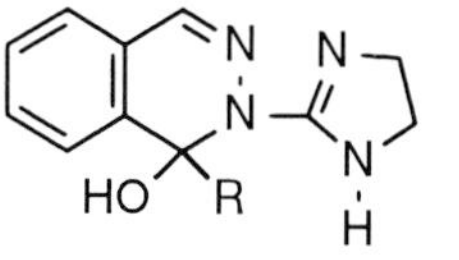 (35) 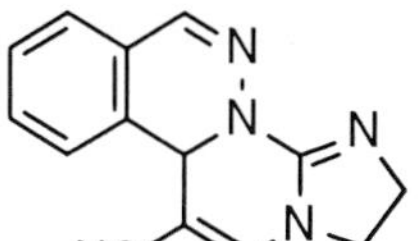 (36)

Quaternization of a nitrogen atom of phthalazine with 2-chloro-imidazoline in the presence of aqueous sodium hydroxide has been used to provide (35) from which tetracyclic phthalazine derivatives such as (36) have been obtained (F. Saczewski and M. Gdaniec, Liebigs Ann., 1996, 1673).

1-Adamantylphthalazine is obtained in 55% yield from phthalazinium trifluoroacetate by reaction with 1-adamantyl chloride, tris-(trimethylsilyl)silane and 2,2′-azobis(isobutyronitrile) (H. Togo, K. Hayashi and M. Yokoyama, Chem. Letters, 1993, 641).

A study of the mechanism by which an oxygen atom is transferred from dimethyldioxirane to (30) to yield phthalazine *N*-oxide (37) has indicated an SN2 process rather than an electron-transfer mechanism (W. Adam and D. Golsch, Angew. chem. int. Ed., 1993, 105, 737).

A new synthesis of 1,6-methano[10]annulenes (38) has been achieved by the [4+2]cycloaddition of electron-rich alkynes ($R^1C{\equiv}CR^2$) with the 4a,8a-methanophthalazine (13) to give an intermediate which spontaneously loses dinitrogen to give (38) (J. Laue and G. Seitz, Liebigs Ann., 1996, 773).

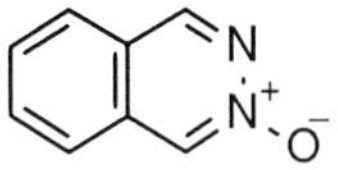

(37) 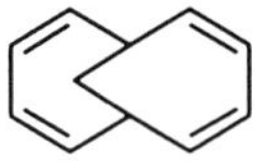 (38)

d. Reactions of substituted phthalazines

The conversion of 1-chlorophthalazine to 1-cyanophthalazine is catalysed by sodium toluenesulfinate but sometimes the better method is to adopt a two stage process including isolation of the intermediate phthalazine 1-sulfinate (A. Miyashita *et al.*, Heterocycles, 1994, 39, 345). 1-Phenoxyphthalazines are useful intermediate products in the conversion of 1-chlorophthalazines into the corresponding amines because the 1-phenoxy derivatives are smoothly converted into the amines in the presence of ammonium acetate at 200°C (V. Colotta *et al.*, Eur. J. med. Chem., 1994, 29, 95). Highly substituted phthalazine derivatives such as (40) are obtained by reaction of chlorophthalazine (39) with a hydrazide, $RCONHNH_2$ (F.A. Yassin *et al.*, Egypt. J. Chem., 1990, 33, 199).

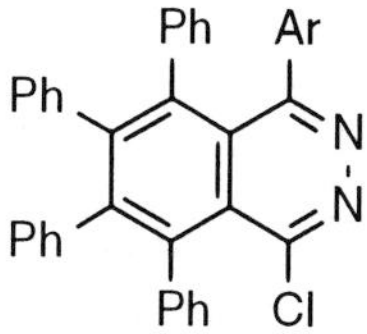

(39)

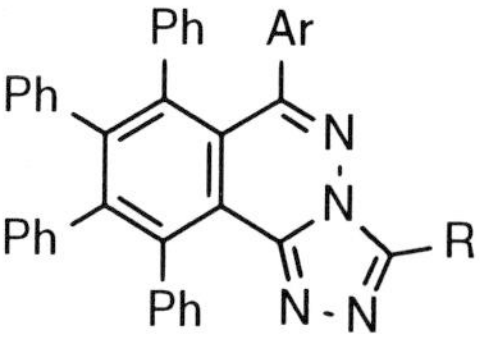

(40)

Interest in 1,4-disubstituted phthalazines has increased because of the development by Sharpless of the very useful process for the asymmetric dihydroxylation of alkenes using osmium tetroxide in the presence of 1,4-bis(cinchona alkaloid) derivatives of phthalazine as the chiral auxillary. These compounds, for example the dihydroquinine derivative (41) can be prepared on a multigram scale by reaction of a 1,4-dichlorophthalazine with the hydroxyl group of the alkaloid in the presence of potassium carbonate and potassium hydroxide in an organic solvent (K.B. Sharpless *et al.*, J. org. Chem., 1992, 57, 2768; W. Amberg *et al.*, J. org. Chem., 1993, 58, 844). These chiral catalysts are effective at concentrations of 1 mole % or less and the enantioselectivity in the dihydroxylation of olefins can be predicted using a mnemonic (H.C. Kolb, P.G. Andersson and K.B. Sharpless, J. Amer. chem. Soc., 1994, 116, 1278). 1,4-Bis(cinchona alkaloid) substitutents on the 6,7-diphenylphthalazine give an exceptionally high ee (H. Becker *et al.*, J. org. Chem., 1995, 60, 3940) and ligand-dependent reversal of facial selectivity in the hydroxylation process has been achieved (K.P.M. Vanhessche and B.K. Sharpless, J. org. Chem., 1996, 61, 7978). A silica gel supported derivative of (41) is a highly efficient catalyst for heterogeneous asymmetric dihydroxylation of olefins giving, in six examples, 88-96% yields with ee greater than 92% (C.E. Song, J.W. Yang and H.J. Ha, Tetrahedron Asymm., 1997, 8, 841).

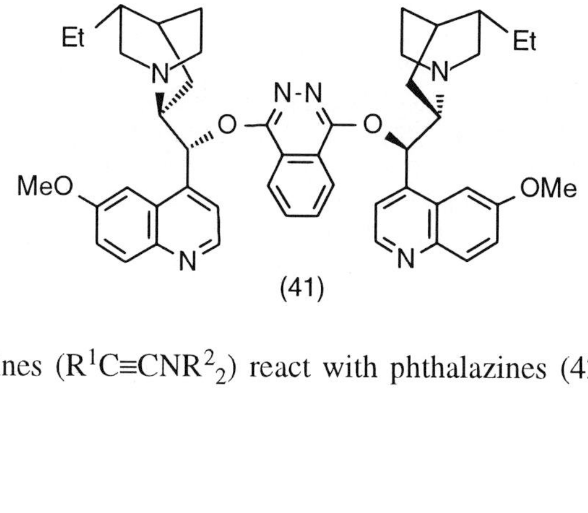

(41)

Ynamines ($R^1C{\equiv}CNR^2_2$) react with phthalazines (42) to give one or more of three products: a [4+2]cycloaddition followed by elimination of dinitrogen yields (43); [2+2]cycloaddition followed by a ring opening process affords (44); and addition of two molecules of the ynamine across the unsubstituted C-N bond and subsequent fission of the *N-N* bond provides the pentasubstituted pyridine (45). The presence of an electron-donating substituent (R^3) disfavours the reactions, electron-withdrawing substituents (e.g. COPh, CN, *p*-toluenesulfonyl) favour formation of (43) but the presence of a chlorine substituent (42, R^3 = Cl) promotes the production of (45) to afford (45) in moderate to good yields (K. Iwamoto *et al.*, Chem. pharm. Bull., 1994, 42, 413; *idem, ibid.*, 1995, 43, 679; K. Iwamoto *et al.*, Heterocycles, 1996, 43, 199).

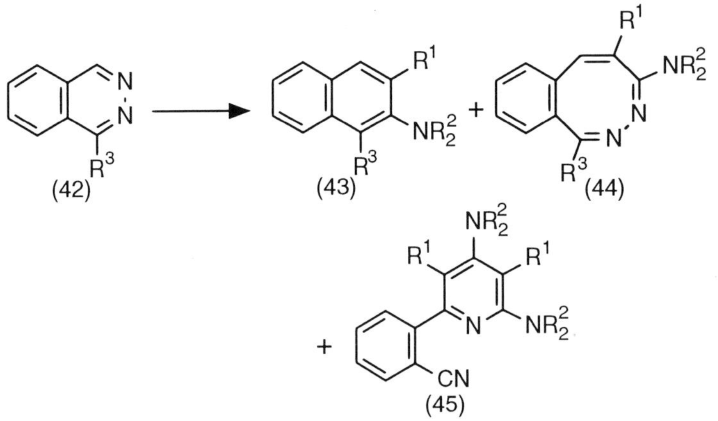

(42) (43) (44) (45)

A 1-hydrazino substituent on phthalazines is converted to a methoxy substituent on treatment with thallium trinitrate in methanol (M. Kocevar, P. Mihorko and S. Polanc, J. org. Chem., 1995, 60, 1466). 1-Hydrazinophthalazine is converted into a tetrasubstituted pyrazole in poor yield using a solid supported 1,3-diketone (A.L. Marzinzile and E.R. Felder, Tetrahedron Letters, 1996, 37, 1003).

The thermolysis at 270°C of cycloalkanone 1-phthalylhydrazones (48) has been shown to give a variety of products which depend on the ring size of the ketone (H. Zimmer *et al.*, J. org. Chem., 1995, 60, 1908). This work corrects results reported by other workers. When cyclopentanone or cycloheptanone derivatives are used the major product is the monocyclic (47) with the tricyclic triazole (46) as the minor component of a mixture. However, when n = 11, the major product is the tricyclic (49) together with a spiro derivative (50) as the minor component.

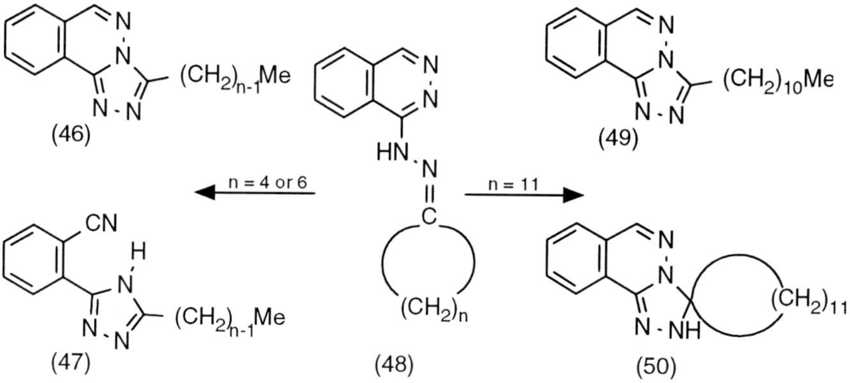

Polyphenylated phthalazines are converted to quinazolines in high yield on pyrolysis at 360°C in a sealed tube. Thus, 1,4,5,8-tetraphenylphthalazine affords 2,4,5,8-tetraphenylquinazoline in 70% yield. Cross-over experiments with two phthalazines having different 1,4-diaryl groups gave no exchanged products, so the reaction appears to be a unimolecular process and the structure for a reaction intermediate is suggested (K.P. Chan and A.S. Hay, J. org. Chem., 1995, 60, 3131). Pyrolysis of poly(aryl ether phthalazines) is used to form polymers (K.P. Chan, H. Yang and A.S. Hay, J. polym. Sci., Part A; Polym. Chem., 1996, 34, 1923).

e. Phthalazine ylides

Treatment of 2-aryl-1-hydroxy-1,2-dihydrophthalazines (51) with perchloric acid gave the cations (52) which were isolated as perchlorate salts. These, on being heated with sodium acetate gave the ylide intermediate (53) which reacted with (52) to yield the biphthalazinylidene (54) (R.N. Butler and C.S. Pyne, J. chem. Soc., Perkin 1, 1992, 3387).

Alkene and alkyne dipolarophiles undergo cycloaddition reactions with the phthalazinium methanide 1,3-dipoles (55, R = H or CN). Addition of *N*-methyl- and *N*-phenyl-maleimide gives the pyrrolo[2,1-*a*]-phthalazine derivatives (56, R^1 = Me or Ph, R^2 = H or CN) and acetylene dicarboxylates afford (57, R^1 = Me or Et, R^2 = H or CN). The endo and exo ratio of the products showed no specificity (R.N. Butler, D.M. Farrell and C.S. Pyne, J. chem. Research (S), 1996, 418).

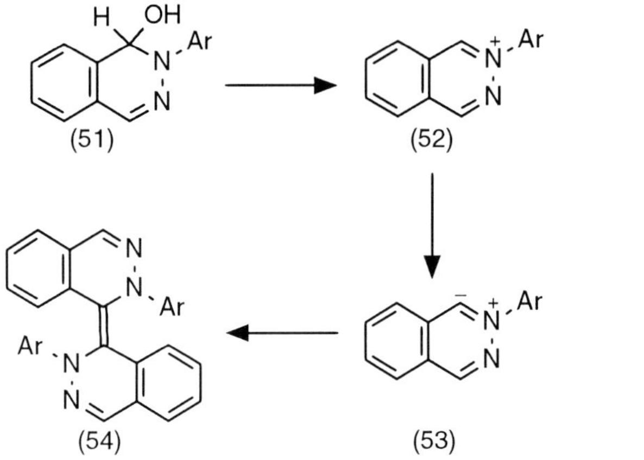

The unsubstituted methanide (55, R = H) is conveniently obtained from 2-(trimethylsilylmethyl)phthalazinium triflate by the action of cesium fluoride. The methanide adds to thioketones (R^1CSR^2) to give mixtures of substituted, partially-reduced thiazolo[4,3-*a*]phthalazine (58) and thiazolo[2,3-*a*]phthalazine (59) (R.N. Butler *et al.*, J. chem. Soc., Perkin 1, 1998, 869).

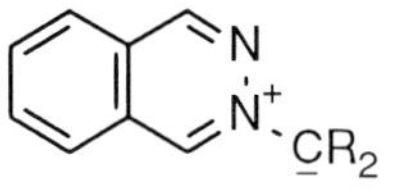

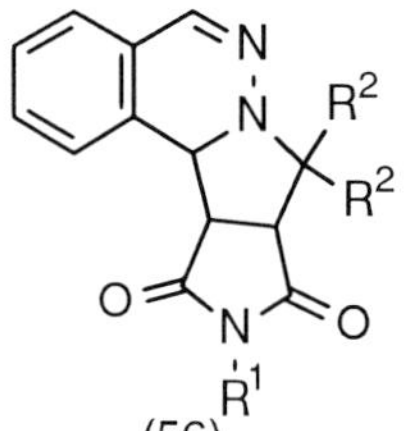

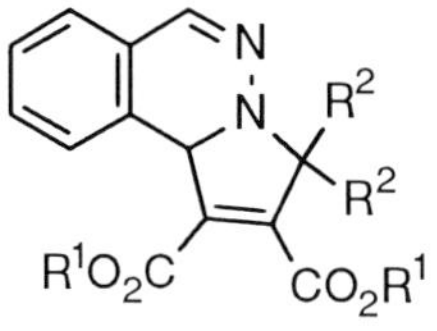

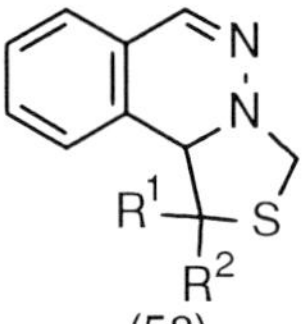

Ring transformations of the 2,3-dihydroimidazo[2,1-*a*]phthalazin-4-ium-6-olates (60, R^1 = H or NO_2) and the substituent dependent asymmetric induction has been investigated (A. Szabo *et al.*, Tetrahedron, 1997, 53, 7021). The action of acetic anhydride on (60, $R^1 = R^2$ = H)

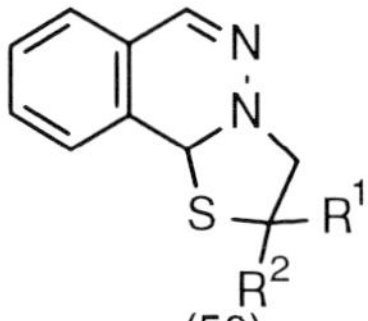

yielded the phthalazinone (61).

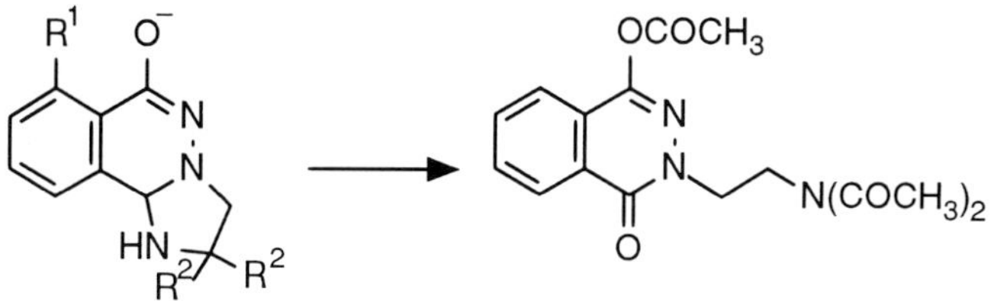

(60) (61)

The quaternary salt (62) undergoes a nucleophilic substitution when heated in *n*-butanol and the initially formed ether (63) rearranges to the 2-butyl-phthalazin-1(2*H*)-one (64). However, when (62) is heated in 1,2-dichloroethane, it rearranges to the pyridyl ether (66) and the intermediate ylide (65) can be trapped by reaction with acrylonitrile (S.A. El-Abbady and S.M. Agami, Indian J. Chem., 1995, 34B, 504).

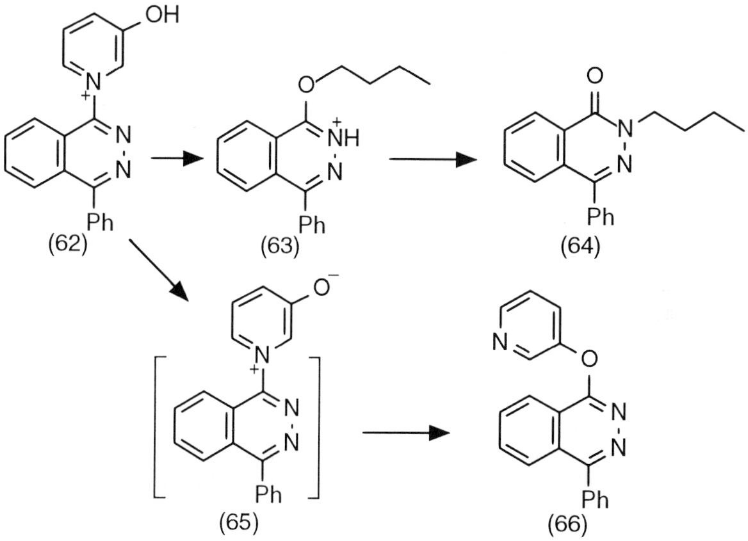

(62) (63) (64)

(65) (66)

*f. Phthalazin-1(2*H*)-ones*

2-Phenylation of phthalazin-1(2*H*)-one is achieved in good yield by the action of iodobenzene in DMF in the presence of copper(I) iodide and potassium carbonate (M. Sugahara and T. Ukita, Chem. pharm. Bull.,

1997, 45, 719). Oxidation of 4-benzyl-2-substituted-phthalazin-1(2*H*)-ones (67) with sodium dichromate in acetic acid converted the benzyl group to a benzoyl function (M.F. Ismail *et al.*, Egypt. J. Chem., 1994, 37, 89). The action of Lawesson's reagent followed by trimethylsilyl azide gave the tetrazalophthalazine (68) in 95% yield (S. Lehnhoff and I. Ugi, Heterocycles , 1995, 40, 801). Treatment of (69, R = H) with acetyl nitrate gave the nitrate (69, R = NO_2).

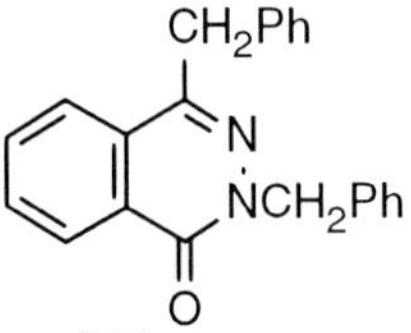

CH_2Ph N NCH_2Ph O (67) N - N N N N (68) Me CH_2OR N NR EtO_2C O (69)

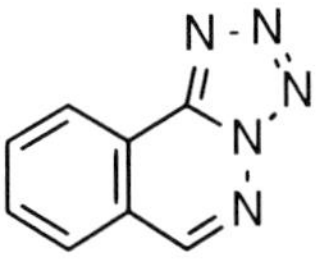

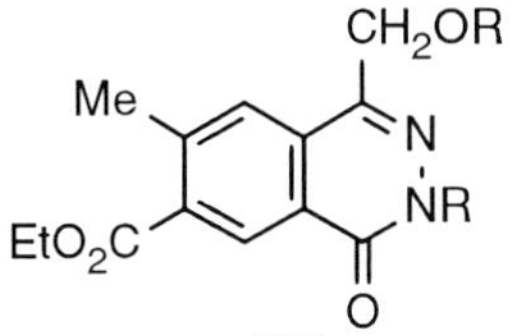

*g. Phthalazin-1,4(*2H,*3*H)-diones and phthalazine quinones

Reaction of 2-substituted phthalazin-1,4(2*H*,3*H*)-diones with lead(IV) dioxide in toluene gives (70) in 80% yield through formation and dimerisation of the free-radical (R. Kluge *et al.*, Chem. Ber., 1992, 125, 2075). Conditions have been established for the formation of either 1-chloro or 1,4-dichloro derivatives in high yield by the action of phosphoryl chloride on benzo[*g*]phthalazin-1,4(2*H*,3*H*)-dione (71) (E.H. Eltamany, J. Indian chem. Soc., 1997, 74, 21).

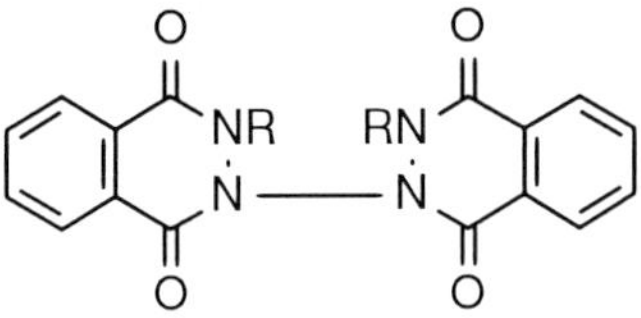

O O NR RN N N O O (70) O NH NH O (71)

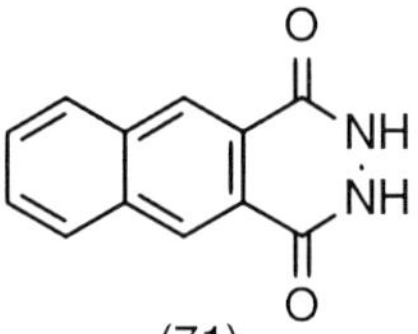

The generation and reactions of phthalazine quinone (72) have been reviewed (S. Radl, Adv. heterocyclic Chem.,1994, 61, 141). Oxidation of (71) to the corresponding quinone followed by [4+2]cyclo-addition with the appropriate methyl-1,3-butadiene gave the benzo[*g*]-phthalazino[1,2-*b*]pyridazin-6,13-diones (73, R^1 = H, R^2 = Me) (M.C. Cano *et al.*, Tetrahedron, 1993, 49, 243) and (73, R^1 = Me, R^2 = H) (*idem*, Canad. J. Chem., 1997, 75, 348). The steric course of reactions involving electrophilic addition of positive halogen to these tetracyclic phthalazin-dione derivatives was studied.

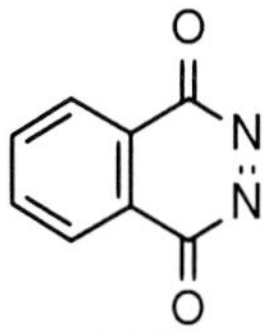

(72) 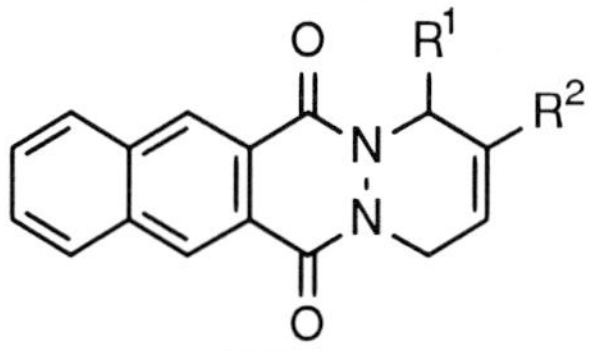 (73)

h. Complexes with metal ions

The oxygenation and relevance to quercitinase of the copper complex (74) is discussed (E.B. Hergovich, J. Kaizer and G. Specier, Inorg. chim. Acta, 1997, 256, 9). Dinuclear copper(II) and nickel(II) complexes bridged by an alkoxide (75, R = Me, Ph or CF_3) are investigated (C. Li *et al.*, Bull. chem. Soc., Japan, 1997, 70, 2429) and the synthesis and spectroscopic properties of copper and silver complexes of phthalazinophane macrocycles (76, R = $(CH_2)_3$ and $(CH_2CH_2)_2S$) are reported (L. Chen *et al.*, Inorg. Chem., 1993, 32, 4063).

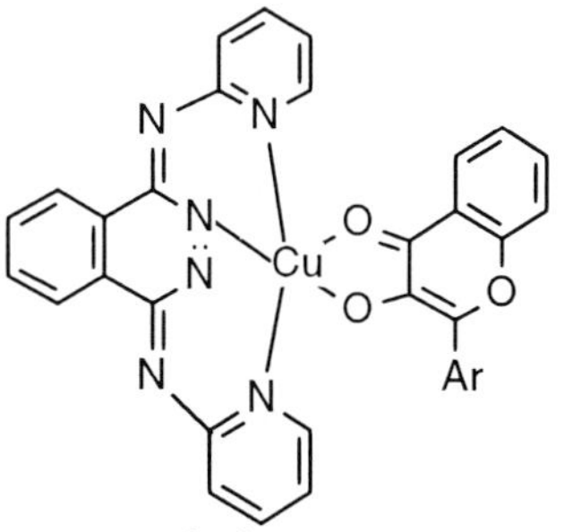

(74) 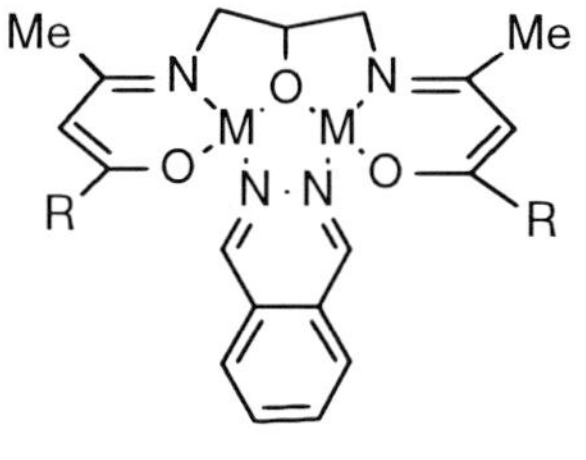 (75)

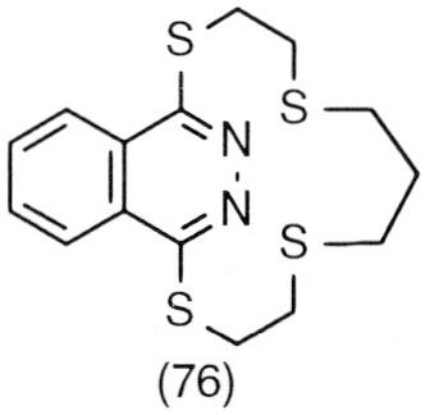

(76)

i. Analytical and environmental aspects

Luminol (77) is still a compound causing considerable interest and activity. The kinetics of its oxidation with ferricyanide in aqueous alkali (I.Y. Vieru, Y.N. Kozlov and A.N. Petrov, Kinetics Catal., 1996, 37, 754) and with horseradish peroxidase catalysed by phenylboronic acids (L.J.

Kricka and X.Y. Ji, Talanta, 1997, 44, 1073) or by heterocycles (F.G. Sanchez, A.N. Diaz and J.A.G. Garcia, J. Photochem. Photobiol. A-Chem., 1997, 105, 11) are reported. The effects of a medium containing a high salt concentration (A.B. Collaudin and L.J. Blu, Photochem. Photobiol., 1997, 65, 303) and of a mixture of water and an organic solvent as a medium are discussed (J.R. Trevithick and T. Dzialoszynski, Biochem. mol. Biol. Int., 1994, 33, 1179; D. Guha *et al.*, Indian J. Chem., 1997, 34A, 307).

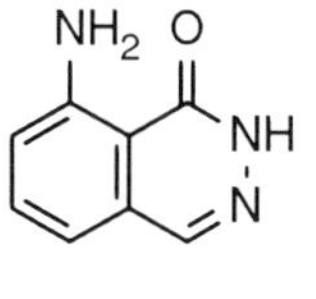

(77) 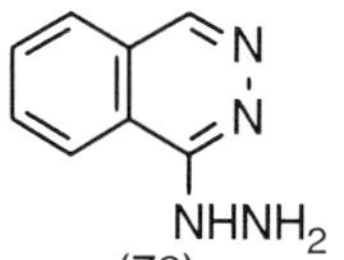 (78) 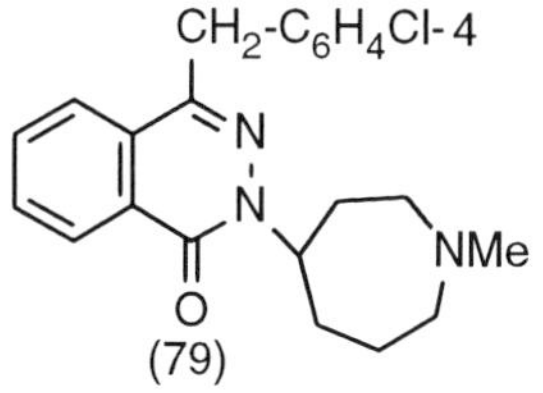 (79)

Applications of the chemiluminescence of luminol include its use to determine hydrogen peroxide concentration at the picomole level (T. Miyazawa *et al.*, J. Chromatrogr. A, 1994, 667, 99), the estimation of catalase activity at physiological hydrogen peroxide concentrations (S. Mueller, H.D. Riedel and W. Stremmel, Anal. Biochem., 1997, 245, 55), and 'reagentless' flow analysis of hydrogen peroxide by electrocatalysed luminol chemiluminescence (J.E. Atwater *et al.*, Anal. Letters, 1997, 30, 21).

The use of luminol in the determination of Cd(II) (G. Li, Diandu Yu Huanbao, 1993, 13, 25; Chem. Abstr., 1994, 121, 212463v) and Cr(VI) in aqueous solution are described (Z. Zhang, W. Qin and S. Liu, Anal. chim. Acta, 1995, 318, 71). The reagent is useful for the determination of cobalt impurity in zinc and cadmium salts (V.K. Zinchuk and M. Makhruka, Zavod. Lab., 1994, 60, 23; Chem. Abstr., 1994, 121, 194515g).

Methods using luminol for the estimation of organic compounds include the HPLC estimation of penicillins (K. Nakashima *et al.*, Biomed. Chromatogr., 1993, 7, 217), titration of thiols (G.I. Lopez, P. Vinas and J.A. Martinez Gil, Fresenius' J. anal. Chem., 1993, 345, 723; Chem. Abstr., 1993, 119, 62082n), enzymatic determination of ethanol (J.E. Atwater *et al.*, Analyt. Letters, 1997, 30, 1145)and the continuous flow assay of total organic carbon in water (R. Liu *et al.*, Fenxi Shiyanshi, 1995, 14, 12; Chem. Abstr., 1995, 123, 295925n).

j. Pharmaceuticals

Some aspects of the chemistry of 1-hydrazinophthalazine (78) have been mentioned already (Section d) but the compound (hydralazine) was introduced in the 1950s as a pre-capillary vasodilator and is still used extensively in experimental work. A crystal structure for hydrazaline hydrochloride is available (N. Okabe, H. Fukuda and T. Nakamura, Acta Crystallogr. C, 1993, 49, 844) and an HPLC assay established (A.M. DiPietra *et al.*, Farmaco, 1993, 48, 1555). The mutagenicity of hydralazine has been studied (B. Chlopkiewicz, A. Ejchart and J. Marczeska, Acta Pol. Pharm., 1995, 52, 219; T. Morita, H. Naitou and E. Oishi, Biol. pharm. Bull., 1995, 18, 363) and its effects on blood perfusion in normal and tumour tissue studied (C.A. Belfi *et al.,* Int. J. radiat. Oncol., Biol. Phys., 1994, 29, 473).

Azelastine (79) a clinically useful antihistamine and its mechanism of action has been investigated (K. Fanous and R.P. Garay, Naunyn-Schmiedeberg's Arch. Pharmacol., 1993, 348, 515; Y. Hamamoto *et al.,* Exp. Dermatol., 1993, 2, 231; H. Hazama *et al.,* Eur. J. Pharmacol., 1994, 259, 143). The drug is an effective scavenger of superoxide and hydroxyl radicals (Y. Ando *et al.,* Pharm. pharmacol. Letters, 1995, 5, 5).

Phthalazin-1(2*H*)-ones are of interest as inhibitors of aldose reductase (N.A. Calcutt *et al.*, Eur. J. Pharmacol., 1994, 251, 27) and as potential antiasthmatic agents (M. Yamaguchi *et al.,* J. med. Chem., 1993, 36, 4061). Some phthalazin-1,4-diones show antineoplastic activity in human and murine cells (I.H. Hall, E.S. Hall and O.T. Wong, Anticancer Drugs, 1992, 3, 55).

Second Supplements to the 2nd Edition of Rodd's Chemistry of Carbon Compounds, Vol.IV IJ, edited by M. Sainsbury

Chapter 43

PYRIMIDINES AND QUINAZOLINES

DEREK T. HURST

1. Introduction

The First Supplement to Vol. IV IJ of the second edition of "Rodd" which contained Chapter 43, on pyrimidines and quinazolines, was published in 1995. That chapter reviewed advances in the chemistry of these compounds from 1989 to 1994. This chapter covers the period from the latter part of 1994 to the early part of 1998. This period has seen the continued use of pyrimidine drugs such as AZT and ddC in the treatment of viral diseases and the development of new analogues of these compounds, for example 3TC (3'-thiocytosine). A variety of pyrimidine derivatives continues to be investigated for a range of biological and other uses.

Regular annual accounts of developments in heterocyclic chemistry, including reports of pyrimidines and quinazolines, have been given in the series "*Progress in Heterocyclic Chemistry*" (Pergamon Press) and a second edition of a major publication covering the topics, "*Comprehensive Heterocyclic Chemistry*" (Eds. A.R. Katritzky, C.W. Rees and E.F.V. Scriven, Pergamon, Elsevier Science Ltd., Oxford, 1996), has been published. A new edition of the encyclopaedic volume "*The Pyrimidines*" in the series on heterocyclic chemistry edited by Taylor/Weissberger (Wiley-Interscience), written by D.J. Brown, has been published (1994). Other reviews published in recent years include those on the chemistry of thiobarbituric acid (112 references) (V. K. Ahluwalia and R. Aggarwal, Proc. Indian Natl. Sci. Acad., Part A, 1996, 62, 369) and the chemistry of 6-substituted pyrimidines and uracils (M. Botta *et al.*, Trends Org. Chem., 1995, 5, 57).

This chapter provides a brief overview of some of the more interesting or important aspects of the chemistry of pyrimidines and quinazolines which have been published since late 1994 but it cannot be as comprehensive as the accounts mentioned above.

2. Pyrimidines

(a) Properties and structural aspects

Many of the recent papers describing syntheses of pyrimidines which are published now contain data of the X-ray crystal structures, however, these data are not reported here.

A study, using high-resolution ^{1}H NMR, semi-empirical PM3 calculations and molecular dynamics, of the conformations of 4-(2-furyl)-2-methylaminopyrimidines (1) has shown that the S-trans conformation is favoured in solution and that there is considerable sp^2 character for the C2-amino bond (J. L. Mokrosz *et al.*, J. Heterocycl. Chem., 1996, 33, 1207).

There have been many reports of different biological and pharmaceutical properties of both pyrimidines and quinazolines during the review period.

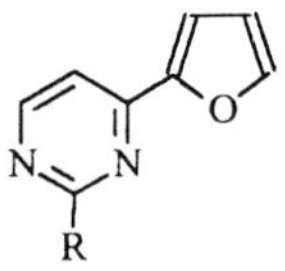

1 R= NHMe, NMe_2

(b) Synthesis

By far the most common synthesis of the pyrimidine ring system is the CCC plus NCN (dicarbonyl plus amidine) approach, the "Principle Synthesis". An example of this method is the reaction of 1,1,3,3-tetramethoxypropane with amidine hydrochloride salts at elevated temperature in a sealed tube to yield 2-alkyl (or aryl)-substituted pyrimidines in a high yielding, one-pot, process (T. Wang and I. S. Cloudsdale, Synth. Commun., 1997, 27, 2521).

N,N-Diethylaminomethylene-1,1,1,5,5,5-hexafluoroacetylacetone (2) reacts with urea derivatives in acetonitrile at room temperature to give pyrimidines 3a which can react further to yield 3b (M. Soufyane, C. Miraud and J. Levy, Tetrahedron Lett., 1993, 34, 7737).

$F_3CCOC(=CHNEt_2)COCF_3$

2

3a R= Me_2N, Et_2N, MeO; R^1= $COCF_3$, CO_2H, COMe

3b X= HO(F$_3$C)... ; F$_3$C...

The reaction of enamines (4) with benzamidine or with *S*-benzylthiourea gives 5 and 6 (E. S. Ratemi, N. Namdev and M. S. Gibson, J. Heterocycl. Chem., 1993, **30**, 1513).

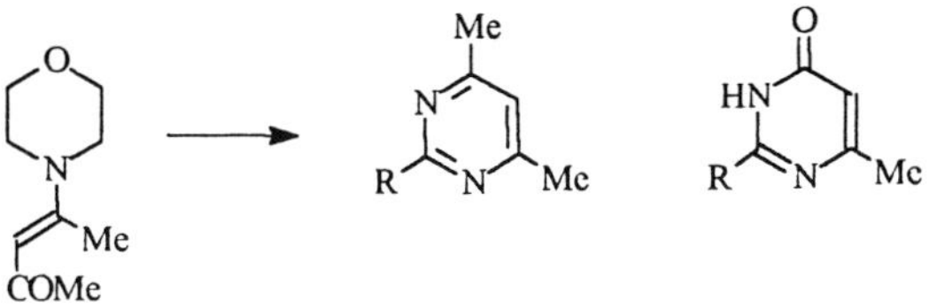

4　　**5** R= Ph, BnS　　**6**

However, there is much interest in other synthetic methods for pyrimidines, many of which have only been developed in recent years. For example an easy, efficient, way to make fluorine-containing pyrimidines is to react ß,ß-bis(trifluoroacetyl)vinylamine with aliphatic, or aromatic, aldehydes in aqueous ammonia under mild conditions to give the dihydropyrimidines (7) which, for R= H, can be aromatised by DDQ (E. Okada *et al.*, Heterocycles, 1997, **44**, 349). Fluorinated pyrimidines (8) are easily obtained by reacting perfluoro-2-methylpent-2-ene with aceto- and trifluoroacetoamidines. The fluorine at position 4 of the pyrimidine ring is very active in reactions with nucleophiles (L. S. German *et al.*, Russ. Chem. Bull., 1997, **46**, 1920; Chem. Abstr., 1998, **128**, 167395).

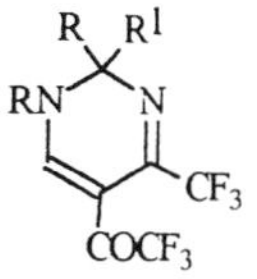

7 R= H, alkyl; R^1= H 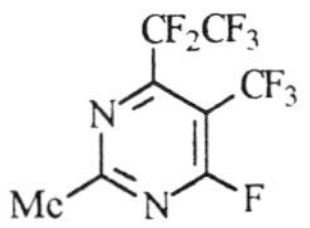　　**8**

Solid-support methodology has also been developed for the synthesis of pyrimidines. For example, polymer-bound thiouronium chlorides react with the alkynyl ester 9 to form, after ester cleavage, polymer-bound pyrimidinecarboxylates which are cleaved by MCPBA/pyrrolidine to give 10 (D. Obrecht *et al.*, Helv. Chim. Acta, 1997, **80**, 65).

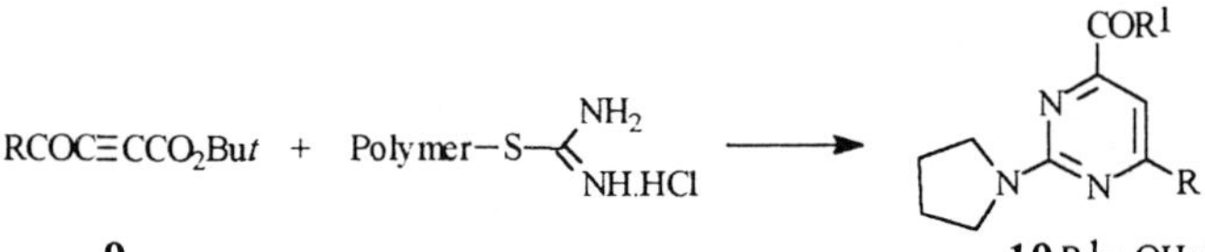

$RCOC{\equiv}CCO_2Bu^t$ + Polymer–S–C(=NH.HCl)NH_2 → **10**

9　　**10** R^1= OH; R= Ar

Pyrimidines (11) can be made by condensing aroyl or cinnamoyl isothiocyanates with enaminones, anilines and cyano(thioacetamides) (M. G. Assy, A. A. Hataba and H. Y. Moustafa, Pol. J. Chem., 1995, **69**, 1018; M. G. Assy and H. Y. Moustafa, Phos. Sulf. Silicon Relat. Elem., 1995, **105**, 213; A. A. Hataba, M. G. Assy and R. M. Fikry, Indian J. Chem., Sect. B: Org. Chem. Incl. Med. Chem., 1996, **35**, 144).

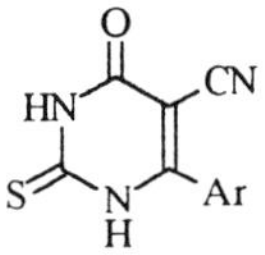

11

5-Arylidene-5,6-dihydrouracil derivatives (13) can be synthesised from methyl 3-aryl-2-isopropylaminomethyl-2-propenoates (12) and phenyl iso(thio)cyanate (F. El Guemmont and A. Foucard, Synth. Commun., 1993, **23**, 2065), whilst aminoenylesters (prepared from methyl propiolate and primary amines) react with primary amines and acetaldehyde to give pyrimidines (14) (T. Koike *et al.*, Chem. Pharm. Bull., 1997, **45**, 1117).

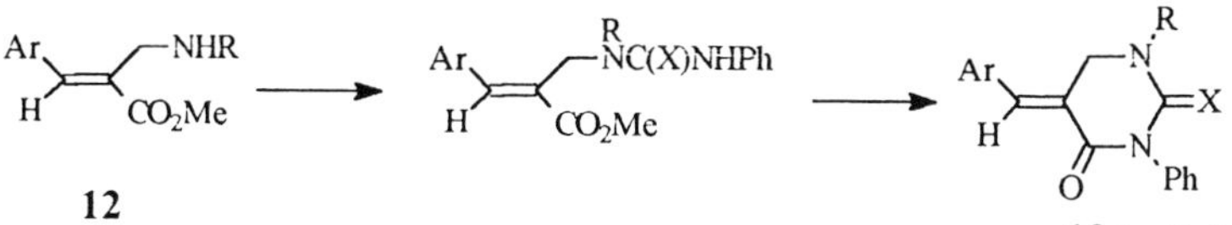

12 → **13** X= O, S

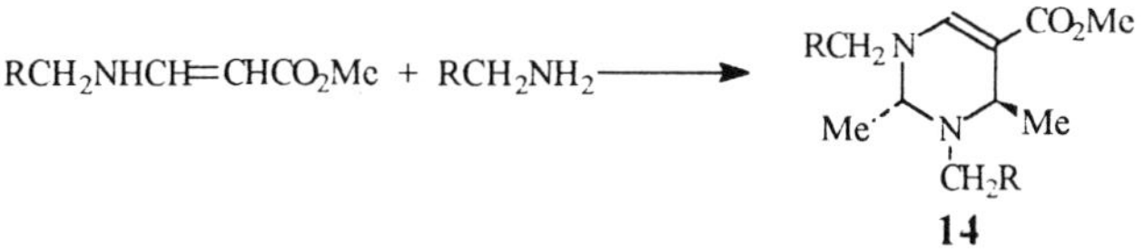

$RCH_2NHCH{=}CHCO_2Me + RCH_2NH_2 \longrightarrow$ **14**

Condensation adducts of N-nucleophiles and ketene dithioacetals or alkoxymethylene reagents undergo regioselective cyclisation to give pyrimidines 15 and 16, a process thought to be governed by stereo-electronic factors and/or the geometry about the C=C bond (Scheme 1) (A. Lorente *et al.*, Heterocycles, 1995, **41**, 71).

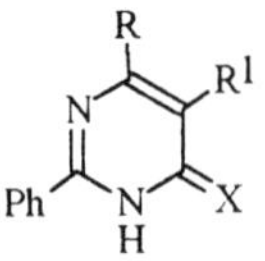

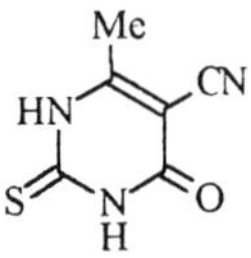

15 R= Ph, MeO, H, MeS
R^1= CO_2Me, CN; X= O, S

16

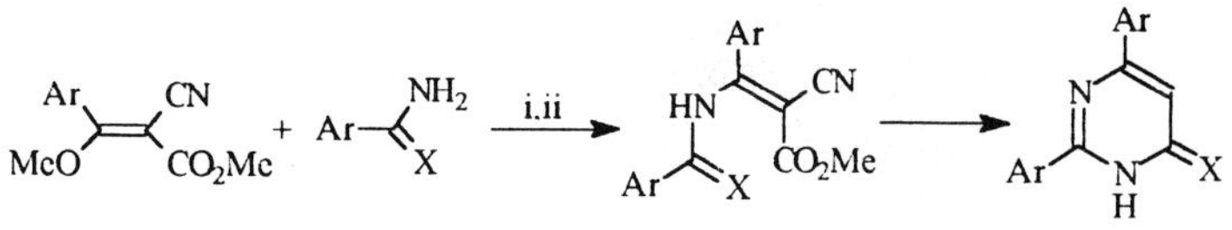

Reagents: i; NaH, DMF ii; H_3O^+

Scheme 1

A phase-transfer catalysed, one-pot, condensation of the dinitrile 17 with the oxalyl ester chloride 18 gives esters which undergo cyclisation with amines to yield pyrimidinones (19) (H. Abdel-Ghany, A. M. El-Sayed and A. K. El-Shafei, Synth. Commun., 1995, 25, 1119).

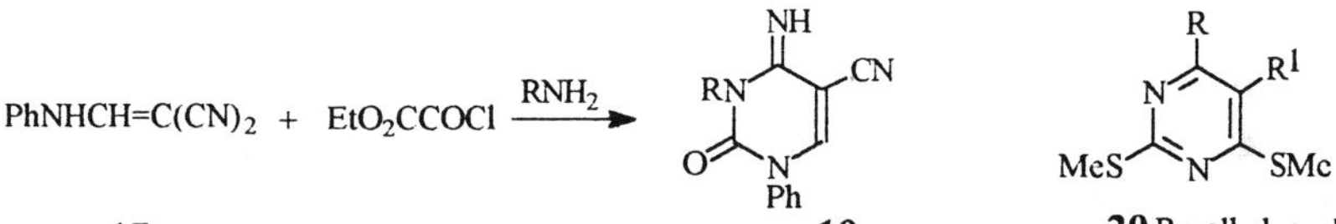

$PhNHCH{=}C(CN)_2 + EtO_2CCOCl \xrightarrow{RNH_2}$

17 **18** **19** **20** R= alkyl, cycloalkyl

The 2,4-bismethylthiopyrimidines 20 are formed from ketones, methyl thiocyanate and triflic anhydride, whilst *N*-tosyl-2- and -3-acetylpyrroles (21), or *N*-tosyl-2-pyrrolidine condenses with cyano compounds in the presence of triflic anhydride to give heteroaryl-pyrimidines (22) and (23) (Scheme 2) (A. G. Martinez *et al.*, Synlett, 1994, 559; Tetrahedron, 1996, 52, 7973).

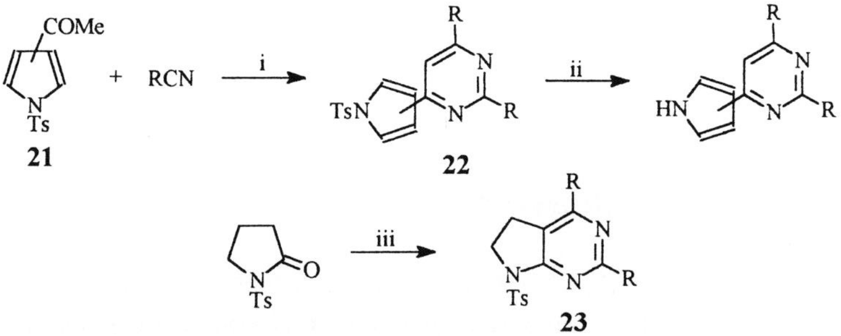

Reagents: i; Tf_2O, CH_2Cl_2 , r.t. 24 h. ii; NaOH, MeOH, reflux 3h. iii; RCN, Tf_2O, CH_2Cl_2 , reflux 8 h.

Scheme 2

1-Aryl-2-phenyl-4-thioalkyl-4-(secondary amino)-1,3-diaza-1,3-buta-dienes (24) react with chloroketenes to give pyrimidines through episulfonium intermediates, whilst 1-aryl-2-phenyl(methylthio)-4-dimethylamino-1,3-diaza-1,3-butadienes give pyrimidinones with haloketenes (S. N. Mazumdar *et al.*, Tetrahedron, 1994, **50**, 7579).

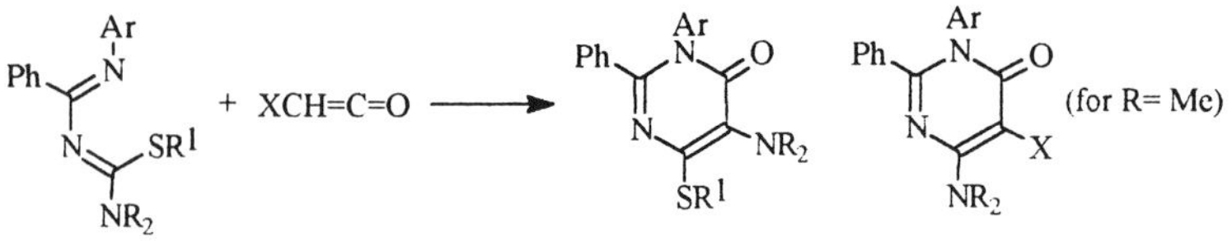

Ar N Ph N SR1 NR2 + XCH=C=O → Ph N Ar O N NR2 SR1 ; Ph N Ar O N X NR2 (for R= Me)

24

A library of 2-thioxo-1,4-dihydropyrimidines has been prepared using the combinatorial approach by reductive alkylation of ß-amino acids with aldehydes, followed by the addition of an isothiocyanate and cyclisation (M. M. Sun, C. L. Lee and A. Ganesan, J. Org. Chem., 1997, **62**, 9358).
Both pyrimidines and quinazolines can be obtained by reacting 2-trimethylsilyloxy- and 2-trimethylsilylthio-1,3-diazabutadienes (25) with enamines derived from aliphatic aldehydes (Scheme 3).
[4+2] Cycloadditions of 4-(*N*-allyl-*N*-aryl)amino-1,3-diaza-1,3-butadienes with vinyl-, isopropenyl- and chloroketenes yield pyrimidinones 26 and 27. The products 27, on refluxing in xylene, give the pyrimido-azepines 28 (J. Barluenga *et al.*, Heterocycles, 1994, **37**, 1109).

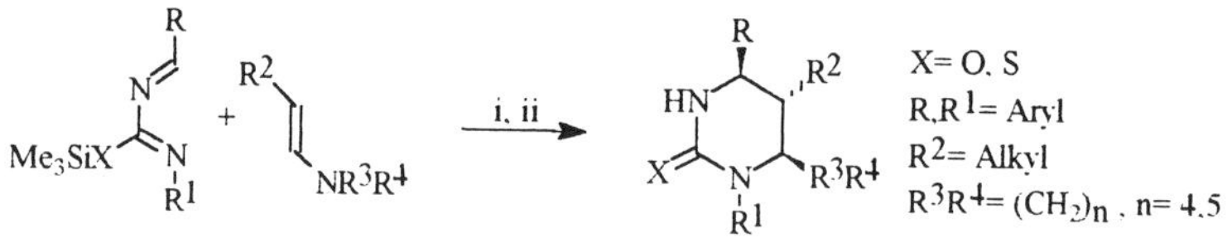

R N Me_3SiX N R1 + R2 NR^3R^4 —i, ii→ R HN R2 X N R1 R^3R^4

25

X= O, S
R,R1= Aryl
R2= Alkyl
R^3R^4= $(CH_2)_n$, n= 4,5

Conditions: i; CH_2Cl_2 , 25 °C, 4 h. ii; H_2O

Scheme 3

The pyrimidinones 27, with DMAD in refluxing toluene, undergo [4+2] cycloaddition, with elimination of *N*-allylaryl-amine functionality to yield quinazolinones 29 (A. K. Sharma and M. P. Mahajan, Tetrahedron Lett., 1997, **53**, 13841).

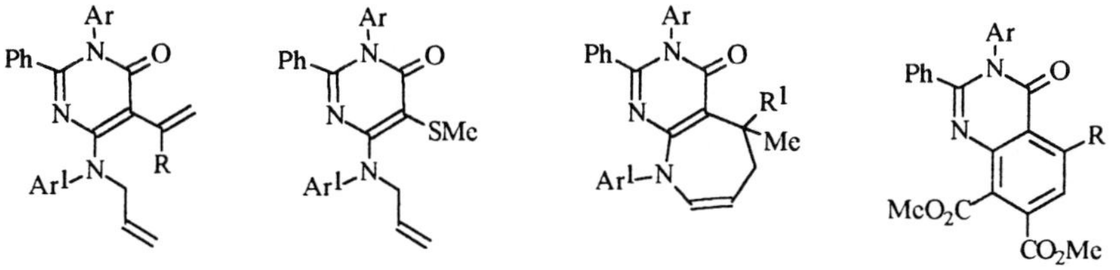

26 R= H, Me **27** **28** R^1= Me, MeO **29** R= H, Me

A simple method of pyrimidine synthesis is to treat phenyltrichloromethane with nitriles in the presence of aluminium chloride to yield 4-chloro-2-phenyl derivatives (30) (I. Somech and Y. Shvo, J. Heterocycl. Chem., 1997, 34, 1639).

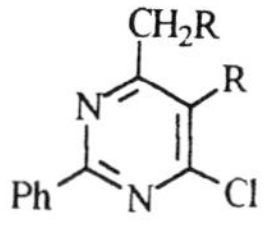

30

Pyrimidines may also be obtained from the reaction of alkyl and benzyl ketones with cyanamide, or *N,N*-dimethylcyanamide, in the presence of phosphorus oxychloride. Chloroformamidine derivatives are formed initially which, in the presence of titanium(IV) chloride, yield 2,4-diaminopyrimidines or 1-aminoisoquinolines (Scheme 4) (W. Zielinski and M. Mazic, Pol. J. Chem., 1993, 67, 1099; Heterocycles, 1994, 38, 375). The use of pyridyl derivatives such as 31 results in the synthesis of pyridylpyrimidines (32) (M. Mazic and W. Zielinski, Monatsh., 1996, 127, 1161).
Pyridylpyrimidines are also obtained by reacting pyridylacetic acid derivatives with 1,3,5-triazine (H. Möhrle and M. Pycior, Arch. Pharm., 1994, 327, 533).

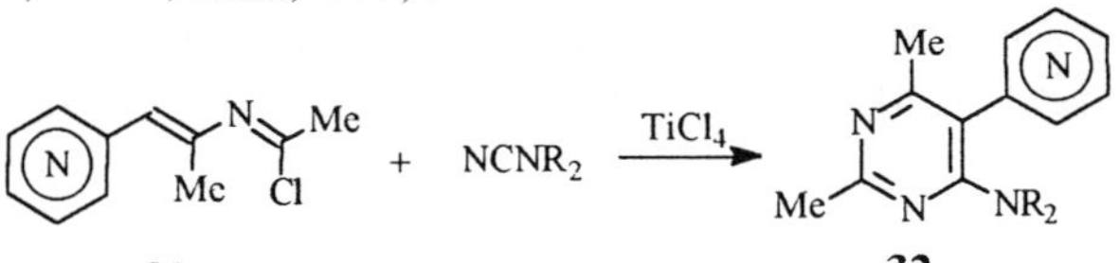

31 + $NCNR_2$ $\xrightarrow{TiCl_4}$ **32**

R= H, Me; R^1= H, alkyl, aryl

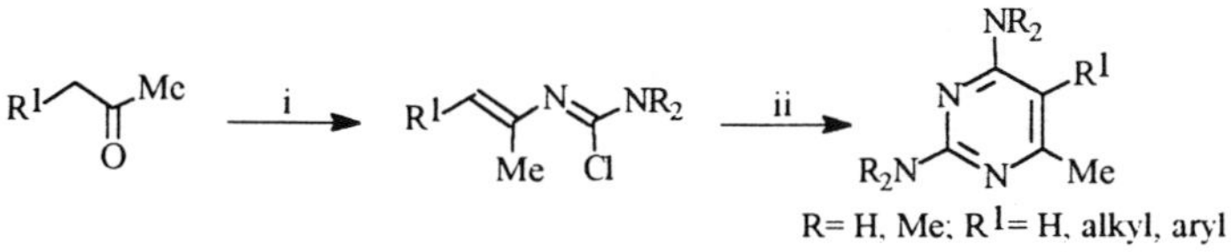

Reagents: i; $POCl_3$, $NCNR_2$ ii; $TiCl_4$, $NCNR_2$

Scheme 4

A one-step synthesis of 2,4-dialkyl (or aryl)-5,6-dihydrocyclobuta[*d*]-pyrimidines, which are precursors for pyrimidine orthoquinodimethanes, is *via* the reaction of cyclobutanone with nitriles. The orthoquinodimethanes can be trapped, for example, by *N*-phenylmaleimide (Scheme 5) (A. Herrera *et al.*, Tetrahedron Lett., 1997, **38**, 4873).

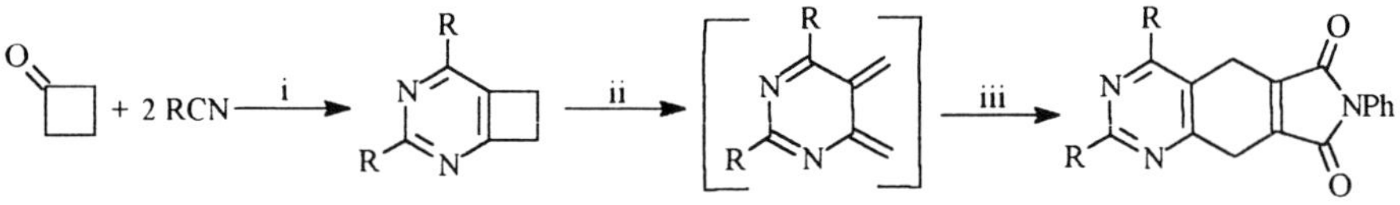

Reagents: i; Tf_2O, CH_2Cl_2, 42 °C, 6 h. ii; 1,2-$Cl_2C_6H_4$, 180 °C iii; N-phenylmaleimide

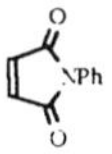

Scheme 5

Several approaches to pyrimidine synthesis involve ring transformations. For example, 2,5-diaryl-4-(chloromethylthio)imidazolines, obtained from *N*-acyl-α-aminoketones, rearrange when treated with sodium hydride in DMF. Spontaneous air oxidation leads to good yields of arylpyrimidines (Scheme 6) (M. Seki *et al.*, J. Org. Chem., 1993, **58**, 6354).

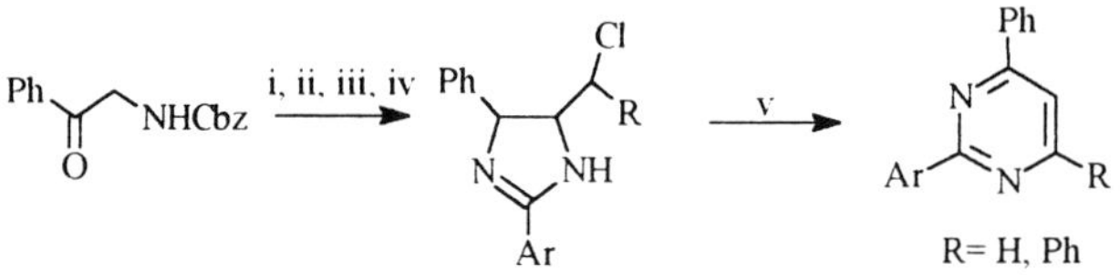

R= H, Ph

Reagents: i; RCHO, $NaN(TMS)_2$ ii; NH_2OH iii; H_2/Pd/C iii; ArC(OEt)NH.HCl iv; $SOCl_2$ v; NaH, DMF

Scheme 6

Ring expansion of 3-amino-1-ethoxycarbonylmaleimides yields *N*-1-substituted orotic acids (33) (M. Shopova *et al.*, Pharmazie, 1993, **48**, 861).

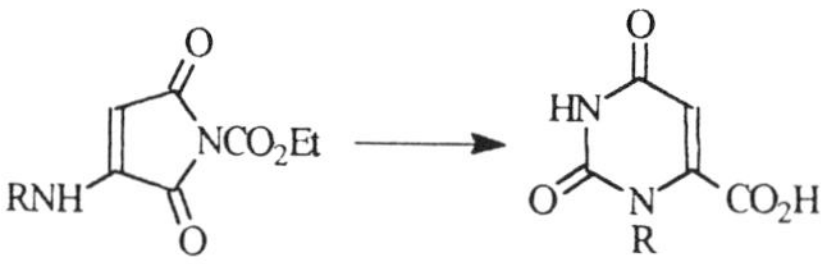

33

Ring transformations of oxazoles to give pyrimidines have also been described. For example, heating 5-benzyloxyacetyl-4-methyloxazole with methanolic ammonia yields 4-benzyloxymethyl-5-hydroxy-6-

methylpyrimidine (34) (S. Ray, S. K. Pal and C. K. Saha, Indian J. Chem., Sect. B: Org. Chem. Incl. Med. Chem., 1995, **34**, 112).

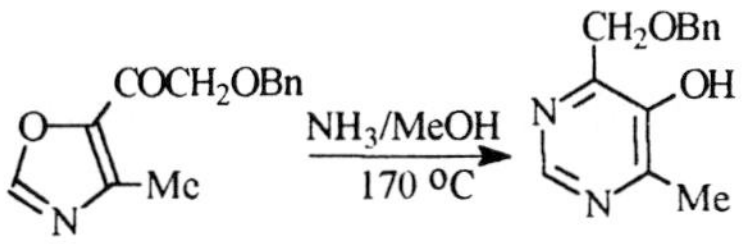

NH_3/MeOH, 170 °C

34

4-Formyl-2*H*-1,2-thiazine-1,1-dioxide (35) acts as a masked 1,3-dicarbonyl compound and undergoes ring transformation with appropriate amidines to give 5-(2-sulfamoylvinyl)pyrimidines (36) (E. Fanghaenel *et al.*, Liebig's Ann., 1997, 2617).

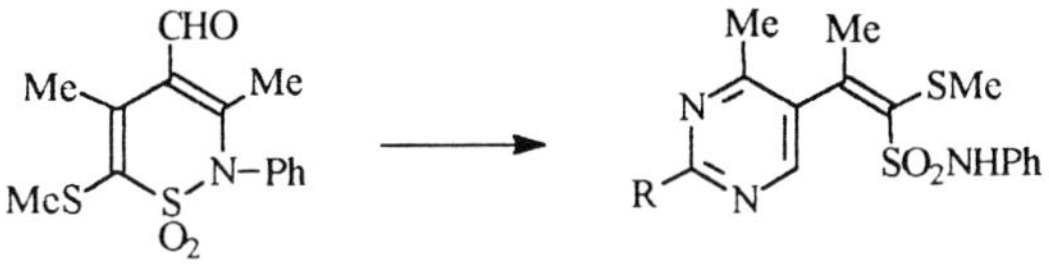

35 **36**

The unusual pyrimidines 37 can be obtained from *N,N,N'*-tris-(trimethylsilyl)amidines and vinamidinium salts (Scheme 7) (S. Brandl, R. Gompper and K. Polborn, J. Prakt. Chem., 1996, **338**, 451).

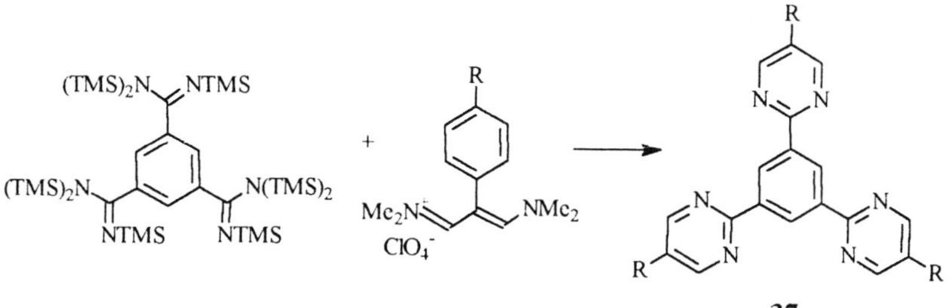

37

Scheme 7

(c) Reactions of pyrimidines

(i) Substitution

A simple synthesis of 5-bromo-2-iodopyrimidine, a very useful reagent for use in a variety of reactions such as Pd-catalysed cross couplings and the synthesis of many 5-substituted pyrimidines using alkylboronic

acids and alkylzinc reagents, is the bromination of pyrimidin-2(1*H*)-one followed by chlorination using phosphorus oxychloride then treatment with hydrogen iodide (Scheme 8) (J. W. Goodby *et al.*, Chem. Commun., 1996, 2719).
In these coupling reactions it is always the 2-iodo substituent which is displaced and many of the products of these coupling reactions have been developed as potential liquid crystal compounds.

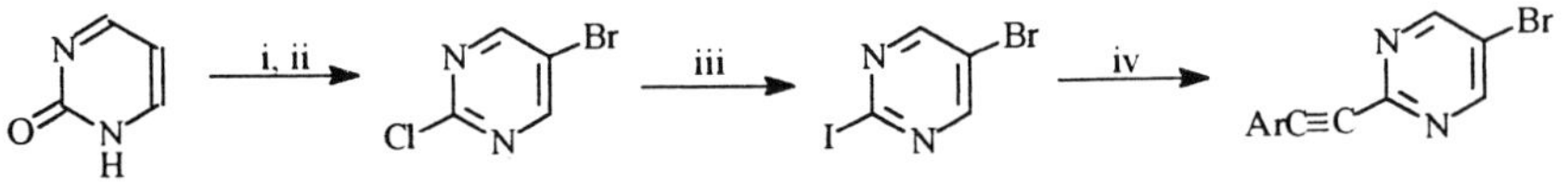

Reagents: i; Br_2/H_2O ii; $POCl_3$ iii; HI/H_2O iv; ArC≡CH, $Pd(PPh_3)_4$, CuI, iPr_2NH (also used $ArB(OH)_2$)

Scheme 8

Selective bromination at the 5-position of *N*-substituted uracils can be carried out using $CHBr_3$/oxygen (C. Moltke-Leth and K. A. Jorgensen, Acta Chem. Scand., 1993, <u>47</u>, 1117).
4,6-Dihydroxypyrimidine can be substituted at the 5-position by the benzylthio group using benzyl sulfenyl chloride. Chlorination using phosphorus oxychloride yields the 4,6-dichloro product (S. H. Thang *et al.*, Synth. Commun., 1993, <u>23</u>, 2363) (Scheme 9).

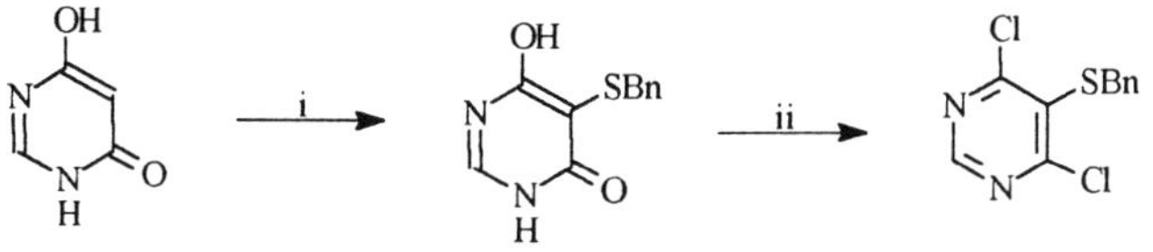

Reagents: i; SO_2Cl_2, $(BnS)_2$, DMF/CCl_4 ii; $POCl_3$

Scheme 9

Nucleophilic substitution of 6-chloropyrimidines (38) with piperidine, morpholine and other secondary amines gives the corresponding 6-amino-substituted pyrimidines. Pyrimidines (38; R= Me, R^1= CHO) react with amines to give barbiturates 39 which are also obtained by the Knoevenagel reaction of 1,3-dimethylbarbituric acid with 2-formylaminohetarenes. Further developments of such reactions yield products such as 40 (A. Schmidt, A Hetzheim and D, Albrecht, Heterocycles, 1996, <u>43</u>, 2153).

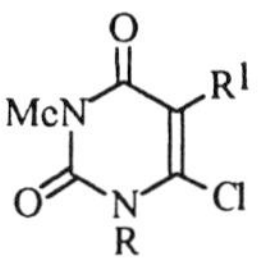

38 R= H, Me; R^1= H 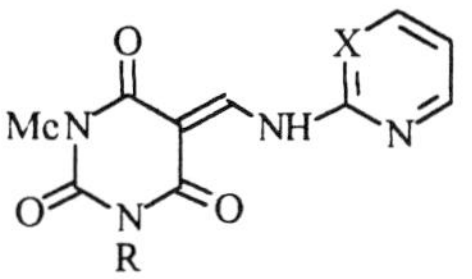 **39** R= Me; X= CH, CMe, N 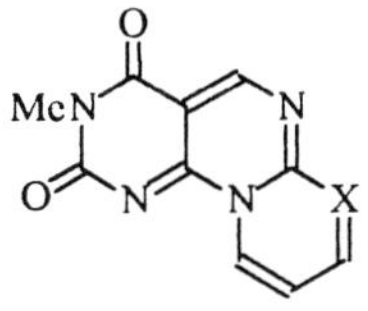 **40** X= N, CR

Reacting 4,6-dichloro-2-methylthiopyrimidine-5-carbonitrile with various amines yields the 4-substituted products, further reactions resulting in the formation of ring-fused compounds, whilst reaction with ethyl mercaptoacetate gives 4,6-bis(ethoxycarbonylmethylthio)-2-methylthiopyrimidine-5-carbonitrile which cyclises to the thieno[2,3-*d*]pyrimidine when treated with sodium ethoxide (S. Timkevicius, Chemija, 1997, 58; Chem. Abstr., 1997, 127, 108894). 4,6-Dichloro-pyrimidine-5-carboxaldehydes can also be used to make thieno[2,3-*d*] pyrimidines (J. Clark *et al.*, J. Heterocyclic Chem., 1993, 30, 1065) (Scheme 10).

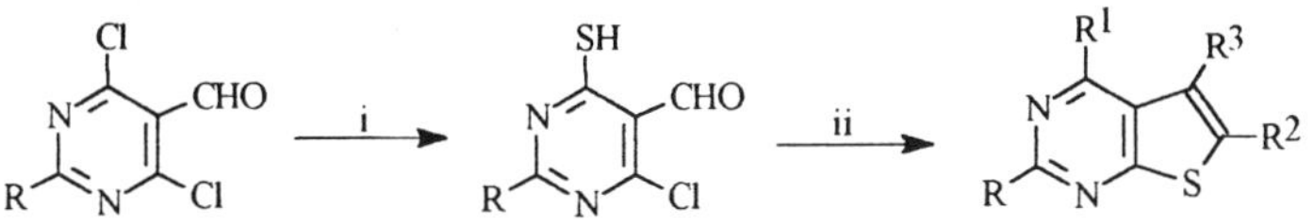

Reagents: i; NaSH or NH_2CSNH_2 ii; $HSCHR^2CO_2Me/Et_3N$

Scheme 10

An unusual and efficient direct nucleophilic C-4 hydroxy substitution on 2-methoxy- and 2-methylthio-4(3*H*)-pyrimidinones carrying a diethylamino moiety on the C-6 position is effected by sodium alkoxides. An unprecedented tandem C-6 side chain Hofmann-like elimination/C-4 pyrimidinone substitution has also been observed when sodium hydride in butanol/dioxane is used which provides a good method for the synthesis of new 6-vinylcytosine derivatives. The reaction probably operates *via* an intermolecular concerted process (Scheme 11) (X. Xia *et al.*, Tetrahedron Lett., 1997, 38, 1111; M. Botta *et al.*, Tetrahedron Lett., 1997, 38, 8249).

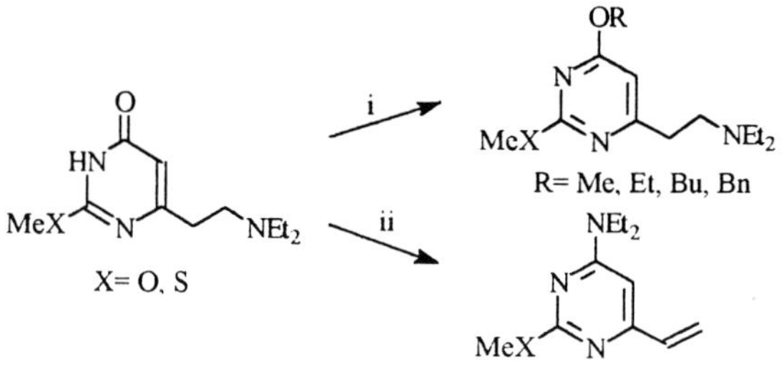

Reagents: i; RONa, ROH, 70 °C ii; NaH, BuOH, dioxane, 70 °C

Scheme 11

Reinvestigation of the chlorination, using phosphorus oxychloride, of *N*-phenylbarbituric acid has shown that 6-chloro-1-phenyluracil can be isolated as well as 6-chloro-3-phenyluracil. Similar results are obtained with *N*-benzylbarbituric acid (H. Goldner, G. Dietz and E. Carstens, Liebig's Ann. Chem., 1966, **691**, 142; T. Nagamatsu, H. Yamasaki and F. Yoneda, Heterocycles, 1994, **37**, 1147) (Scheme 12).

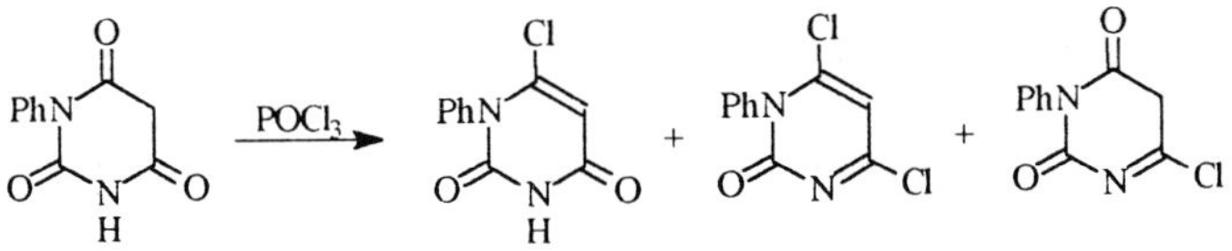

Scheme 12

2,4,6-Trichloropyrimidine reacts with pyridines and with other heterocycles to give tris(hetarenium) salts, *e.g.* 41 and 42 (A. Schmidt and A. Hetzheim, Tetrahedron, 1997, **53**, 1295).

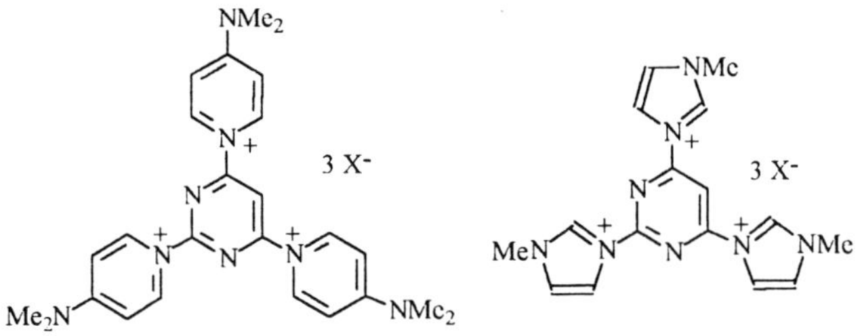

41 **42**

Barbituric acid reacts with 4,6-dihydroxypyrimidine-5-carboxaldehyde in water at 20 °C to 60 °C to yield 43. In acetic acid at 120 °C (44) is obtained. The reaction of 4,6-dihydroxypyrimidine with 4,6-dihydroxy-pyrimidine-5-carboxaldehyde gives the pyranopyrimidinecarboxamides (45) (A. V. Moskvin, N. M. Petrova and B. A. Ivin, Zh. Obshch. Khim., 1996, **66**, 1748; Chem. Abstr., 1997, **126**, 171544).

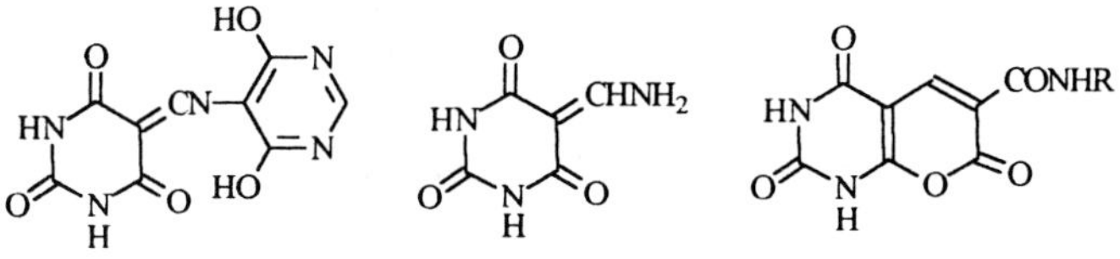

43 **44** **45** R= H, CHO

The electrophilic condensation of pyrimidines with cyclic ketones to give 5-vinyl-substituted pyrimidines is controlled by the degree of ionisation of the pyrimidine in acidic media and the electron density at the 5-C position (O. Lee and H. Wang, Tetrahedron Lett., 1997, **38**, 6401).
The reaction of 2-aminopyrimidine with a variety of heteroaromatic aldehydes to yield imines has been carried out. The products are readily reduced to give the 2-(substituted)aminopyrimidines (G. Musumarra and G. Sergi, Heterocycles, 1994, **37**, 1033).
Reacting (uracil-6-ylamino)phosphorane (46), isocyanates and pyridines yields pyrido[1',2':3,4]pyrimido[4,5-*d*]pyrimidines (47). Similar reactions have been used to obtain pyrimido[4',5':4,5]pyrimido[6,1-*a*]-isoquinoline (48, X= CH) and -phthalazine (48, X= N) (H. Wamhoff and A. Schmidt, J. Org. Chem., 1993, **58**, 6976).

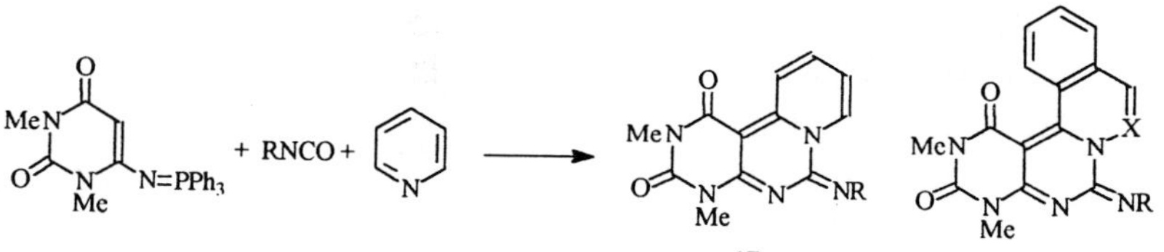

46 + RNCO + pyridine → **47** **48**

5-Iodo-2,4-dimethoxypyrimidine undergoes a heterogenous Heck reaction with methyl acrylate in aqueous tetra-*n*-butylammonium hydrogen sulfate and potassium carbonate at room temperature to afford (E)-5-(2-methoxycarbonylvinyl)-2,4-dimethoxypyrimidine (49) in high yield. A similar reaction using the 5-bromo derivative and thiophene under reflux gives 5-(2-thienyl)-2,4-dimethoxypyrimidine (I. Basnak, S. Takatori and R. T. Walker, Tetrahedron Lett., 1997, **38**, 4869).

Thymine is alkylated by 4-bromobutyl acetate at 100 °C to give the N-1 substituted compound 50 (55% after 48 hours) as the main product with small amounts of the bis N-substituted and O-alkylated products also being formed (P. Grandjean *et al.*, Tetrahedron Lett., 1997, **38**, 6185). 2,4-Bis(trimethylsilyloxy)pyrimidines react with aldehydes to give 1-silyloxymethyluracils (51) (T. Lukevics and M. Trusule, Khim. Geterotsikl. Soedin., 1996, 1653; Chem. Abstr., 1997, **126**, 186049). Palladium(0) catalysed allylation of 2-hydroxypyrimidines with cyclopentadiene monoepoxide yields the nucleoside analogue 52 (N. Arnau *et al.*, J. Heterocycl. Chem., 1997, **34**, 233).

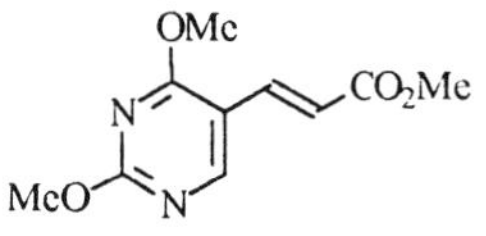

49 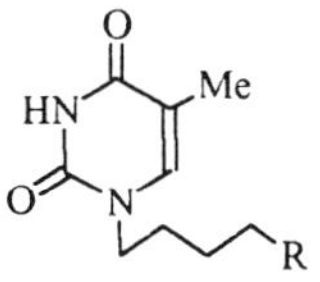 **50** R= H, OAc 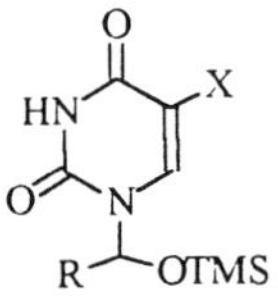 **51** X= H, F 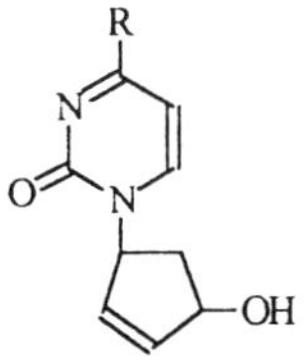 **52** R= H, Me

(ii) Metallation reactions

Extensive studies of metallation reactions of pyrimidines have been carried out during the period of this review and the topic of coupling reactions of π-deficient heterocycles has been reviewed (K. Undheim and T. Benneche, Adv. Heterocyclic Chem., 1995, **62**, 305). For example, pyrimidine-5-carboxylic acid and its 2-methyl-, 4-methyl-, 2,4-dimethyl- and 4,6-dimethyl- derivatives have been obtained by lithiation of the 5-bromo derivatives with subsequent carboxylation, or by using alternative methods. These pyrimidine-5-carboxylic acids were studied for their covalent hydration characteristics using ^{1}H NMR in dilute acid solution. The parent pyrimidine-5-carboxylic acid gives an equilibrium mixture of hydrates at both 2- and 4- ring positions. The 2-methyl derivative hydrates exclusively at 4- whilst the 4-methyl derivative hydrates exclusively at the 2-position. The dimethyl derivatives do not hydrate (T. J. Kress, Heterocycles, 1994, **38**, 1375) (Scheme 13).

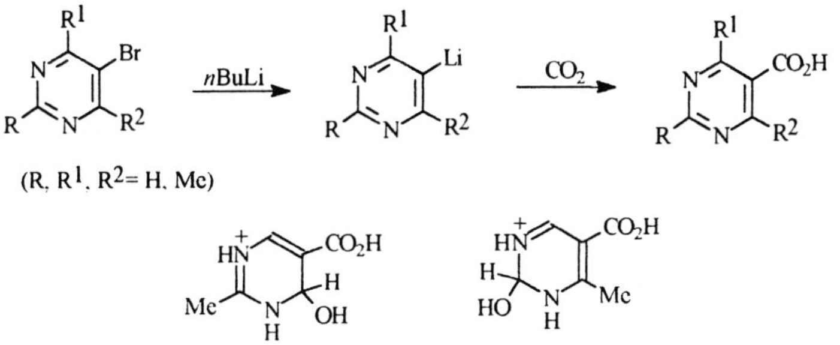

(R, R^1, R^2= H, Me)

Scheme 13

Both 2- and 4-pyrimidinyl triflates readily couple with stannanes in the presence of lithium chloride to give alkylated pyrimidines. The triflates show reactivity which is comparable with the corresponding chlorides and can show stepwise coupling to give disubstituted products (Scheme 14) (J. Sandosham and K. Undheim, Heterocycles, 1994, **37**, 501).

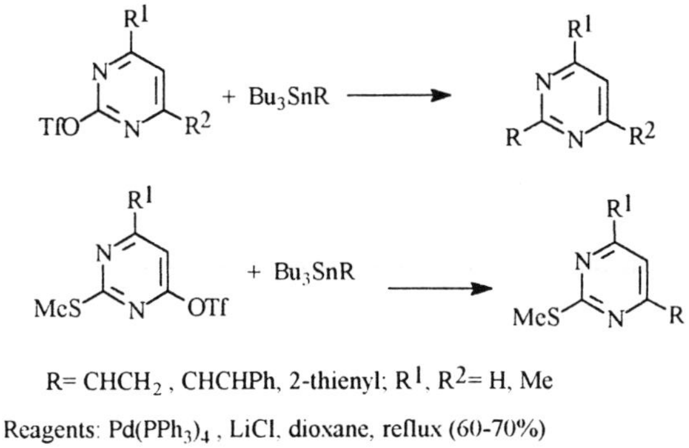

R= $CHCH_2$, CHCHPh, 2-thienyl; R^1, R^2= H, Me

Reagents: $Pd(PPh_3)_4$, LiCl, dioxane, reflux (60-70%)

Scheme 14

2- or 4-Stannylated pyrimidines are highly reactive in coupling reactions at low temperature in the absence of a catalyst and will react with acyl halides to yield ketones (J. Sandosham and K. Undheim, Tetrahedron, 1994, **50**, 275). However the 2- and 4-trimethylstannyl-pyrimidines react even more smoothly with acylformyl chlorides to give the corresponding ketones (Y. Yamamoto *et al.*, 1995, **41**, 1275). 5-Bromo-2-chloropyrimidine is selectively stannylated in the 5-position using stannyllithium or copper reagents. The reaction of 2,4-dibromo-pyrimidine with trimethylstannylsodium results in regioselective substitution at the 2-position to yield 4-bromo-2-trimethylstannyl-pyrimidine. A similar reaction using stannyllithium fails (J. Sandosham and K. Undheim, Tetrahedron, 1994, **50**, 275).

In addition, 2- or 4-trimethylstannylpyrimidines react more readily with acylformyl chlorides than with acyl chlorides to yield acylpyrimidines (Y. Yamamoto *et al.*, Heterocycles, 1995, **41**, 1275).
Lithiation of 2-methylthio-4-trifluoromethylpyrimidine, obtained from 4-iodo-2-methylthiopyrimidine, has been achieved using 2,2,6,6-tetramethylpiperidide. Subsequent treatment with a variety of electrophiles yields 6-substituted products (Scheme 15) (N. Ple *et al.*, J. Heterocycl. Chem., 1997, **34**, 551).

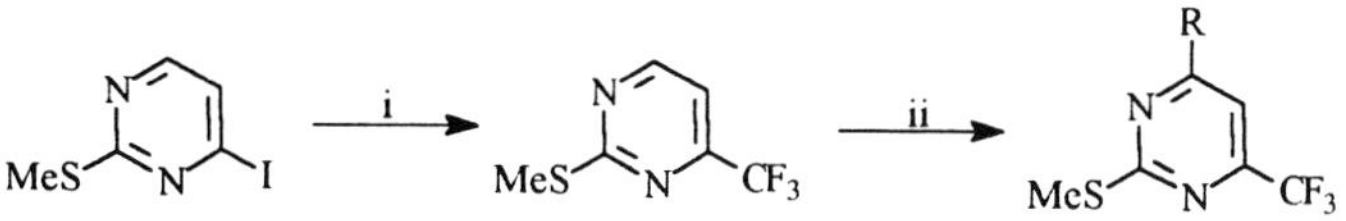

R= Cl, I, SPh, MeCH(OH), PhCH(OH), $Ph_2C(OH)$

Reagents: i; $ClCF_2CO_2Me$/KF/CuI, 115 °C, 10 h, DMF
ii; 3-4 equiv. LiTMP, -100°C, THF, 1h, then electrophile 2h.

Scheme 15

Other reactions of pyrimidines, involving organometallic reagents, include the formation of 2-chloro-5-methylpyrimidine from the 5-bromo derivative using trimethylaluminium in the presence of $Pd(PPh_3)_4$ (Q. Lu *et al.*, Acta Chem. Scand., 1996, **51**, 302), and the reaction of 2-methylthio-4,6-pyrimidinyl ditriflate with arylzinc or stannane reagents to give dicoupled products (Scheme 16) (J. Sandosham and K. Undheim, Heterocycles, 1994, **37**, 501).

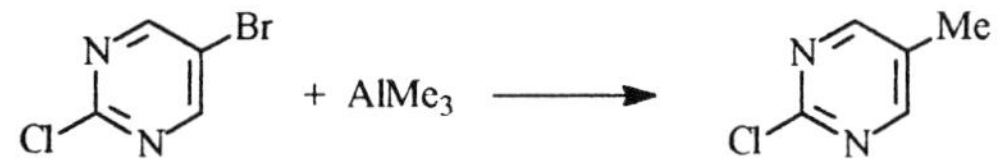

+ $AlMe_3$

Reagents: $Pd(PPh_3)_4$, THF, reflux (65%)

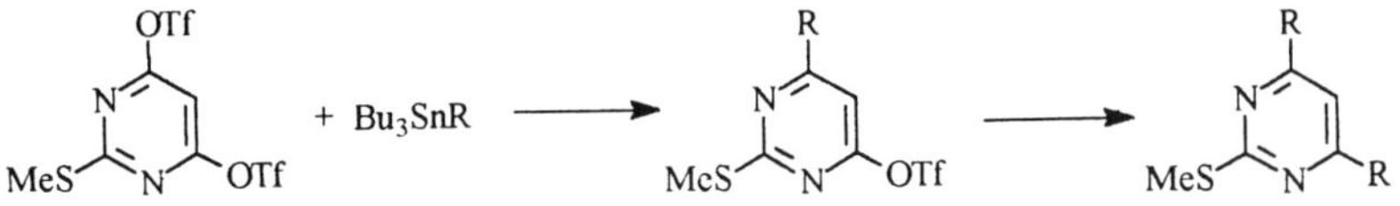

+ Bu_3SnR

Scheme 16

(iii) Ring transformations and miscellaneous reactions of pyrimidines

1-Benzyloxy-2(1*H*)-pyrimidinones (53) undergo ring transformation when treated with hydroxylamine to give 5-*N*-(benzyloxy)urea derivatives of 2-isoxazolines and isoxazoles (A. Katoh *et al.*, Heterocycles, 1994, **37**, 1141).

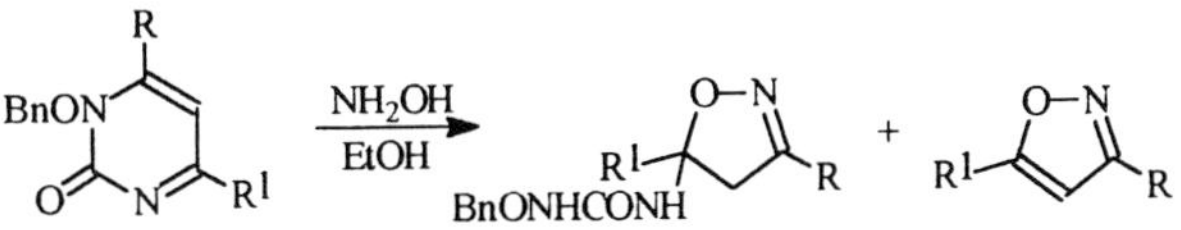

53 R, R^1= H, Me, Et

The ease of O to N rearrangement of 2- and 4-alkoxypyrimidines is shown by the fact that 2,4-bis(benzyloxy)-6-methylpyrimidine converts to the N-substituted isomer by crystallisation from methanol, heating in benzene with a catalytic amount of benzyl chloride, by chromatography on alumina, or by heating above 110 °C (G. I. Podzigdin *et al.*, Russ. Chem. Bull., 1997, **42**, 384; Chem. Abstr., 1997, **127**, 161781).

Treating 5-acylamino-6-hydroxy(or -benzyloxy)methyl-3-phenyl-pyrimidin-4(3*H*)-ones (54) with 5% ethanolic sodium hydroxide yields 2-alkyl-5-hydroxymethyl-4-phenylcarbamoyl-1*H*-imidazoles (T. Ueda *et al.*, J. Heterocycl. Chem., 1997, **34**, 761).

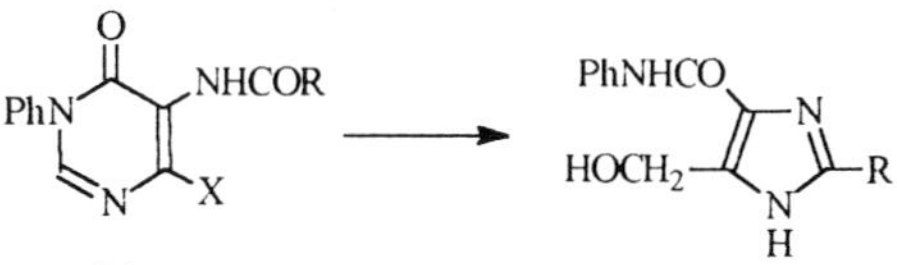

54

3-Methyl-5-nitropyrimidin-4(3*H*)-one is transformed into 4,5-disubstituted pyrimidines (55) and disubstituted 3-nitro-2-pyridinones when reacted with ketones in the presence of ammonia. The nitro-pyrimidine acts as an activated diformylamine in the first case and as a synthetic equivalent of α-nitroformyl acetate in the second (N. Nishiwaki *et al.*, Heterocycles, 1994, **38**, 249; J. Chem. Soc., Perkin Trans. 1, 1997, 2261).

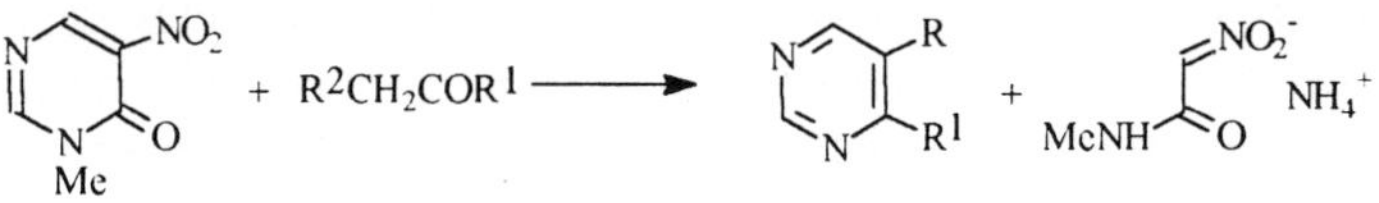

55

Reagents: NH_3 in MeOH (or MeCN), 65-130 °C, 3 h.

The conversion of 4-chloro-5-nitropyrimidines to 1,3,5-triazines by alkylisothioureas (M. P. Nemeryuk, Chem. Heterocyclic Compd. Engl. Transl., 1992, **27**, 804) has been shown to proceed via the intermediate 56 (Scheme 17) (D. T. Hurst, J. Heterocycl. Chem., 1995, **32**, 697).

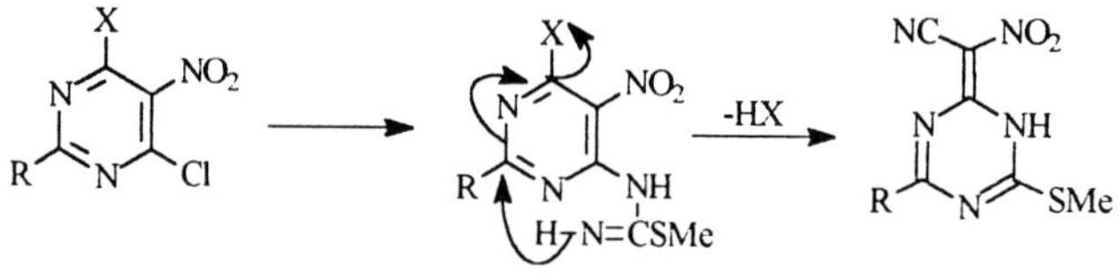

X
NO_2
N
R N Cl
X
NO_2
-HX
N
R N NH
$H_2N{=}CSMe$
NC NO_2
N NH
R N SMe

56

Reagents: $MeSC(NH_2)NH$, EtOH, then NaOH/EtOH, warm.

Scheme 17

Oximes of 6-(alk-2-enyl)amino-1,3-dimethyl-2,4-dioxo-1,2,3,4-tetrahydropyrimidine-5-carboxaldehyde (57) undergo thermally induced nitrone cyclo-additions to yield isoxazolo[3',4':3,4]pyrido[2,3-*d*]-pyrimidines (58) (M. Noguchi *et al.*, Heterocyclic Commun., 1996, **2**, 417; Chem. Abstr., 1997, **126**, 31325).

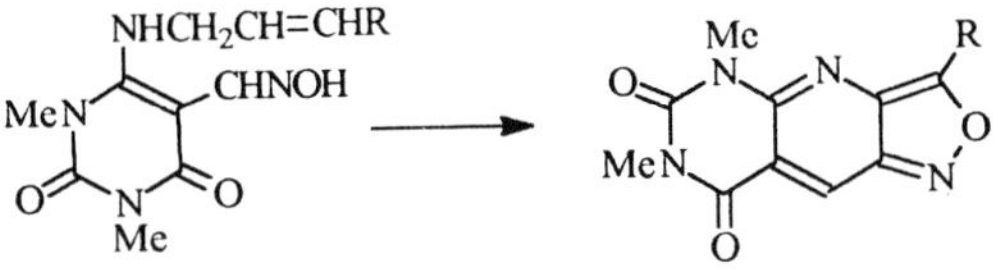

$NHCH_2CH{=}CHR$
CHNOH
MeN
O N O
Me
Me R
O N N
O
MeN N
O

57 **58**

The pyrimidinediones 59 have been shown to undergo intramolecular ene reactions to give pyrimido[4,5-*b*]azepines (60) (T. Inazumi *et al.*, J. Chem. Soc., Perkin Trans. 1, 1994, 565).

O
CHO
MeN
O N N R
Me Bn
$H_2NCH_2CO_2Et$
O $NHCH_2CO_2Et$
R
MeN
O N N
Me Bn

59 **60**

A second way of generating pyrimidine orthoquinodimethanes (*vide infra*) is the thermal extrusion of sulfur dioxide from 61. The resulting intermediates are trapped by dienophiles or nucleophiles to give products such as 62 (A. C. Tome *et al.*, Synlett, 1993, 347).

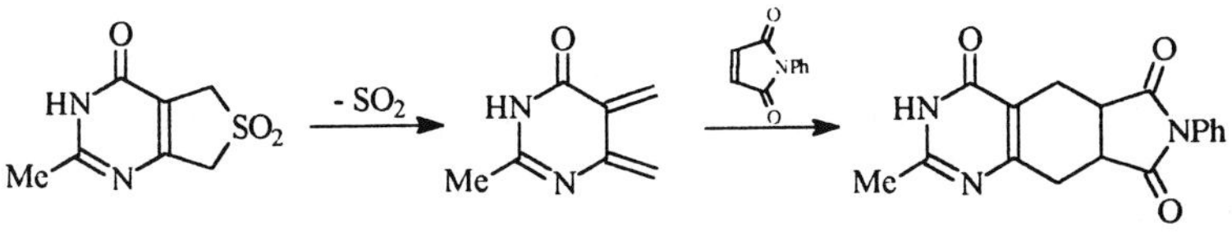

61 62

The 2'-substituted 2-amino-6-[(aminoethyl)amino]-4,5-pyrimidinediones 63 undergo ready intramolecular Schiff base condensation to yield the labile 4a-hydroxytetrahydropterins 64. The mechanism of the nonenzymic dehydration of the tetrahydropterins (64) to quinoid dihydropterins has been studied (S. W. Bailey *et al.*, J. Amer. Chem. Soc., 1995, **117**, 10203).

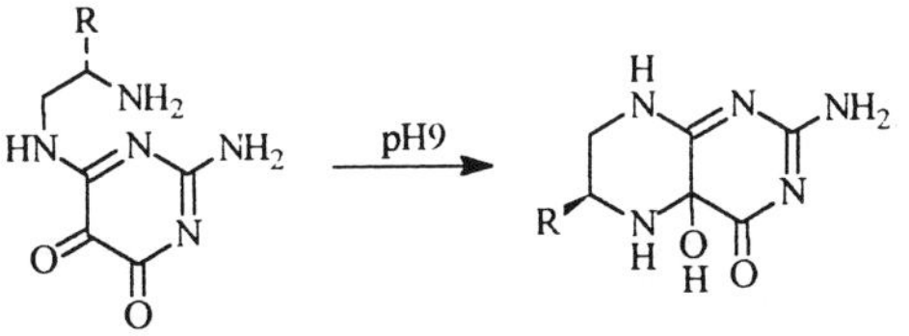

63 64

The 5-thioformyluracil 65 reacts readily with amines to give imines and with a variety of other nucleophiles to give comparable products. With enamines 66 pyrido[2,3-*d*]pyrimidines (67) are obtained (K. Hirota, K. Kubo and H. Sajiki, Chem. Pharm. Bull., 1997, **45**, 542; K. Hirota *et al.*, J. Org. Chem., 1997, **62**, 2999).

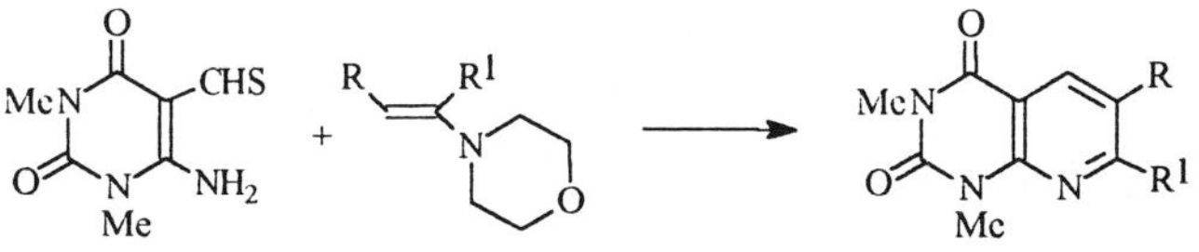

65 66 67

Upon irradiation the thiobarbiturates 68, having an alkenyl *N*-sidechain, give bi- and tricyclic fused pyrimidine derivatives through regioselective [2+2] cycloaddition (Scheme 18) (H. Takeuchi and M. Machida, Heterocycles, 1996, **42**, 117).

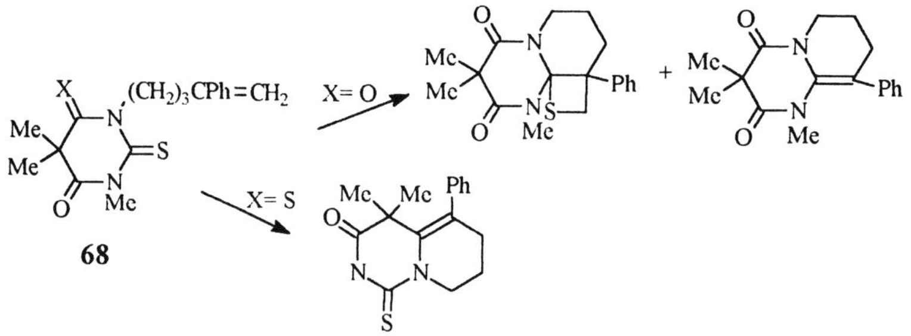

X, $(CH_2)_3CPh=CH_2$, Me, Me, N, S, N, Me, O, **68**; X= O; X= S; Ph; Me; Ph; Me, Me, Ph, O, N, N, S

Scheme 18

Tetrabutylammonium fluoride in dry THF cleaves silylpyrimidinyl ethers to yield pyrimidinones (J. Sandosham and K. Undheim, Acta Chem. Scand., 1994, **48**, 279).

Other developments of pyrimidine reactions include a solid-phase synthesis of pyrimidinamines by attaching ethyl 2-chloro-4-trifluoromethylpyrimidine-5-carboxylate to NovaSyn TG thiol resin, followed by displacement with amines (Scheme 19) (L. M. Gayo and M. J. Suto, Tetrahedron Lett., 1997, **38**, 211).

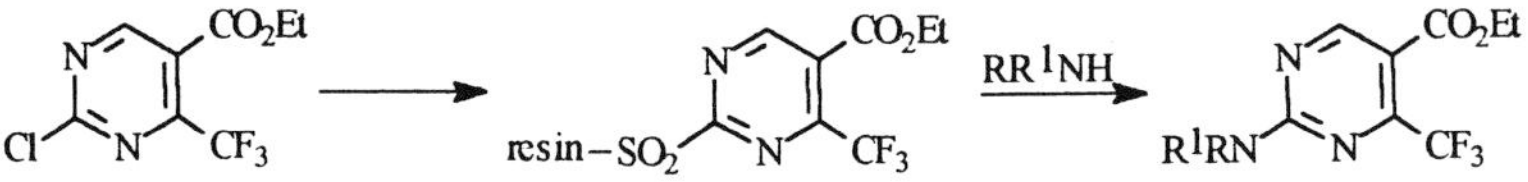

CO_2Et, Cl, N, CF_3 → resin–SO_2, CO_2Et, CF_3 —RR^1NH→ R^1RN, CO_2Et, CF_3

Scheme 19

Stille-type C-C coupling has been used to obtain oligo tridentate ligands based on alternating pyridines and pyrimidines, *e.g.* 69 (G. S. Hanan *et al.*, Can. J. Chem., 1997, **75**, 169) and the pyrimidine 71 has been synthesised from 70 for use in non-linear optics (R. Gompper and P. Walther, Tetrahedron, 1996, **52**, 14607).

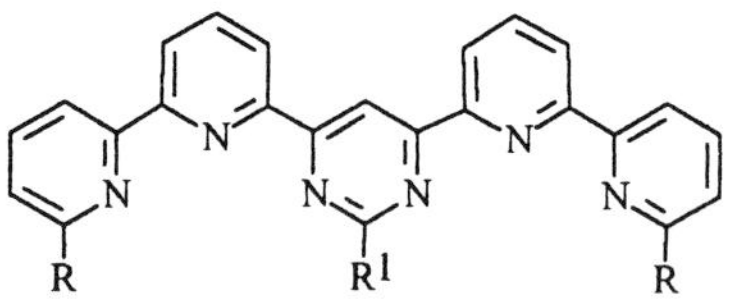

R, R^1, R

69 R= H, Ph; R^1= H, Me, CH_2Br

Me_2N, Me_2N, + BF_4^- —$RCNH_2NH.HX$→ SMe

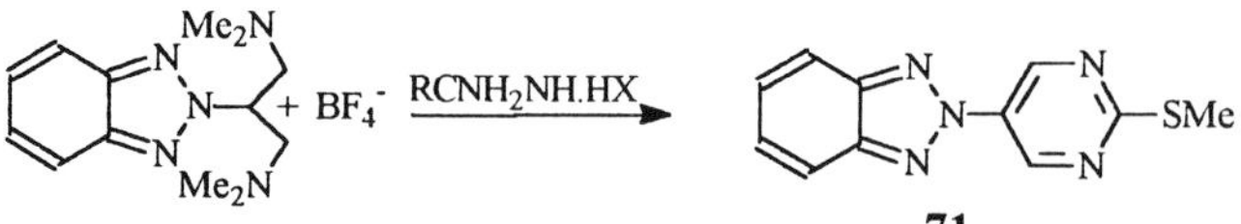

70 **71**

Pyrimidine derivatives continue to find uses in a variety of commercial applications, particularly so as agricultural chemicals. New pyrimidines having herbicidal action include 72 (D. B. Kamme and M. P. Prisbylla, USP 5 714 438 (1998); Chem. Abstr., 1998, **128**, 140721) and 73 (M.-W. Drewes, R. Andree and M. Dollinger, Ger. Offen. DE 19 632 005 (1998); Chem. Abstr., 1998, **128**, 167437).

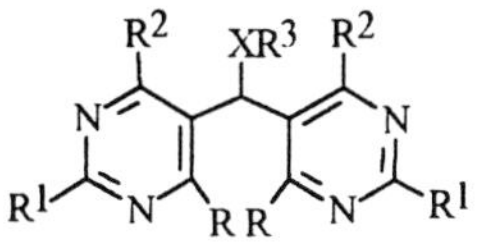

72 RS= h, var. alkyl, aryl; X= OR, NR 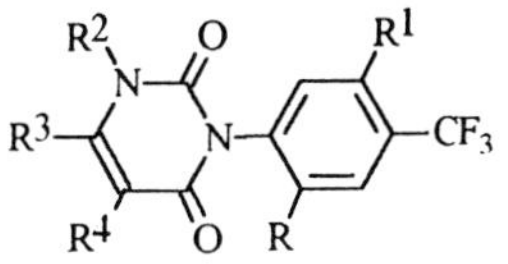 **73** RS= H, halogen, alkyl, haloalkyl

3. Quinazolines

(a) Properties and structural aspects

Comprehensive data on the structure and properties of quinazolines have been reported in "*Comprehensive Heterocyclic Chemistry*" (*vide infra*) and there have not been significant developments in the last few years which need to be recorded here.

(b) Synthesis

Hydroxyglycine reacts with 2-aminobenzophenones to give 1,2-dihydro-4-phenylquinazoline-2-carboxylic acids in high yield and under mild conditions. These products are readily converted into the corresponding 3,4-dihydro isomers and into the quinazoline derivatives *via* rearrangement and air oxidation respectively (Scheme 20) (A. J. Hoefnagel, Tetrahedron, 1993, **49**, 6899).

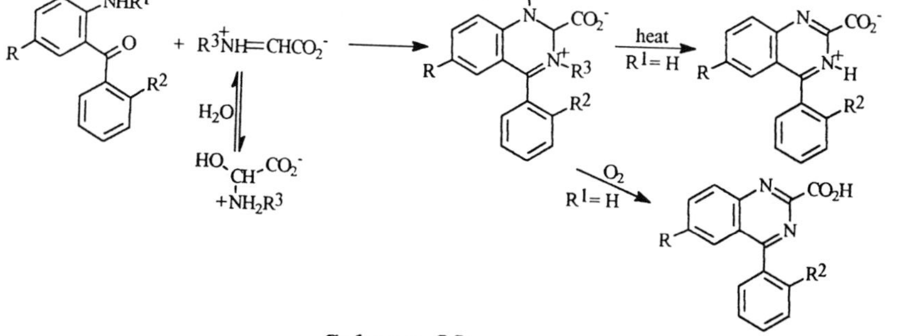

Scheme 20

Aldehydes, converted *in situ* to *N*-(1-chloroalkyl)pyridinium chlorides using thionyl chloride and pyridine, react with 2-aminobenzylamine under mild conditions to afford 1,2,3,4-tetrahydroquinazolines which can be aromatised by 2,3-dichloro-5,6-dicyano- or 2,3,4,5-tetrachloro-benzoquinone in benzene (Scheme 21) (J. J. Vanden Eynde, Synthesis, 1993, 867).

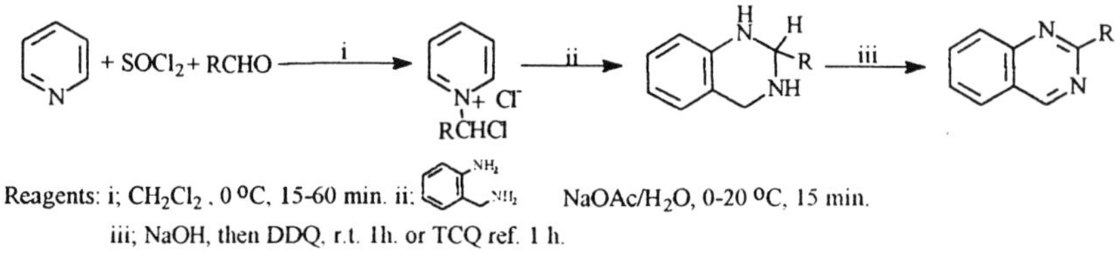

+ $SOCl_2$ + RCHO —i→ N+ Cl^- RCHCl —ii→ —iii→

Reagents: i; CH_2Cl_2, 0 °C, 15-60 min. ii; $NaOAc/H_2O$, 0-20 °C, 15 min.
iii; NaOH, then DDQ, r.t. 1h. or TCQ ref. 1 h.

Scheme 21

Quinazolin-4(3*H*)-ones (74) are obtained in a one-pot reaction by the sodium perborate oxidation of *o*-amidobenzonitriles with concomitant cyclisation (B. Baudoin, Y. Ribeill and N. Vicher, Synth. Commun., 1993, <u>23</u>, 2833) whilst 1,2-dihydroquinazolinones (75) are obtained by the ring closure of anthanilic amides with oxo compounds (J. Lessel, Arch. Pharm. (Weinheim, Ger.), 1994, **327**, 571).

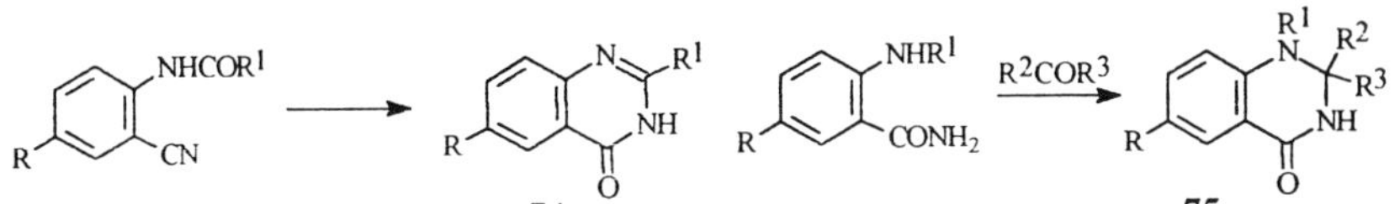

NHCOR[1] CN R → 74 NHR[1] $CONH_2$ R —R^2COR^3→ 75

Another synthesis of quinazolinones is the reaction of ketene dithioacetals with anthanilamide to yield 2-substituted-methylene-1,2-dihydroquinazolinones (76) (Scheme 22) (Y. Cheng *et al.*, Synth. Commun., 1996, <u>26</u>, 475).

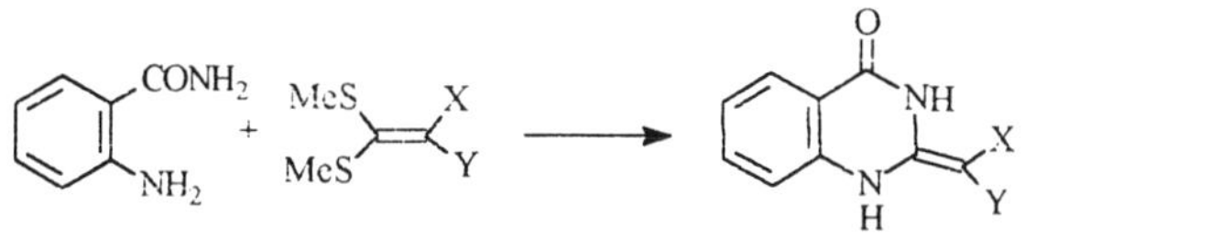

$CONH_2$ NH_2 + MeS MeS X Y →

76 (X, Y= CN, H, ArCO, CO_2R)

Reagents: i; THF, NaH, ref. 1 d. or ii; ref. in xylene 1 d. or iii; KOH, DMSO, 90 °C, 2 d.

Scheme 22

A new synthesis of 4-aminoquinazolines (78) is the cyclisation of *N*-(2-cyanophenyl)amidines (77) with arylamines (E. Erba and D. Sporchia, J. Chem. Soc., Perkin Trans. 1, 1997, 3021) and the treatment of a variety of substituted 2-aminobenzonitriles with formic acid under strong acid catalysis gives quinazolin-4(1*H*)-ones (79) in good yield (G. A. Roth and J. J. Tai, J. Heterocycl. Chem., 1996, **33**, 2051) whilst quinazolin-4(3*H*)-ones can be obtained by reacting *o*-amidobenzonitriles with urea and hydrogen peroxide (B. P. Bandgar, Synth. Commun., 1997, **27**, 2065). A series of 4-amino-8-cyano-quinazolines (80) has also been obtained by reacting 2-aminobenzene-1,3-dicarbonitrile with formamide or guanidine (P. J. Victory *et al.*, Heterocycles, 1993, **36**, 2273) and 2-aminoquinazolines are produced by reacting guanidine carbonate with various *o*-fluorobenzaldehydes in *N,N*-dimethylacetamide (J. B. Hynes and J. P. Campbell, J. Heterocycl. Chem., 1997, **34**, 385).

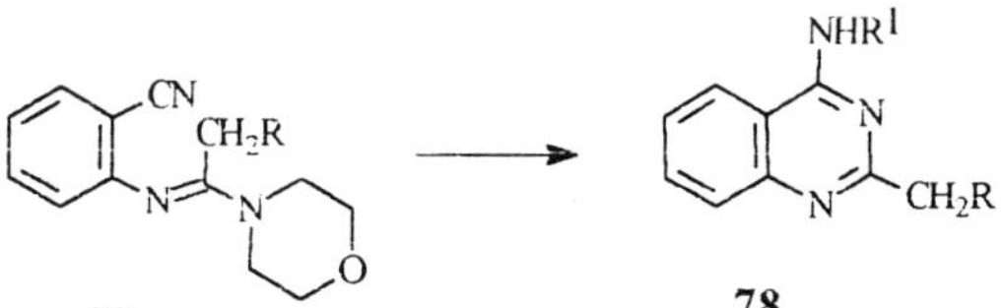

CN CH$_2$R N N O **77** → NHR^1 N N CH$_2$R **78**

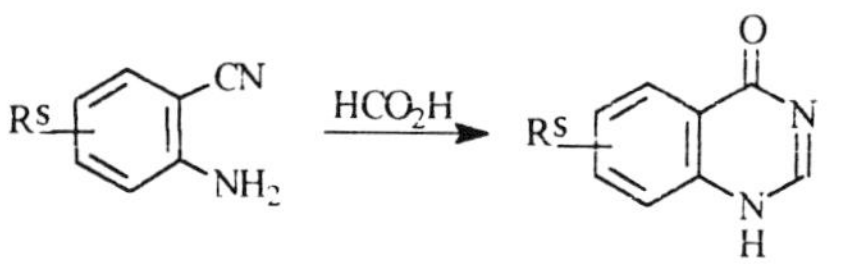

RS CN NH$_2$ HCO$_2$H → RS O N N H **79** CN RS N R N NH$_2$ **80**

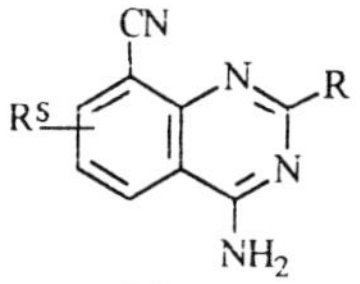

Bis(triphenylphosphine) palladium(II) chloride and molybdenum pentachloride catalyse the intramolecular reductive cyclisation of 2-nitrobenzaldehyde or 2-nitrophenyl ketones by formamide under carbon monoxide to give quinazolines. A nitrene intermediate, generated by carbon monoxide deoxygenation of the nitro group, is a possible intermediate (M. Akazome *et al.*, J. Organometal. Chem., 1995, **494**, 229).

2-(Benzotriazol-1-yl)enamines (81) undergo regiospecific thermal rearrangement at 110 °C in toluene to yield 2,4-diarylquinazolines (82) (A. R. Katritzky *et al.*, J. Org. Chem., 1995, **60**, 246). However lithiated *N*-(α-alkoxyalkyl)benzotriazoles (83) in contrast undergo ring opening and concomitant nitrogen extrusion at -78 °C to give *o*-imino-phenyl anions which can lead to 2,3,4-triaryl-3,4-dihydroquinazolines, 2-phenyl-4-arylquinazolines and 3-aryl-3-phenyl-3,4-quinazoline-4-thiones (A. R. Katritzky *et al.*, J. Org. Chem., 1995, **60**, 7625) (Scheme 23).

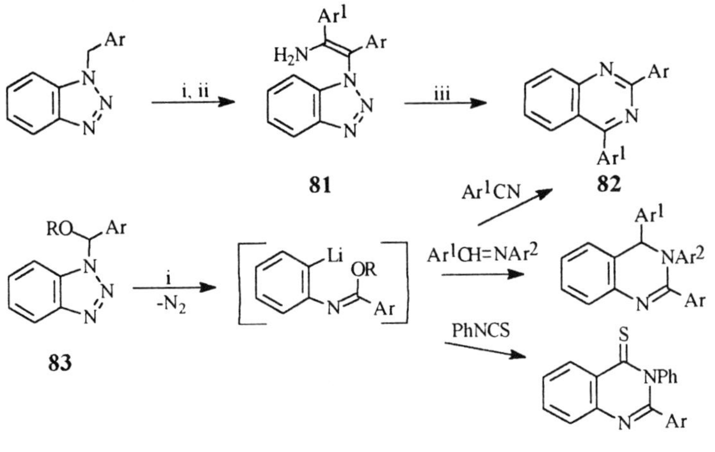

Reagents: i: BuLi/THF, -78 °C ii: RCN, -78 °C to r.t. iii: Toluene, reflux

Scheme 23

Deprotonation of *N*-phenyldiethylketenimine with strong bases (LTMP, *t*-BuOK) in *t*-BuOMe, followed by the addition of another equivalent of the ketenimine and subsequent introduction of *t*-butylcarbonyl chloride, gives the quinazoline 84 (R. Gertzmann *et al.*, Tetrahedron, 1995, 51, 9031).

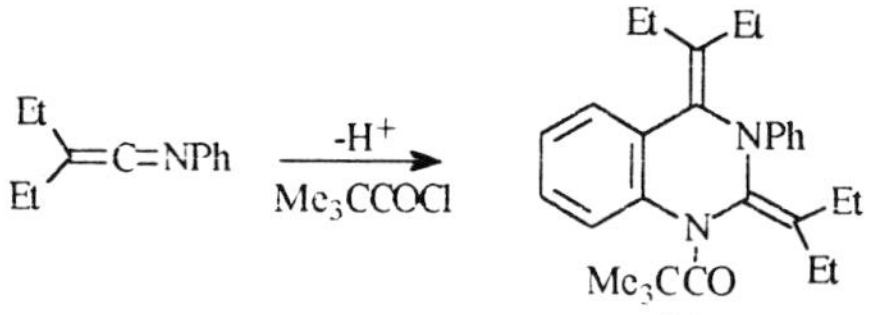

Quinazolinium salts (85) can be obtained from the phosphoryl chloride mediated Vilsmeier formylation of *N,N*-dimethyl *p*-toluidine with *N*-formylated dialkylamines (O. Meth-Cohn and D. L. Taylor, J. Chem. Soc., Chem. Commun., 1995, 1463).

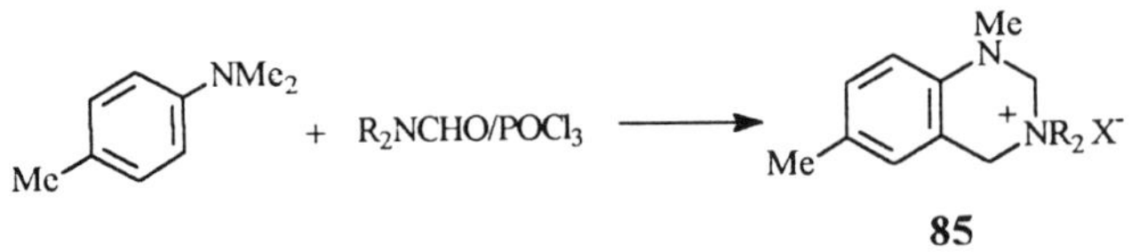

3-Hydroxyquinazolines (86) are formed by the reaction of *o*-aminobenzohydroxamic acid with aldehydes (O. A. Luk'yanov and P. B. Gordeev, Russ. Chem. Bull., 1997, **46**, 110; Chem. Abstr., 1997, **127**, 121690).

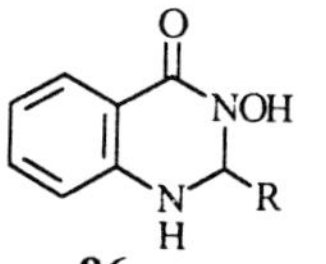

86

A novel, easy and efficient method for the synthesis of quinazolines using solid phase techniques involves solid supported, substituted, cinnamyl iminophosphoranes treated with aryl isothiocyanates to generate solid supported carbodiimides which, when exposed to secondary amines, undergo 1,2-addition followed by an intramolecular Michael addition to give 3,4-dihydroquinazolines (Scheme 24) (F. Wang and J. R. Hanske, Tetrahedron Lett., 1997, **38**, 8651).

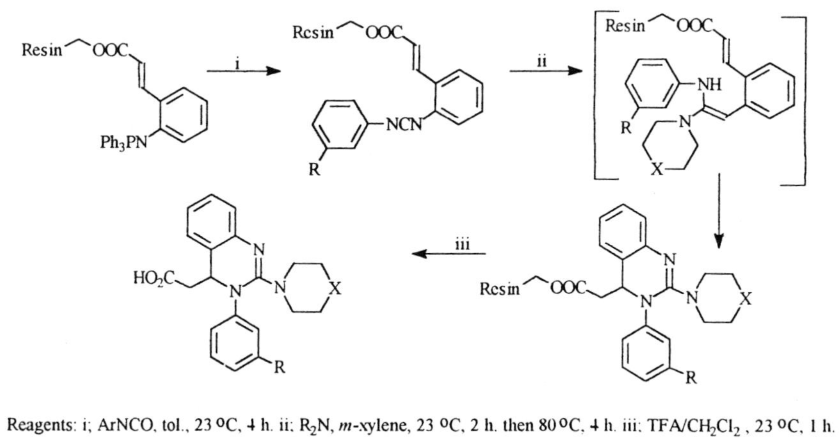

Reagents: i; ArNCO, tol., 23 °C, 4 h. ii; R_2N, *m*-xylene, 23 °C, 2 h. then 80 °C, 4 h. iii; TFA/CH_2Cl_2, 23 °C, 1 h.

Scheme 24

(c) Reactions

Most of the reactions of pyrimidines can, similarly, be carried out using quinazolines. For example, metal-catalysed carbosubstitutions of chloroquinazolines using Grignard reagents and other organometallic reagents including those derived from tin, aluminium, palladium and zirconium (Scheme 25) (I. Mangalagiu, T. Benneche and K. Undheim, Acta Chem. Scand., 1996, **50**, 914; Tetrahedron, 1996, **37**, 1309).

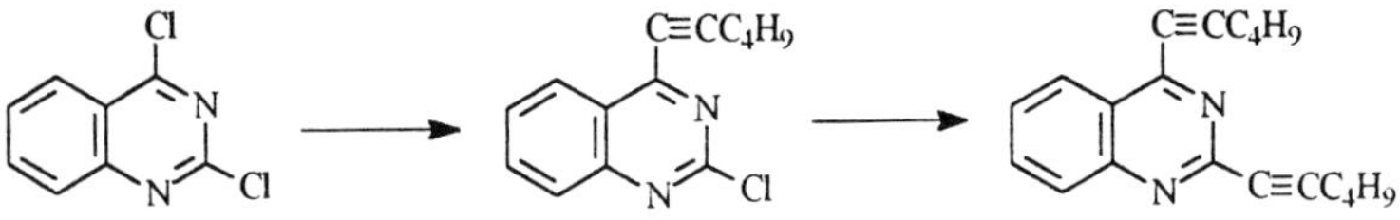

Reagents: either $Cp_2ZrClCH{=}CHC_4H_9$. THF, reflux, or $PdCl_2(PPh_3)_2$/CuI, NEt_3 , 20 °C, 20 h.

Scheme 25

Quinazolinethiones are converted to the corresponding thiocyanato derivatives by *p*-tosyl cyanide and sodium hydride in THF (A. Miyashita *et al.*, Heterocycles, 1997, **45**, 745).
The lithiation of 3-amino- and 3-acylaminoquinazolin-4(3*H*)-ones has been studied and has been shown to result in the formation of 2-substituted quinazolin-4(3*H*)-one derivatives *via* regioselective lithiation at C-2 (K. Smith *et al.*, J. Org. Chem., 1996, **61**, 647, 656).
Queguiner and his coworkers have carried out extensive studies of the lithiation of heterocycles including a report of the first functionalisation of the benzene moiety of quinazolines. According to the nature and position of the directing group(s) and the nature of the metallation reagent lithium can be introduced into different positions. For example, 4-methoxy-2-phenylquinazoline undergoes metallation at C-8. Without C-2 substitution C-2 metallation is favoured. The use of LTMP instead of butyllithium favours metallation in the benzene moiety and reaction can occur here even if the C-2 is unsubstituted (N. Ple *et al.*, Tetrahedron, 1997, **53**, 2871). Some examples of these reactions are shown in Scheme 26.

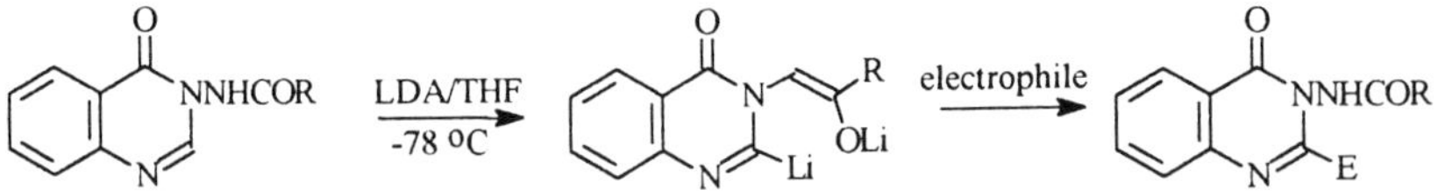

LDA/THF -78 °C

electrophile

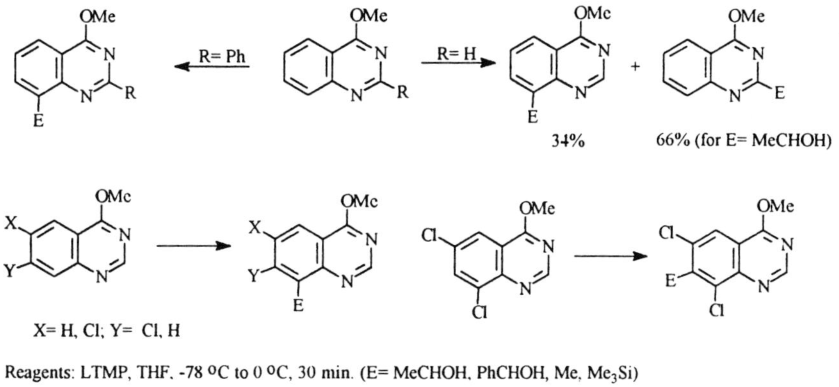

Scheme 26

A study of the reaction between benzaldehyde and 3-amino-2-methyl-3,4-dihydroquinazolin-4-one has shown that it condenses in two stages, first with the 3-amino group, then with the methyl at C-2 (K. B. Lal, Asian J. Chem., 1998, **10**, 35).
The treatment of 2-mercaptoquinazolin-4(3*H*)-one with hydrazonyl halides affords 1,3-disubstituted-1,2,4-triazolo[4,3-*a*]quinazolin-5-ones (87) (Scheme 27) (H. A. Abdelhadi, T. A. Abdallah and H. M. Hassaneen, Heterocycles, 1995, **41**, 1999).

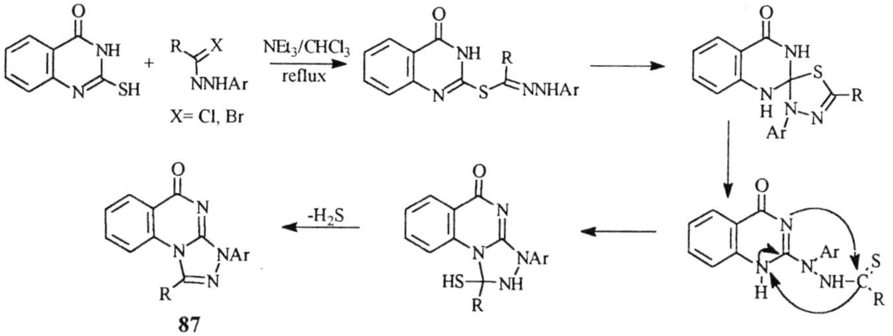

Scheme 27

4-(α-Benzyl-α-hydroxybenzyl)quinazolines (88) undergo retro-benzoin condensation catalysed by cyanide ion, most efficiently using tetrabutylammonium cyanide, to give deoxybenzoin and quinazolines (Scheme 28) (Y. Suzuki *et al.*, Chem. Pharm. Bull., 1998, **46**, 199).

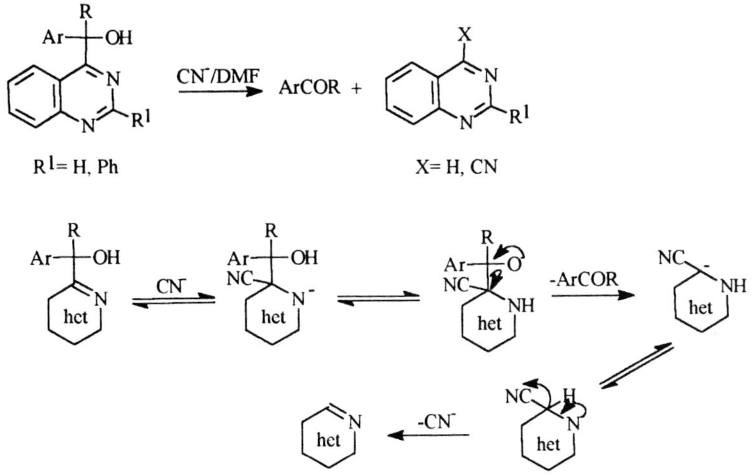

Scheme 28

Like other *N*-heterocycles quinazolines show a variety of useful biological activities including acting as hyperlipemics (Y. Kurogi *et al.*, J. Medicin. Chem., 1996, **39**, 1433), as antimicrobials and anticancer agents (L. F. Kuyper *et al.*, J. Medicin. Chem., 1996, **39**, 892; M. A. Aziza *et al.*, Indian J. Heterocycl. Chem., 1996, **6**, 25) and in many other applications (K. Undheim and T. Benneche in "*Comprehensive Heterocyclic Chemistry*", 2nd Edn., Eds. A. R. Katritzky, C. W. Rees and E. F. V. Scriven, Pergamon, Elsevier Science Ltd., Oxford, 1996, Chap. 6.02, 93).

Second Supplements to the 2nd Edition of Rodd's Chemistry of Carbon Compounds, Vol.IV IJ, edited by M. Sainsbury

Chapter 44

PYRAZINES AND RELATED RING STRUCTURES

Kevin J. McCULLOUGH

1. Pyrazines, 1,4-diazines

(a) Introduction

This second supplement is intended to cover advances in the chemistry of pyrazine during the period 1994 until June 1998.

Two families of dimeric steroid-pyrazines, a) the cephalostatins (17 members), e.g. cephalostatin 1 (1), isolated from the Indian Ocean hemichordate *Cephalodisus gilchristi*, (G. R. Pettit *et al.*, Bioorg. med. Chem. Letters, 1995, 5, 2027), and b) the structurally related ritterazines (13 members), e.g. ritterazine A (2), metabolites of the Japanese tunicate *Ritterella tokioka* (S. Fukuzawa, S. Matsunaga and N. Fusetani, J. org. Chem., 1994, 59, 6164), have attracted considerable attention because of their potent cytotoxicity. Strategies for the synthesis of such compounds have been reviewed (A. Ganesan, Angew Chem., Intern. Ed., 1996, 35, 611).

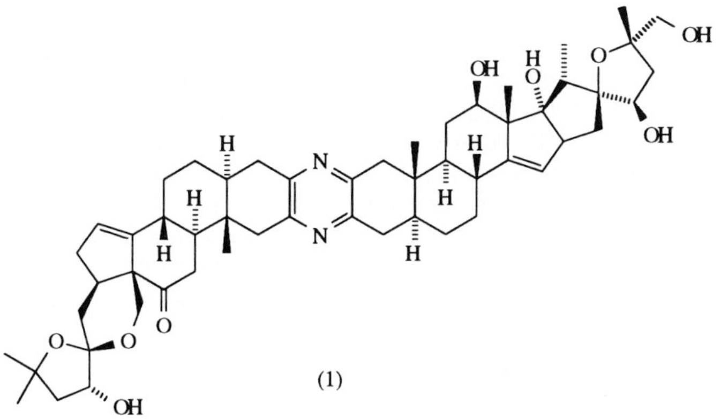

(1)

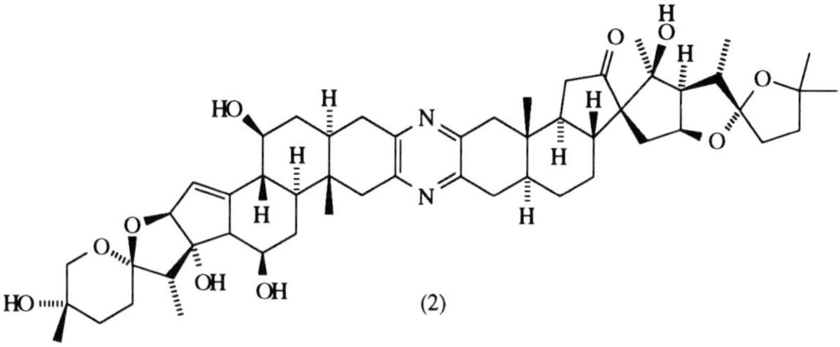

(2)

Syntheses of several cephalostatin analogues have been reported (J. U. Jeong and P. L. Fuchs, J. Amer. chem. Soc., 1994, 116, 773; C. H. Heathcock and S. C. Smith, J. org. Chem., 1996, 59,6828; C. Guo, S. Sudhakar and P. L. Fuchs, J. Amer. chem. Soc., 1996, 118, 10672; M. Drögemüller, R. Jautelat and E. Winterfeldt, Angew Chem., Intern. Ed., 1996, 35, 1572).

A novel pyrazine (3) has been obtained from chitin by the action of an enzyme isolated from *V. alginolyticus* TK-24 (T. Yanai *et al.*, J. Ferment. Bioeng., 1995, 80, 406).

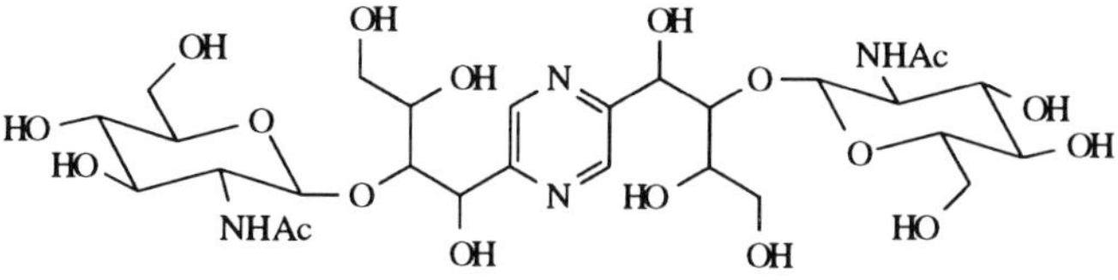

(3)

A general review on the chemistry of pyrazines and related compounds, primarily covering the period 1980-1994, has been published (N. Sato "Comprehensive Heterocyclic Chemistry II", Vol 6, ed. A. R. Katritzky, C. W. Rees and E. F. V. Scriven, Pergamon, Oxford, 1997, pp. 233). Other recent reviews on pyrazine chemistry include (a) the reactions and syntheses of pyrazines, particularly 2,5-disubstituted derivatives (A. Ohta and Y. Aoyagi, J. pharm. Soc. Japan, 1997, 117, 1), (b) the synthesis of naturally occurring pyrazines (*idem*, *ibid.*, 1997, 117, 32), (c) metallation reactions of pyrazine derivatives (A. Turck, N. Plé and G. Quéguinier, Heterocycles, 1994, 37, 2149), and (d) the chemistry of pyrazine *C*-nucleosides (M. A. E. Shaban and A. Z. Nasr, Adv. heterocyclic Chem., 1997, 68, 379).

(b) Physical and spectroscopic properties

Ab initio SCF-MO calculations have shown that, in common with other diazines, the HOMO energy correlates well with pKa and first π-electron ionization energy (H. J. S. Machado and A. Hinchcliffe, Theochem - J. mol. Structure, 1995, 339, 255). A force field has also been developed which accurately predicts the structure and thermodynamic properties of pyrazine and related compounds (W. L. Jorgensen and N. A. McDonald, *ibid.*, 1998, 424, 145).

A comprehensive set of ^{15}N nmr data has been compiled for pyrazine, methylpyrazine derivatives and related mono- and di-*N*-oxides (C. Sakuma *et al.*, Mag. Reson. Chem., 1996, 34, 567).

Studies of the enolization of phenacylpyrazine and related compounds by ^{1}H and ^{13}C nmr spectroscopic techniques indicate that the pyrazines exist preferentially in the enol form (R. A. M. O'Farrell and B. A. Murray, J. chem. Soc. Perkin II, 1994, 2461; A. R. E. Carey *et al.*, *ibid.*, 1994, 2471). Deuterium isotope effects on ^{13}C nmr chemical shifts of phenacylpyrazines also provide similar results (A. R. Katrizky *et al.*, *ibid.*, 1997, 2605)

(c) General methods of synthesis

General strategies for the synthesis of pyrazines have been discussed in detail elsewhere (K. J. McCullough, Rodd's Chemistry of Carbon 2nd Ed. Vol IV, Part IJ, 1989, pp. 241; 1st Supplement, 1995, pp. 93; N. Sato "Comprehensive Heterocyclic Chemistry II", Vol 6, ed. A. R. Katritzky, C. W. Rees and E. F. V. Scriven, Pergamon, Oxford, 1997, pp. 233).

Condensation of dialkynyl-1,2-diones with 2,3-diaminomaleonitrile (DAMN) affords the tetra-substituted pyrazine (4), except for R = R' = Ph (R. Faust *et al.*, Tetrahedron, 1997, 43, 14655).

R–C≡C–CO–CO–C≡C–R' + H_2N(CN)C=C(CN)NH_2 → (AcOH / RT) → (4)

The reaction of active methylene compounds with *O*-(*p*-toluenesulfonyl)isonitrosomaleodinitrile (OTMD) affords azapropenide salts which can be cyclised under either acidic or basic conditions to provide tetra-substituted pyrazines such as (5) and (6) respectively (A. K. El-Shafei *et al.*, Synthetic Comm., 1994, 24, 1895).

Tosyl
O
N
NC CN
OTMD

CN CN pyridine, 0 °C
[NC N CN CN CN] HCl 80 °C
NC N CN H$_2$N N Cl
(5)

CO$_2$Et CN pyridine, 0 °C
[NC N CO$_2$Et CN CN] MeONa 5 °C
NC N CO$_2$H CH$_3$O N NH$_2$
(6)

Treatment of dipeptidyl chloromethyl ketones with hydrochloric acid at reflux temperature affords the pyrazinone (hydroxypyrazine) derivatives (7) in good to excellent yield (60-90%) rather than the expected amino acid hydrolysis products (Y. Okada *et al.*, Chem. Comm., 1994, 247; Tetrahedron Letters, 1995, 35, 1231).

tBuO H N O R^1 O R^2 N H O Cl
6M HCl Δ
Me N R^1 R^2 N H O
(7)

R^1, R^2 = H, Me, iBu, PhCH$_2$

(d) Pyrazine, its homologues and derivatives

(i) Pyrazine, alkyl-, alkenyl-, alkynyl- and aryl-pyrazines

Dehydrodeamination of ethylenediamine using Al/Pt catalysts doped with either Ir_2O_3 or Re_2O_7 gives pyrazine in around 60% yield together with some alkylpyrazines. Under similar conditions, *N*-(2-hydroxypropyl)ethylenediamine affords methylpyrazine (*ca.* 40% yield) (K. M. Gitis, G. E. Neumoeva and G. V. Isagulyants, Khim. Geter. Soedinenii, 1993, 1516; C. A., 1994, 122, 9991). *N*-(2,3-Dihydroxy-propyl)ethylenediamine is transformed into mixtures of pyrazine and methylpyrazine on reaction over chromite-based catalysts (M. Subrahmanyam *et al.*, Ind. J. Chem., 1995, 34B, 573).

Direct metallation of pyrazine can be successfully achieved using a 4-fold excess of lithium 2,2,6,6-tetramethylpiperidide (LTMP) at -75 °C. The resulting pyrazinyllithium reacts readily with electrophiles such as aldehydes or iodine yielding the corresponding monosubstituted pyrazine derivatives (N. Plé *et al.*, J. org. Chem., 1995, 60, 3781).

LTMP (4 equiv.), THF, -75 °C; RCHO

R = Me (64%), Ph (65%)

Reaction of pyrazine with excess allyltibutyltin in the presence of chloroformate affords a mixture of the tetrahydropyrazine derivatives (8) and (9) (T. Itoh *et al.*, Heterocycles, 1994, 37, 709; Chem. pharm. Bull., 1994, 42, 1768).

$Bu_3SnCH_2CH=CH_2$, $ClCO_2CH_2CF_3$

$CO_2CH_2CF_3$ (8) + (9)

Pyrazine readily participates in 1,3-dipolar cycloaddition reactions with (a) nitrilimines to give bis(cycloadducts) (10) in 60-70% yield (L. Grubert *et al.*, Ann., 1994, 1005); (b) benzonitrile oxide to give a mixture of the *anti* bis(cycloadducts) (11) and (12) (G. Grassi, F. Risitano and F.

(10) R = Ar, CO_2Et (11) (12)

Foti, Tetrahedron, 1995, 51, 11855; A. Corsaro, G. Perrini and V. Pistarà, *ibid.*, 1996, 52, 6421).

A series of trialkylpyrazines (13), identified as components of ant pheromones (see T. H. Jones *et al.*, J. chem. Ecology, 1998, 24, 125), has been synthesised (yields >70%) by Ni-catalysed cross-coupling reactions between 2-chloro-3,6-dialkylpyrazines with the appropriate dialkylzinc. Trialkylpyrazines (13) undergo acylation under Minisci radical conditions to give the corresponding 2-acylpyrazines (14) in high yield though in certain cases (R ≠ Me) the product is contaminated by significant quantities of tetralkylpyrazines (N. Sato and T. Matsuura, J. chem. Soc., Perkin Trans. 1, 1996, 2345)

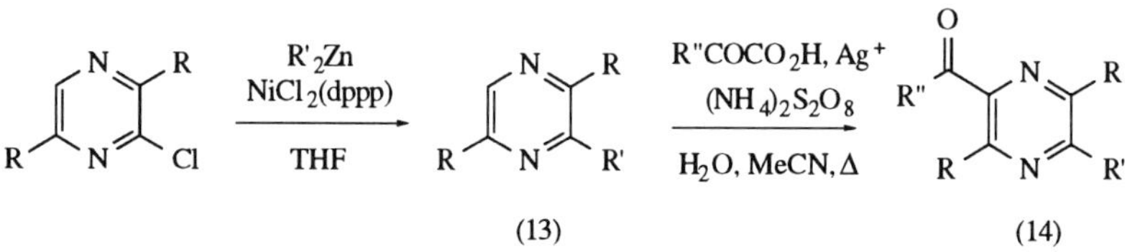

R'_2Zn, $NiCl_2(dppp)$, THF; R''$COCO_2H$, Ag^+, $(NH_4)_2S_2O_8$, H_2O, MeCN, Δ

(13) (14)

R = Me, iPr, iBu, sBu; R' = Me, Et, isopentyl; R'' = Et, iBu

Photochemical irradiation of propyl and higher alkyl pyrazines promotes an intramolecular hydrogen abstraction process which results in cleavage of the side-chain to give methylpyrazine (A. Mukherjee, S. A. M. Duggan and W. C. Agosta, J. org. Chem., 1994, 59, 178).

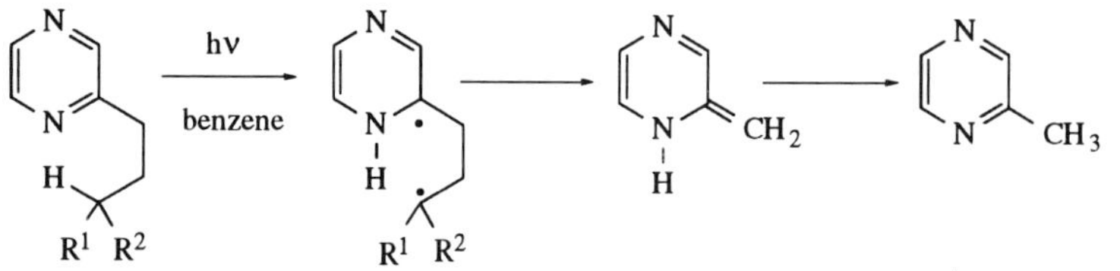

hν, benzene; CH_2; CH_3

R^1, R^2 = H, Me

Homolytic acetylation of 2,5-dimethylpyrazine results in the formation of a mixture 2,5-diacetyl-3,6-dimethylpyrazine and 2-acetyl-3,5,6-trimethylpyrazine (V. Opletalová *et al.*, Coll. Czech. chem. Comm., 1995, 60, 1551).

Exhaustive chlorination of 2,5-dimethylpyrazine using $POCl_3$ in excess PCl_5 results in the formation of the hitherto unreported 2,5-bis(trichloromethyl)pyrazine in moderate yield (D. Cartwright *et al.*, J. chem. Soc., Perkin Trans. 1, 1995, 2595).

Pyrazines (15) with a 5-membered-ω-alkenyl side-chains are transformed into the corresponding tricyclic compounds (16), via intramolecular Diels-Alder reactions, on heating in trifluoroacetic acid at reflux. Subsequent heating of (16) at higher temperatures results in almost complete regeneration of the original pyrazine (15) (B. Chen, C.-Y. Yang and D.-Y. Ye, Tetrahedron Letters, 1996, 37, 8205).

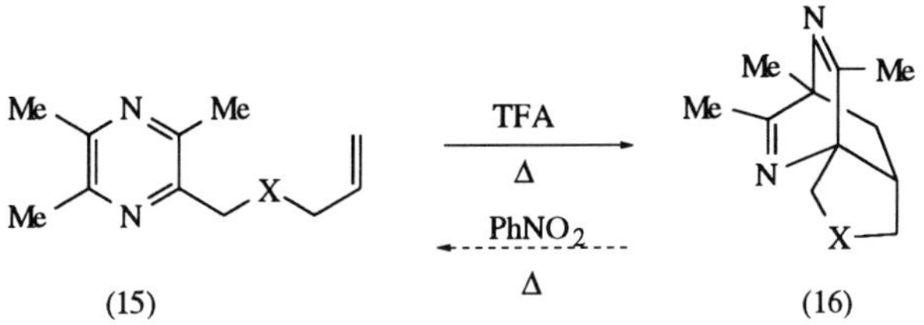

TFA, Δ; $PhNO_2$, Δ

(15) (16)

X = O, S, SO, SO_2

Self-condensation of α-amino-4-hydroxyacetophenone yields 2,5-bis(4-hydroxyphenyl)pyrazine (17) which can be selectively nitrated to produce either the coresponding di- or tetra-nitro derivatives (18). Such compounds have considerable potential as environmentally friendly alternatives to pigments based on inorganic heavy metals (O. S. Fruchey *et al.*, New J. Chem., 1996, 20, 249).

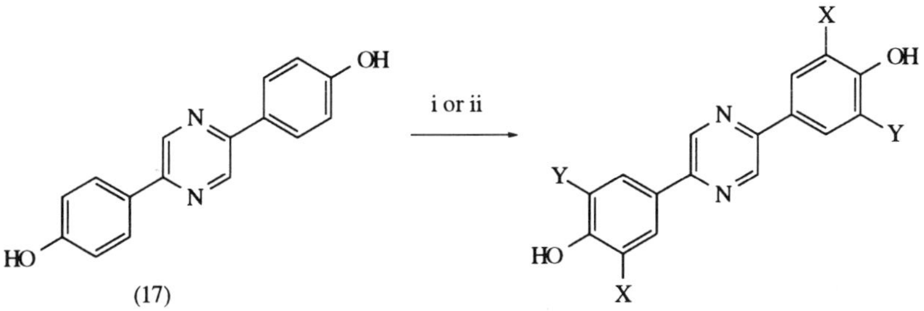

i or ii

(17)

i. c. HNO_3, RT → (18a), X = H, Y = NO_2

ii. c. HNO_3 / c. H_2SO_4, RT → (18b), X = Y = NO_2

Pyrazinium dicyanomethylide (19) undergoes dipolar cycloaddition with the strained cyclooctyne, followed by loss of hydrogen cyanide to give the polyfused system (20) in 96% yield (K. Matsumoto *et al.*, J. heterocyclic Chem., 1995, 32, 367; 1997, 34, 203).

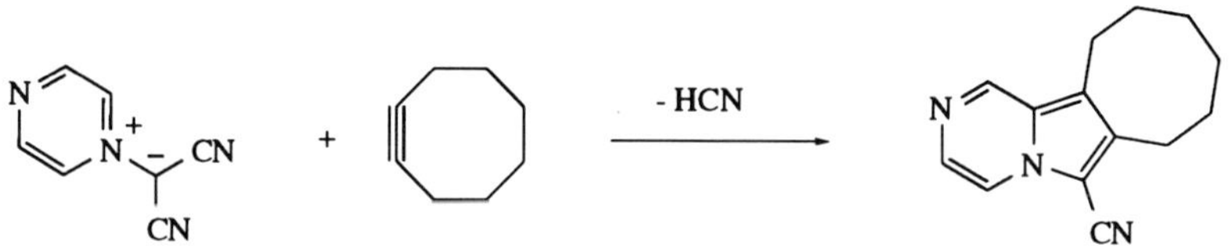

(19) (20)

(ii) Halopyrazines

A variety of polyhalogenated pyrazines can be obtained by tandem electrophilic halogenation and aromatisation of tetrahydropyrazine-2,5-dione with phosphoryl chloride, phosphorus pentachloride, phosphorus pentabromide or their mixtures in the presence of either chlorine or bromine as indicated below (I. L. Yudin *et al.*, Mendeleev Comm., 1995, 196).

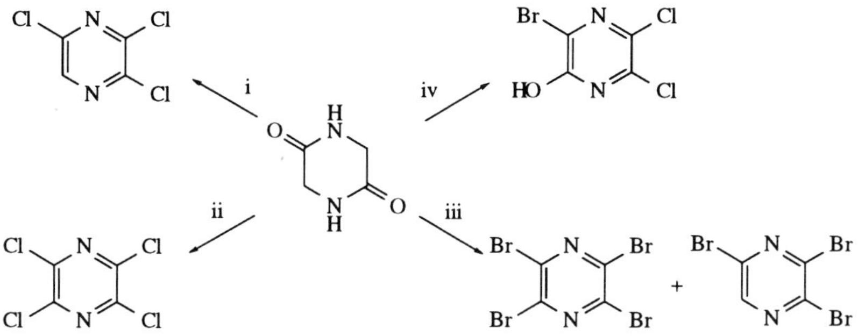

i. $POCl_3/PCl_5/Cl_2$, rt; ii. $POCl_3/PCl_5/Cl_2$, 20 → 140 °C;
iii. $PBr_3/Br_2/DMSO$, 40 → 80 °C; iv. $POCl_3/PCl_5/BrI_2$, rt

Treatment of 3-substituted pyrazine *N*-oxides with phosphorus oxychloride in the represence of an amine affords 3-substituted 2-chloropyrazines regioselectively whereas with chloroacetyl chloride, the corresponding 6-substituted 2-chloropyrazines are obtained instead (N. Sato and M. Fujii, J. heterocyclic Chem., 1994, 31, 1177).

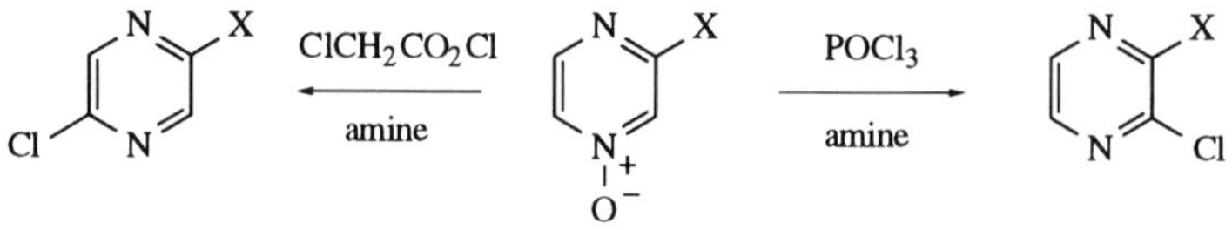

X = CO_2Me, OMe, Cl

2-Trifluoromethylpyrazines derivatives can be prepared in moderate yield by selective *ipso*-substitition of the iodine in the corresponding iodopyrazines (N. Plé *et al.*, J. heterocyclic Chem., 1997, 34, 551).

$ClCF_2CO_2Me$ / KF / CuI

115 °C / 3h / DMF

X = Cl, 50%
X = SPh, 63%

Lithiation of fluoropyrazine with LTMP followed by quenching with electrophiles yields a series of 2,3-disubstituted pyrazines (21). The metallation process could be further extended to allow access to 2,3,6-trisubstituted and 2,3,5,6-tetrasubstituted pyrazines such as (22) and (23) respectively (N. Plé *et al.*, Tetrahedron, 1998, 54, 4899). Directed metallation of iodopyrazine, reported for the first time, provides convenient entry to a range of new 3-substituted 2-iodo-pyrazine derivatives (24) (*idem, ibid.*, 1998, 54, 9701).

E = RCHOH, R_2COH, SPh, halogen, RCO

(21) (22) (23)

i. LTMP (1.2 equiv.)/THF/-78 °C/5 min ii. electrophile/ -78 °C/1 h
iii. LTMP (1.2 equiv.)/THF/-78 °C/5 min iv. TMSCl
v. LTMP (3 equiv.)/THF/-78 °C/5 min vi. I_2 (2 equiv.)/-78 °C/ 1 h
vii. LTMP (2.1 equiv.)/THF/-78 °C/1 h

E = Me, RCHOH, R_2COH, SPh, halogen, RCO

(24)

i. LTMP (1.2 equiv.)/THF/-78 °C/5 min ii. electrophile / -78 °C/1 h

Bipyrazine (25) is conveniently prepared in moderate yield by Ni-catalysed homocoupling of chloropyrazine (Y. Fort, S. Becker and P. Caubère, Tetrahedron, 1994, 50, 11893).

tBuONa / $Ni(OAc)_2$ / PPh_3

65 °C / 10h / DME

(25)

A coupling reaction involving a halopyrazine often provides the key step in the synthesis of several pyrazine-containing natural products or their analogues.

The trisubstituted pyrazine (26), a key intermediate in the synthesis of the antiarythmic agent, arglecin, was efficiently prepared from 2-iodo-6-methoxypyrazine in a sequence involving a directed metallation reaction and a Pd-catalysed cross-coupling of an iodopyrazine with propargyl alcohol (A. Tuck *et al.*, J. heterocyclic Chem., 1994, 31, 1449)

i. LDA / THF / -70 °C ii. iPrCHO / -70 °C

iii. p-TosylOH / Δ

iv. $HC{\equiv}CCH_2OH$ / $Pd(PPh_3)_2Cl_2$ / CuI / NEt_3

v. Pd.C(10%) / H_2 / EtOH

(26)

A series of 5-substituted and 2,5-disubstituted 2-aminopyrazines (27) and (28) respectively, precursors of imidazopyrazinone possessing bioluminescent or chemiluminescent properties, has been prepared in variable yield (35-99%) by the Pd-mediated Stille coupling of 3-bromo- or 3,5-dibromo- 2-amino-pyrazines with the appropriate stannyl compounds in DMF (H. Nakamura, D. Takeuchi and A. Murai, Synlett, 1996, 1227).

Aminopyrazines (29) with similar substitution patterns to (27) and (28) above, can also be prepared in comparable or, in many cases, superior yield by the analogous Suzuki coupling of arylboronic acids to bromo-aminopyrazines (K. Jones, M. Keenan and F. Hibbert, Synlett, 1996, 509).

N NHR Br N X R = H, Ac X = H, Br

(27) R^1 N X

(28) R^1 N R^1

R^1 = Ph, Ar, allyl, 2-furyl, 2-thienyl, nBu

i. $R_1Sn(nBu)_3$ (1 equiv.) / $Pd(PPh_3)_4$ / $(iPr)_2EtN$ / LiCl

ii. $R_1Sn(nBu)_3$ (3 equiv.) / $Pd(PPh_3)_4$ / $(iPr)_2EtN$ / LiCl

Fluorinated semi-synthetic aequorins, required for bioluminescent studies, have also been synthesised via similar Suzuki coupling reactions (T. Hirano *et al.*, Tetrahedron Letters, 1998, 39, 5541).

$ArB(OH)_2$ / (1,4-dppb)$PdCl_2$

(29) 82-96%

An efficient synthesis of pyrazinopsoralen (31) has been achieved by a Suzuki coupling involving the benzofuran derivative (30) and methyl 3-iodo-2-pyrazinoate (D. J. Yoo *et al.*, Bull. Korean chem. Soc., 1995, 16, 1212).

(30)

i. $Pd(PPh_3)_4$ / $NaCO_3$ / $PhCH_3$ / EtOH

ii. BBr_3 / CH_2Cl_2

iii. H^+ / Δ

(31)

(iii) Aminopyrazines

Azidopyrazines, prepared from either the corresponding *N*-oxides or chloropyrazines, can be converted into the corresponding aminopyrazines by either Pd-catalysed hydrogenolysis, or reduction with tin(II) chloride in methanolic hydrochloric acid (N. Sato, T. Matsuura and N. Miwa, Synthesis, 1994, 931).

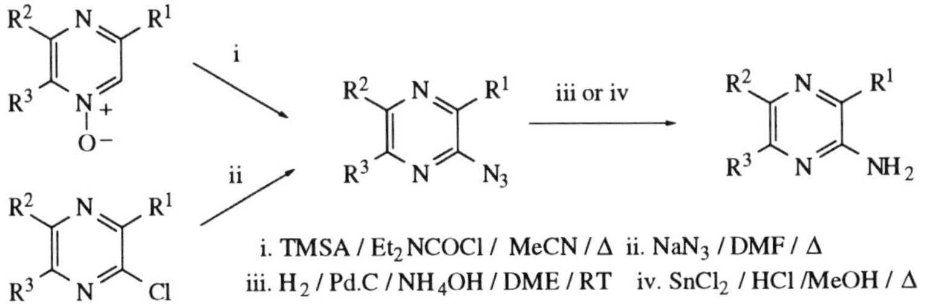
i
ii
iii or iv
i. TMSA / Et_2NCOCl / MeCN / Δ ii. NaN_3 / DMF / Δ
iii. H_2 / Pd.C / NH_4OH / DME / RT iv. $SnCl_2$ / HCl /MeOH / Δ

Direct alkylation of 2-aminopyrazine using organolithium reagents affords the corresponding 3-alkyl substituted products (32, R^1 = H) in variable yield. With the corresponding acylamino derivatives, however, improved yields of (32, R^1 = Ac, Piv) are generally obtained (H. Nakamura, M. Aizawa and A. Murai, Synlett, 1996, 1015).

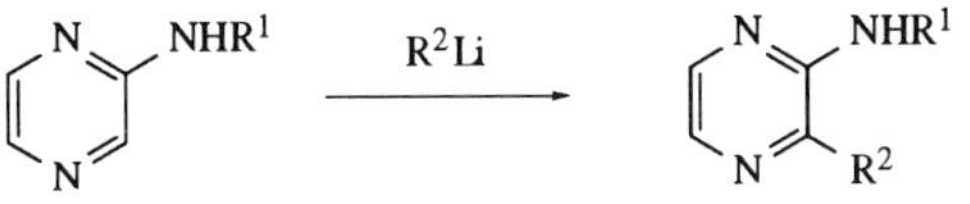
R^2Li

(32)

The iminophosphorane (33), derived from methyl 3-amino-2-pyrazinoate, is transformed into 2,3-disubstituted pteridinones (34) and

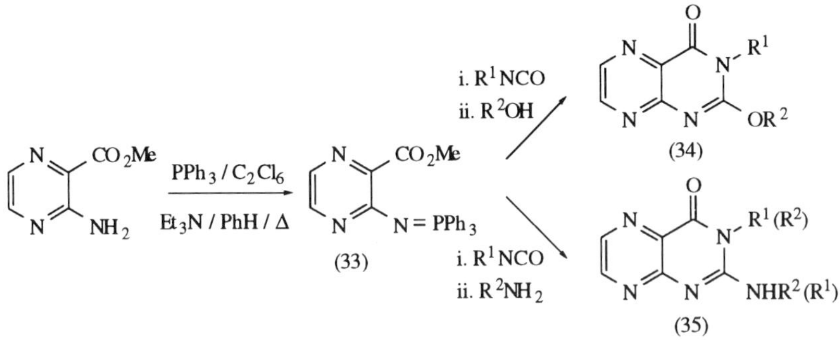
PPh_3 / C_2Cl_6
Et_3N / PhH / Δ
(33)
i. R^1NCO
ii. R^2OH
(34)
i. R^1NCO
ii. R^2NH_2
(35)

(35) in a one-pot reaction procedure involving treatment of (33) with isocyanates followed by heterocylisation on the addition of either alcohols or amines (T. Okawa, S. Eguchi and A. Kakehi, J. chem. Soc., Perkin Trans 1, 1996, 247).

A three-component condensation reaction between 2-aminopyrazine, an aldehyde and an isonitrile catalysed by scandium triflate gives imidazolo[1,2-*a*]pyrazines (36) which are of pharmaceutical interest. This procedure can be readily adapted to combinatorial methods for the synthesis of libraries of analogues of (36) (C. Blackburn *et al.*, Tetrahedron Letters, 1998, 39, 3635).

NH_2 R^1CHO / R^2NC $Sc(OTf)_3$ NHR^2

(36)

Cyclocondensation of 2-amino-3-ethoxypyrazine with α-bromoacetophenone affords the imidazolopyrazine (37) which can be transformed into the hypnotic agent, zolpidem (38). Attempts to prepare analogues of (38) more directly by condensing aminopyrazines with ethyl 3-benzoyl-3-bromopropionate were largely unsucessful (L. Avallone, M. G. Rimolli and E. Abignente, Monat. Chem., 1996, 127, 947).

EtO NH_2 Ph Br EtOH Δ EtO Ph EtO Ph CH_2CO_2H

(37) (38)

(iv) Pyrazinols (Pyrazinones), Alkoxypyrazines, Alkylthiopyrazines and Related Derivatives.

As mentioned above, dipeptidyl chloromethyl ketones can be readily transformed into a variety of pyrazinone derivatives (7) on treatment with hydrochloric acid at reflux temperature (Y. Okada *et al.*, Chem. Comm., 1994, 247; Tetrahedron Letters, 1994, 35, 1231; Tetrahedron, 1995, 51, 7361). This procedure can be adapted to allow the introduction of carboxylate groups using Boc-dipeptidyl chloromethyl ketones containing Asp or Glu residues as substrates (H. Taguchi, T. Yokoi and Y. Okada, Chem. pharm. Bull., 1996, 44, 2037). Optically active deoxyaspergillic acid (39), a natutrally occurring secondary metabolite, has been conveniently synthesised from H-L-Leu-L-Ile-H.HCl

(Y. Okada, H. Taguchi and T. Yokoi, Tetrahedron Letters, 1996, 37, 1231).

HCl.H$_2$N ... OH $\xrightarrow[\text{rt}]{\text{MeCN}}$ (39)

Procedures for the regiospecific enzymatic hydroxylation of pyrazinoic acid and cyanopyrazine have been reported (A. Kiener *et al.*, Synlett, 1994, 814; M. Wieser, K. Heinzmann and A. Kiener, Appl. microbiol. Biotechnol., 1997, 48, 174).

Treatment of 2,6-dimethoxy-3,5-diphenylpyrazine (40) with iodotrimethylsilane yields a mixture of 2,6-dihydroxy- and 2-hydoxy-6-methoxypyrazine (39% and 30% respectively). Under similar conditions, the 3-methyl-5-phenyl compounds gave only the monodemethylated products (*ca.* 20% of each isomer). From UV spectral analysis, 2,6-dihydroxy- and 2-hydroxy-6-methoxypyrazines exist mainly as hydroxypyrazine tautomers (N. Sato *et al.*, J. chem. Soc., Perkin Trans 1, 1997, 3167).

(40) $\xrightarrow[\Delta]{\text{Me}_3\text{SiI / MeCN}}$

Phenylthio- and phenylsulfonylpyrazines (41) are selectively *ortho*-metallated, though product yields are variable, whereas the corresponding phenylsulfinyl derivative appears to undergo nucleophilic displacement. Unlike methylthiopyrazine, the corresponding methylsulfinyl and methylsulfonyl derivatives (42) are metallated initially on the methyl group and then on the ring (A. Turck *et al.*, J. heterocyclic Chem., 1997, 34, 621). Low to moderate yields of *ortho*-substituted products are also obtained from metallation reactions of 2-t-butylsulfinyl- and 2-t-butylsulfonyl-pyrazine (A. Turck *et al.*, J. heterocyclic Chem., 1998, 35, 429).

The photochemical properties of pyrazinones has been reviewed (T. Nishio and C. Kashima, Rev. heteroatom Chem., 1995, 13, 149).

(41)
X = SPh, SO$_2$Ph

i. LTMP
ii. RCHO

X = SPh, 74%
X = SO$_2$Ph, 33%

(42) n = 1,2

(v) Pyrazine N-oxides

Coupling of the indolylmagnesium bromide (43) with the chloromethylpyrazine (44) *N*-oxide to give (45) which is subsequently elaborated into astechrome (46), a naturally occurring hydroxamic acid-iron complex (A. Ohta *et al.*, Chem. pharm. Bull., 1994, 42, 277). The structurally related indolylmethylpyrazine N-oxide OPC-15161 (47), which inhibits superoxide anion generation in biological systems, has been prepared in four steps from tryptophan methyl ester followed by selective methylation of the 5-hydroxy group in (47) (Y. Kita *et al.*, 1994, J. chem. Soc., Perkin Trans. 1, 1994, 875). Cyclocondensation of methyl amino-

(43) (44) (45)

(46)

cyanoacetate and 2-hydroxyimino-4-methylpentanoic acid yields the the highly functionalised *N*-oxide (48) which is further transformed into (49), a useful precursor of (47) (S, Hashizume, A. Sano and M. Oka, Heterocycles, 1994, 38, 1581).

i. DCC, HOSu; ii. NaOH;
iii. NaOAc; iv. DBU then $Me_3O^+BF_4^-$

(47, R = H)
(47, R = Me)

(48) (49)

i. DCC, DMF; ii. 4-dimethylaminopyridine; iii. $PhCH_2Br$, K_2CO_3, DMF
iv. i-$C_5H_{11}ONO$, $CuCl_2$, CuCl, MeCN; v. NaOMe, MeOH; vi. $LiAlH_4$

3-Chloropyrazine 1-oxides undergo Ni-catalysed cross-coupling with dialkylzinc reagent to give the corresponding 3-alkylpyrazine-1-oxides (50) in moderate yield (N. Sato and T. Matsuura, J. heterocyclic Chem., 1996, 33, 1047).

R_2Zn, THF
$NiCl_2[Ph_2(CH_2)_3Ph_2]$

(50)

Pyrazine *N*-oxides with electron-donating substituents in the 3-position afford the 2-azidopyrazines (51) on reaction with trimethylsilyl azide in the presence of diethylcarbamoyl chloride in acetonitrile (N. Sato, N. Miwa and N. Hirokawa, J. chem. Soc., Trans. Perkin 1, 1994, 885; N. Sato, T. Matsuura and N. Miwa, Synthesis, 1994, 931). Treatment of a series of 3-substituted pyrazine-*N*-oxides with acetic anhydride, in some cases with added zinc bromide and triethylamine, yielded predominantly the corresponding 2-acetoxy-6-substituted pyrazines (52) (N. Sato *et al.*, J. heterocyclic Chem., 1994, 31, 1229).

Me_3SiN_3, MeCN; Et_2NCOCl

(51)

R^1 = OMe, NH_2, Ph; R^2 = H, OMe, Ph
R^3 = H, Me, Ph

Me_3SiN_3, MeCN; Et_2NCOCl

(52)

X = CO_2Me, OMe, Cl, Ph, Me

Pyrazine mono- and di-*N*-oxides, apart from 2-hydroxy-substituted derivatives, are efficiently deoxygenated using zinc and aqueous ammonium chloride in THF (Y. Aoyai, T. Abe and A. Ohta, Synthesis, 1997, 891).

The sterically hindered tetramethylpyrazine N-oxide readily participates in the Ru-catalysed oxidation of styrene to styrene oxide (H. Ohtake, T. Higuchi and M. Hirobe, Heterocycles, 1995, 40, 867). From thermochemical studies, the N–O bond dissociation energy of pyrazine 1,4-dioxide is estimated to be 254.0±2 kJ mol^{-1}, similar to that of quinoxaline 1,4-dioxide (W. E. Acree *et al.*, J. org. Chem., 1997, 62, 3724).

(vi) Pyrazinoic acids and Related Derivatives

Hydrocarbons are oxidised, via the corresponding hydroperoxide, to alcohols using pyrazinoic acid and hydrgen peroxide in the presence of vanadium complexes (G. B. Shulpin and co-workers, J. chem. Soc., Perkin Trans 2, 1995, 1459; Tetrahedron, 1996, 52, 13051; Russ. chem. Bull., 1998, 47, 247; J. mol. Catal. A, 1998, 130, 163).

A procedure for the synthesis of methyl 3-aminopyrazinoate (53) from 2,3-pyrazinedicarboxylic acid in multi-gram quantities has been reported; the 3-amino group was efficiently introduced via a Curtius rearrangement. (L. B. Townsend *et al.*, Synthetic Comm., 1996, 26, 617).

i. Ac_2O / Δ; ii. MeOH, rt; iii. $SOCl_2$
iv. NaN_3/acetone/H_2O/ 0°C; v. PhH/Δ

(53)

A series of pyrazinoic acid esters have been found to have broad-spectrum antimycobacterial activity; several esters are significantly more active than pyrazinamide (M. H. Cynamon *et al.*, J. med. Chem., 1995, 38, 3902). Several pyrazinamide derivatives have been prepared and their antituberculotic and antiinflammatory properties evaluated (M. Dolezal *et al.*, Coll. Czech. chem. Comm., 1994; 59, 2562; 1995, 60, 1236; 1996, 61, 1093; 1996, 61, 1102; 1996, 61, 1109). Novel pyrazinoic acid C-nucleosides such as (55), which are potential nucleic acid analogues, have been prepared in good yield by metallation of (54) with LTMP at low temperature, followed by treatment with excess ethyl cyanoformate (J. A. Walker *et al.*, J. med. Chem, 1998, 41, 1236).

1. LTMP/ THF/ -94 °C
2. $NCCO_2Et$ (5 equiv.)

(54) R = Bn (55)

Pyrazinoic acid is regiospecifically hydroxylated by *P. acidovarans* to give 5-hydroxypyrazinoic acid which can be subsequently transformed by conventional chemical procedures into a variety of 5-substituted pyranzoic acid derivatives (A. Kiener *et al.*, Synlett, 1994, 814).

(vii) Cyanopyrazines

2-Cyanopyrazine is converted into the versatile synthetic intermediate 5-hydroxypyrazinoic acid by *Agrobacterium sp.* DSM 6336 in high yield (M. Wieser, K. Heinzmann and A. Kiener, Appl. Microbiol. Biotechnol., 1997, 48, 174).

2,5-Diamino-3,6-dicyanopyrazine (56), a precursor for a series of new fluorescent dyes, has been synthesised in 75% yield by the oxidative coupling of 2,3-diamino-3-(phenylthio)acrylonitrile (M. Matsuoka *et al.*, Dyes and Pigments, 1998, 39, 49).

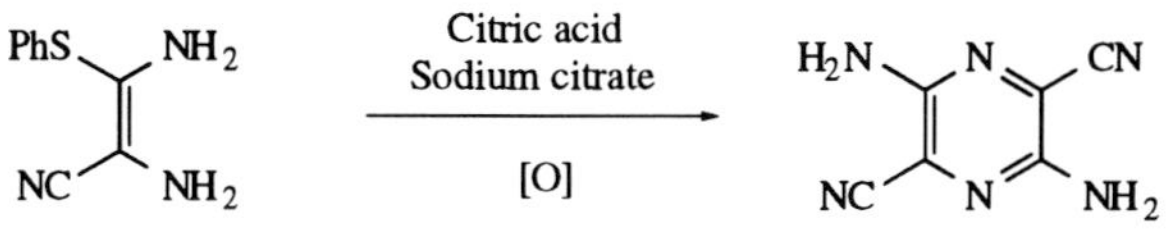

Citric acid
Sodium citrate
[O]

(56)

Alkylation of 5-methylpyrazine-2,3-dicarbonitrile with (*N*-acylanilino)alkyl or hydroxymethyl radicals, generated by the Minisci procedure, yields the corresponding tetrasubstituted pyrazines (57) and (58), which can be further elaborated into pteridine derivatives related to folic acid (M. Tada, Y. Asawa and M. Igarashi, J. heterocyclic Chem., 1997, 34, 973).

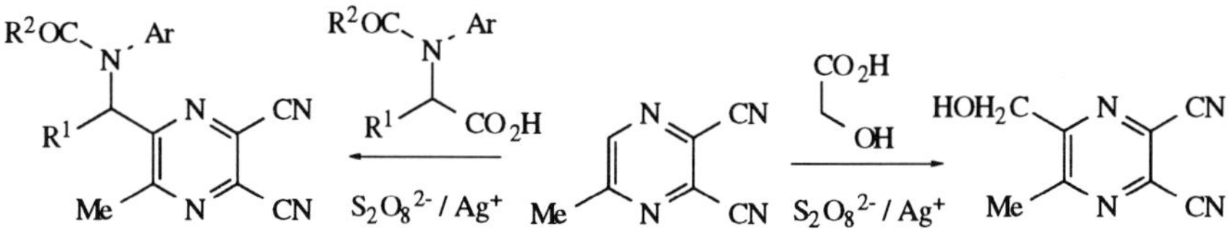

$S_2O_8^{2-}$ / Ag^+

(57) (58)

Enamines, derived from either pyrrolidine or morpholine, and also N-methylindole and 1,2,5-trimethylpyrrole, react readily with 2,3-dichloro-5,6-dicyanopyrazine by nucleophilic displacement of one chlorine to provide the aminovinylpyrazines, e.g. (59) (D. Hou, A. Oshida and M. Matsuoka, J. heterocyclic Chem., 1993, 30, 1571). At elevated temperatures, the pyrrolo[2,3-*b*]pyrazines (60) and the tetrahydropyrazinoindoles (61) are obtained in good yield from the analogous reactions with imines derived from ketones and primary aliphatic amines. (J. Jaung, M. Matsuoka and K. Fukunishi, J. chem. Research-S, 1998, 284). Aromatisation of (61) using NBS gives (62) which, on subsequent treatment with DBN or DBU in ethanol, undergoes cyclotetramerisation to afford tetrapyrazino[2,3-*b*]indoloporphyrazines (J. Jaung, M. Matsuoka and K. Fukunishi, Synthesis, 1998, 1347).

The reaction of 2,3-dichloro-5,6-dicyanopyrazine with ammonia or primary amines affords the 2-amino compounds (63), which cyclodimerise in the presence of base to give the corresponding dihydrodipyrazinopyrazines (64) (J. Jaung, K. Fukunishi and M. Matsuoka, J. heterocyclic Chem., 1997, 34, 653).

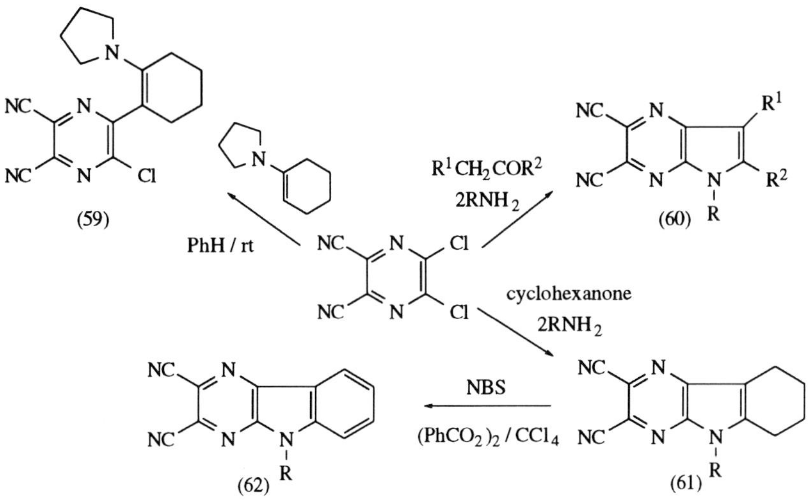

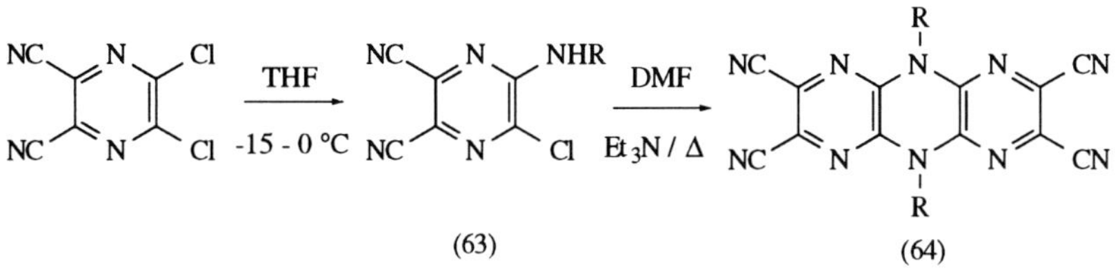

Irradiation of acetonitrile solutions of 2,3-dichloro-5,6-dicyano-pyrazine in the presence of *N*-acyl-*N*-trimethylsilylmethylanilines results in the displacement of a chloro substituent with concomitant formation of (65). Thee reactions are considered to proceed via photo-induced electron-transfer processes (M. Igrashi and M. Tada, Heterocycles, 1994, 38, 2277).

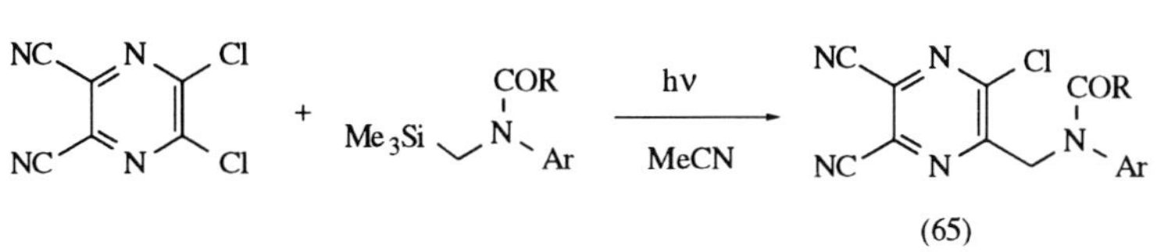

Photochemical reaction of 2,3-dicyanopyrazines with either allylsilanes, or benzylsilane, or ketene silyl acetal was found to give the corresponding mono-substituted products (66-68) respectively in high yield (K. Mizuno *et al.*, Chem. Letters, 1995, 1078). Conversely, the related photoreaction between 5,6-dimethyl-2,3-dicyanopyrazine and allylic silanes in benzene produces the novel tricyclic compounds (69), formation of which can be rationalised by an initial [2+2] photocycloaddition, followed by rearrangement of the resulting adduct (K. Mizuno *et al.*, Tetrahedron Letters, 1997, 38, 5313).

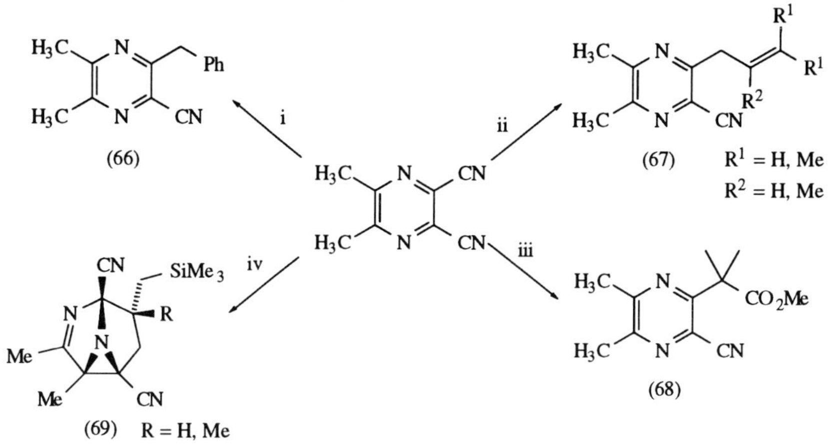

(66) (67) R^1 = H, Me; R^2 = H, Me (68) (69) R = H, Me

i. $PhCH_2SiMe_3$ / hν / MeCN; ii. $(R^1)_2C{=}CR^2CH_2SiMe_3$ / hν / MeCN
iii. $Me_2C{=}C(OMe)OSiMe_3$ / hν / MeCN; iv. $H_2C{=}C(R)CH_2SiMe_3$ / hν / MeCN

(viii) Pyrazine C-nucleosides

The chemistry of pyrazines *C*-nucleosides has been reviewed (M. A. E. Shaban and A. Z. Nasr, Adv. heterocyclic Chem., 1997, 68, pp. 379-384). Recent interest in pyrazine *C*-nucleosides, particularly those compounds which are isosteres of nucleosides commonly found in DNA or RNA, stems from the expectation that these compounds have considerable potential as chemotherapeutic agents.

The isocytidine isostere (70) has been synthesised via a lengthy sequence in which the 6-aminopyrazin-2-one ring was constructed by the cyclocondensation of the amino nitrile (71) with pyruvic aldehyde oxime (J. J. Voegel and S. A. Benner, J. Amer. chem. Soc., 1994, 116, 6929).

Alternatively, in more direct approaches to pyrazine *C*-nucleosides, the pyrazine ring has been introduced as a single structural unit. Thus, the regiospecific Pd-catalysed cross-coupling reaction between the iodopyrazine aglycon (72) and the ribofuranoid glycal (73) yields the novel 2'-deoxypyrazine *C*-nucleoside (74) after selective deprotection and stereoselective hydride reduction (L. B. Townsend *et al.*, Tetrahedron Letters, 1995, 36, 8363; Nucleosides and Nucleotides, 1997, 16, 1999).

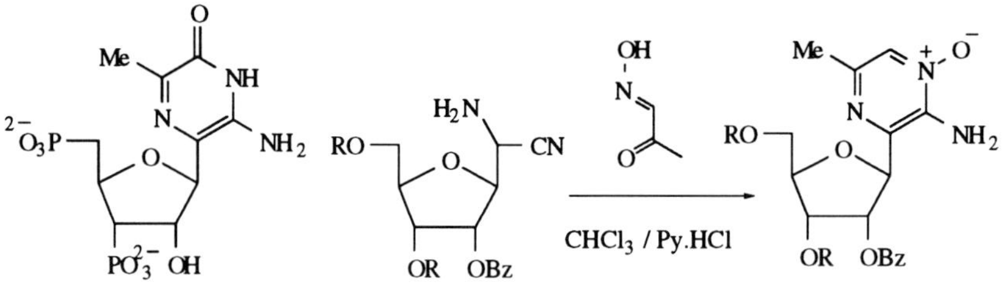

(70) (71)

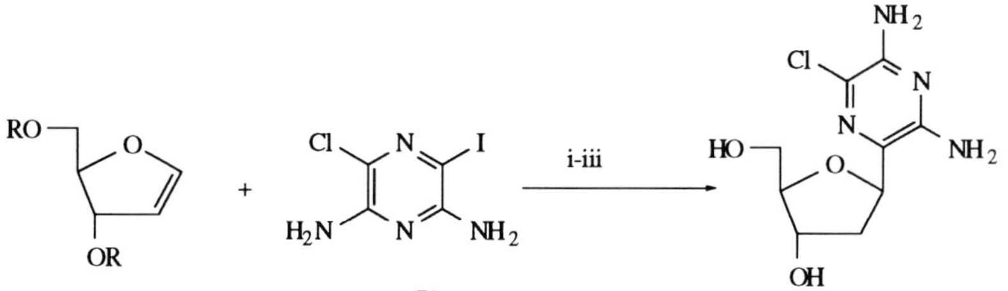

CHCl$_3$ / Py.HCl

(72) R = TBDMS, TBS (73) (74)

i. Pd(OAc)$_2$ / AsPh$_3$ / Et$_3$N / MeCN / 50 °C; ii. TBAF / THF / 25 °C
iii. NaB(OAc)$_3$H / MeCN / 0 °C

The reaction of selectively lithiated pyrazines with 2,3,5-tri-*O*-benzyl-D-ribono-1,4-lactone followed by reductive dehydroxylation of the resulting hemiacetals and removal of the benzyl protecting groups provides an efficient, versatile route to the pyrazine *C*-nucleosides (75) (L. B. Townsend *et al.*, Tetrahedron Letters, 1996, 37, 5325).

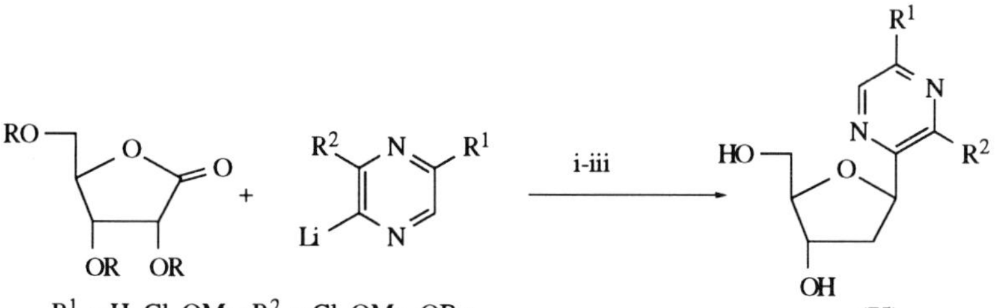

R^1 = H, Cl, OMe; R^2 = Cl, OMe, OBn (75)

i. THF / -78 °C; ii. BF$_3$-OEt$_2$ / Et$_3$SiH / CH$_2$Cl$_2$ / -78 °C→rt
iii. BCl$_3$ or BBr$_3$ / CH$_2$Cl$_2$ / -78 °C→rt or Pd.C / H$_2$ / MeOH

On treatment with acetic acid, the pyrazine *C*-nucleosides (76) rearrange, via polycyclic intermediates, into the pyrido[2,3-*b*]pyrazines (77) (J. J. Chen, D. S. Wise and L. B. Townsend, J. Amer. chem. Soc., 1996, 118, 8953)

(76)

(77)

(e) Reduced pyrazines

(i) Dihydropyrazines

The base-catalysed condensation between ethylenediamine and 2,3-butanedione yields mainly the novel spiro compound (78) (71%) rather than the expected dihydropyrazine (79) (5%) (T. Yamaguchi *et al.*, Chem. pharm. Bull., 1996, 44, 1977). The reaction of the 2,3-dihydropyrazine (80), derived from the condensation of 2,3-butanedione and 1,2-diamino-2-methylpropane, provides a useful route to unsymmetrical α-diones by selective deprotonation at the less hindered methyl group using LDA, alkylation of the resulting anion with alkyl iodides or activated bromides followed by *in situ* hydrolysis (D. Gopal, D. V. Nadkarni and L. M. Sayre, Tetrahedron Letters, 1998, 39, 1877). The 2,3-dihydropyrazine (80), even on storage at -15 °C, dimerises to give the crystalline solid (81) (L. M. Sayre *et al.*, Acta chem. Scand., 1997, 51, 938; T. Yamaguchi *et al.*, *loc. cit.*).

(78)

(79)

(80)

(81)

Reaction of the electron-rich 1,4-bis(triisopropylsilyl)-1,4-dihydropyrazine (82a) with the fullerene C_{60} results in electron exchange with concomitant formation of the radical ion pair $[(82a)^{+\cdot},C_{60}^{-\cdot}]$ indicating that the oxidation potential of (82a) is similar in magnitude to the reduction potential of C_{60}. Although (82a) has a planar structure, the dihydropyrazine ring in the related dimethyl derivative (82b) has a pronounced boat conformation (W. Kaim *et al.*, Chem. Ber., 1995, 128, 745). Carbon dioxide and carbonyl sulphide react with 1,4-bis-(trimethylsilyl)-1,4-dihydropyrazine (83) to give the corresponding insertion products (84); with carbon disulfide, the structurally novel 1:2 cycloadduct (85) is obtained as major isolable product (W. Kaim *et al.*, J. organomet. Chem., 1995, 501, 283).

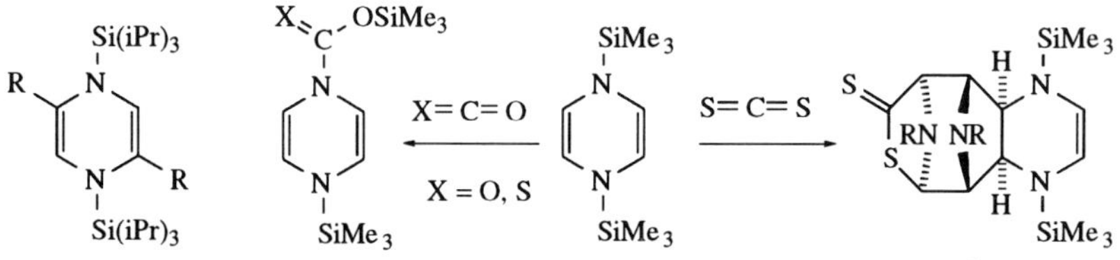

(82) a: R = H
b: R = Me

(84) $X=C=O$; $X = O, S$ (83) $S=C=S$ (85)

A series of *N*-substituted pyrazinone analogues (86) and (87) of the anti-virus agent, acyclovir, have been prepared by either (a) treatment of the appropriate pyrazinone with NaH followed by an alkyl halide, or (b) Lewis acid catalysed coupling of the 2-silyloxypyrazine derivative with α-chloroethers or 1-*O*-acetyl sugars (P. Krauz *et al.*, Nucleosides and Nucleotides, 1998, 17, 875;1489)

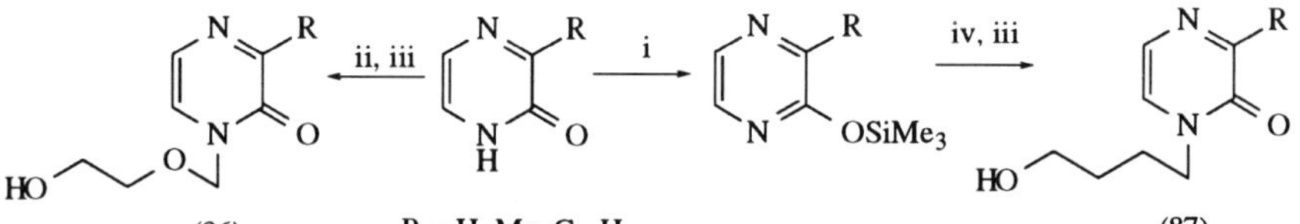

(86) $R = H, Me, C_{10}H_{21}$ (87)

i. HMDS / TMSCl / $C_2H_4Cl_2$; ii. $SnCl_4$ / $ClCH_2O(CH_2)_2OAc$ / CH_2Cl_2

iii. NaOMe / MeOH; iv. NaH / $Br(CH_2)_4OAc$ / DMF

The synthesis and reactions of 3,5-dichloro-2(1*H*)-pyrazinones have been briefly reviewed (G. Hoornaert, Bull. Soc. chim. Belg., 1994, 103, 583). Selective introduction of phenyl and benzyl groups into the 3-position of the 3,5-dichloropyrazinones (88) was accomplished by Pd-catalysed coupling reactions with either $SnPh_4$ or benzyl bromide (via the stannane derivative (89, X = $SnBu_3$)) respectively to yield (89). Subsequent side-chain bromination of (89) followed by treatment with one or two equivalents of sodium methoxide affords the 6-alkylidene /benzylidene derivatives (90) and (91)which are analogues of several bacterial and fungal piperazine-2,5-diones (G. J. Hoornaert *et al.*, J. chem. Soc., Perkin Trans 1, 1996, 231).

(88) R = H, Ph, $CHMe_2$

(89) X = Ph, CH_2Ph

(91)

(90)

i. $(Ph_3P)_4Pd$ / $SnPh_4$ / $PhCH_3$ / Δ
ii. (a) $(Ph_3P)_4Pd$ / $(SnBu_3)_2$ / $PhCH_3$ / Δ, then
(b) $(Ph_3P)_4Pd$ / $PhCH_2Br$ / $PhCH_3$ / Δ
iii. NBS / $(PhCO_2)_2$ / CCl_4 / Δ
iv. NaH / MeOH / THF / rt
v. NaH (2 equiv) / MeOH / THF / rt

Cyclocondensation of diastereomerically pure aminonitriles (92) with oxalyl chloride affords the corresponding atropisomeric pyrazinones (93) with a high degree of stereoselectivity when the *ortho*-substituent is X = I or Br (R. B. Gammill *et al.*, Tetrahedron Letters, 1995, 36, 2017).

$(COCl)_2$, $PhCH_3$ / 70 °C

(92) (S,S)

(93) (aS,S)

Self-condensation of the chloromethylpyrazinone (94) yields the stabilised 1,4-dihydropyrazine analogue (95). The related fused system (96a) can be converted into the corresponding thioxo derivatives (96b) on treatment with phosphorus pentasulfide in pyridine (D. J. R. Brook, B. C. Noll and T. H. Koch, J. chem Soc., Perkin Trans. 1, 1998, 289). Autoxidation of (96a) results in the formation of mixtures of the *trans*-diol (97) and the cleavage product (98); the product ratio varies with solvent (*idem*, J. org. Chem., 1997, 62, 6767).

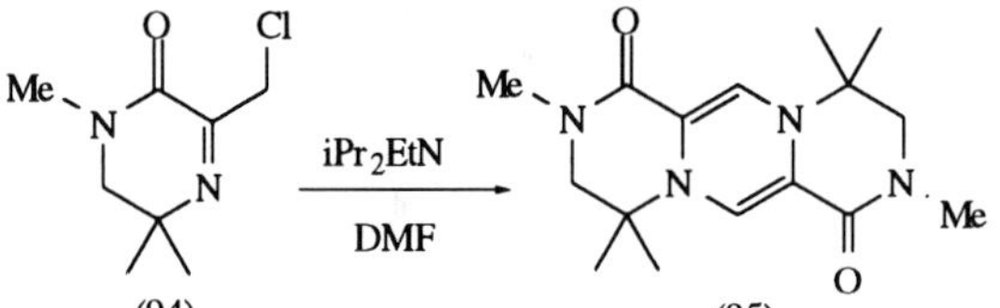

(94) (95)

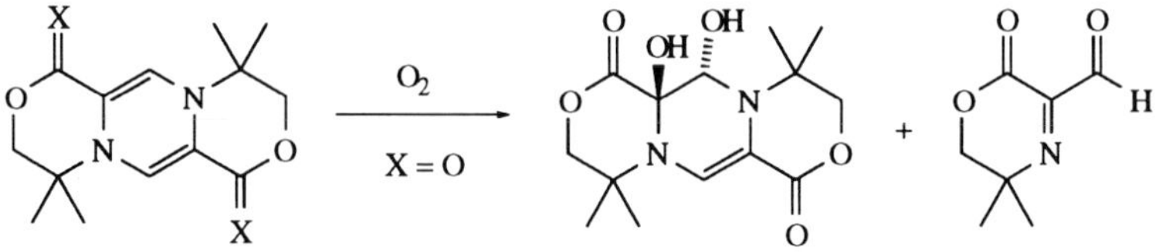

(96) a: X = O
b: X = S
(97) (98)

N-Hydroxypyrazines have been incorporated into a variety of hexadentate ligands which form stable chelates with Fe(III) and Ga(III) ions (J. Ohkanda and A. Katoh, J. org. Chem., 1995, 60, 1583; Tetrahedron, 1995, 51, 12995; Chem. Letters, 1996, 423; J. Ohkanda, H. Shibui and A. Katoh, Chem. Commun., 1998, 375). Photolysis of 1-benzyloxypyrazinones results in extensive N–O bond cleavage; small amounts of rearrangement of the benzyloxy group to C(3) of the ring and [2 + 2] cycloaddition were also observed (A. Katoh *et al.*, Heterocycles, 1996, 43, 883).

The Diels-Alder reaction between the substituted pyrazinones (99) and 1,2,4-triazoline-2,5-dione proceeds readily at room temperature to yield stable cycloadducts (100) (T. Nishio, J. heterocyclic Chem., 1998, 35, 655). Conversely, the analogous reactions involving 3-amino-5-chloropyrazinones and a variety of olefinic dienophiles (e.g. ethene, methyl methacrylate, styrene and *N*-phenylmaleimide) result in

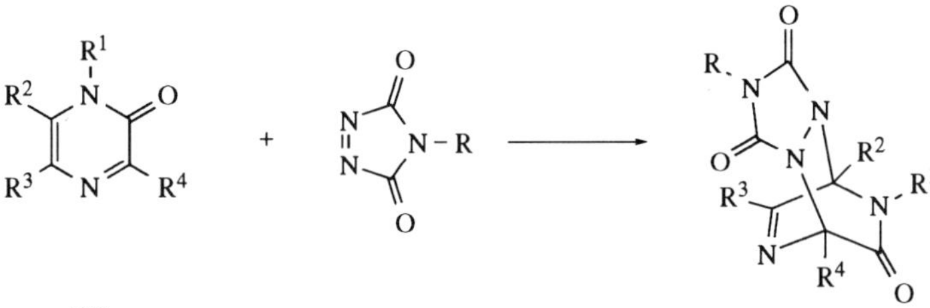

(99) (100)

cycloaddition, followed by elimination of HCl to form tetrahydropyridine derivatives such as (101) (G. J. Hoornaert *et al.*, J. org. Chem., 1996, 61, 304).

Ph N O Cl N NHR + CO_2Me toluene Δ [MeO_2C Cl N NHR N Ph O] -HCl N≡C Ph N O NHR CO_2Me

(101)

Pyrazin-2(1*H*)-ones bearing dienophilic side-chains at the 3-position undergo intermolecular [4 + 2] cycloaddition providing the novel fused pyridinones (102) and (103) depending on the nature of the substituents (D. M. Vanderberghe and G. J. Hoornaert, Bull. Soc. chim. Belg., 1994, 103, 185; G. J. Hoornaert *et al.*, Tetrahedron, 1995, 51, 12463). With 6-(2-propynylaminomethyl) or 6-(2-butynylaminomethyl) pyrazinone derivatives, the intermediate bicyclic adducts can be isolated in favourable instances (K. J. Buysens, D. M. Vanderberghe and G. J. Hoornaert, Tetrahedron, 1996, 52, 9161).

R R^1 N O Cl N O– $(CH_2)_n$ C≡C R^2 Δ [R N O R^1 R^2 Cl N O $(CH_2)_{n-1}$]

n = 2, 3

R^1 R N R^2 O $(CH_2)_{n-1}$ O

(102)

R^1 Cl R^2 N $(CH_2)_{n-1}$ O

(103)

Michael addition of saturated heterocyclic amines to methyl 2-nitro-3-ethoxyacrylate followed by reduction/cyclisation of the nitro group of the resulting intermediate yields bicyclic pyrazines (104) which are structurally related to several natural products (M. A. Brimble and A. D. Johnston, Tetrahedron, 1994, 50, 4887).

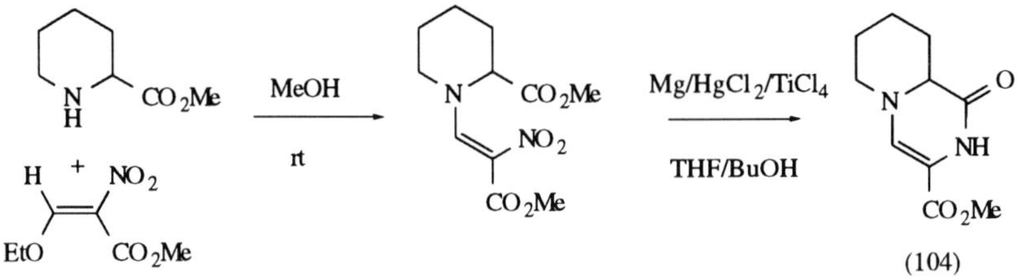

A procedure for the multi-gram synthesis of (3S)-3,6-dihydro-2,5-dimethoxy-3-isopropylpyrazine (105), used as a chiral auxiliary in the synthesis of non-proteinogenic α-amino acids, has been published (S. D. Bull, S. G. Davies and W. O. Moss, Tetrahedron Asymmetry, 1998, 9, 321).

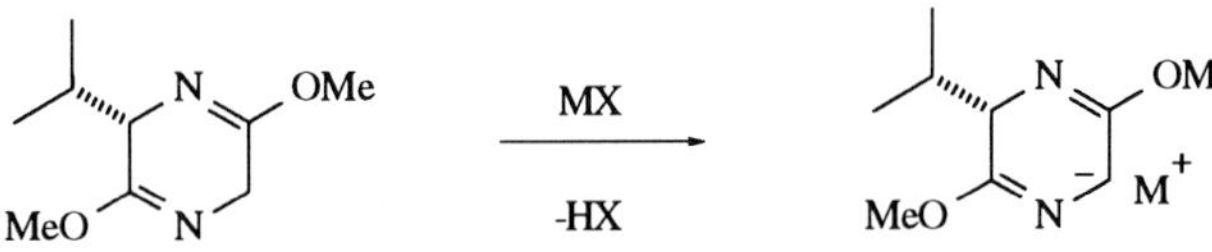

Lithiated anions obtained from (105) participate in a variety of enantioselective reactions including (a) alkylation in the synthesis of the isoxazole containing amino acid (S)-AMPA (P. Pevarello *et al.*, Synthesis, 1996, 1177) and S-*o*-carboranylalanine (W. Karnbrock, H.-J. Musiol and L. Moroder, Tetrahedron, 1995, 51, 1187), (b) aldol condensations featured in the synthesis of the amino sugars elsaminose (M. Ruiz, V. Ojea and J. M. Quintela, Tetrahedron Letters, 1996, 37, 5743) and D-fucosamine (V. Ojea, M. Ruiz and J. M. Quintela, Synlett, 1997, 83), (c) 1,2-addition to diethyl vinylphosphonate (A. Schick *et al.*, Tetrahedron, 1995, 51, 11207) and (d) 1,6-addition to butadienylphosphates (J. M. Quintela et al., Tetrahedron Letters, 1997, 38, 4311). Cuprates derived from (105) also undergo (a) alkylation (J. Zhu *et al.*, J. org. Chem., 1996, 61, 771; 9309) and (b) conjugate 1,6-addition to dienones (H. Wild, J. org. Chem., 1994, 59, 2749).

Novel α-amino-β-hydroxy acids have been prepared from the (R)-enatiomer of (105) by a sequence involving alkylation, aldol condensation, Ru-catalysed ring closure metathesis and hydrolysis (K. Hammer and K. Undheim, Tetrahedron, 1997, 53, 5925).

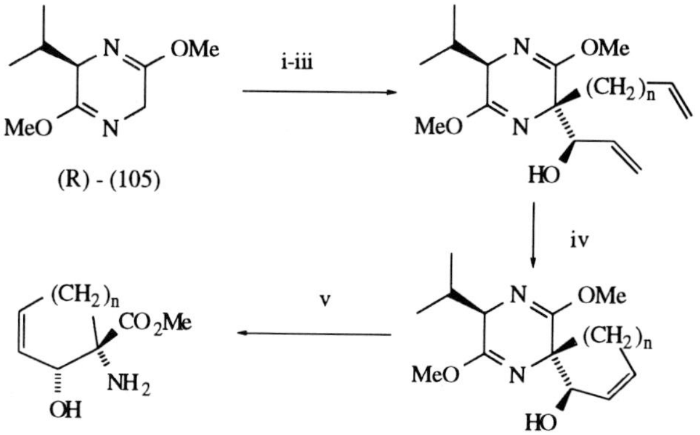

i. (a) BuLi/THF/-78 °C, (b) $CH_2=CH(CH_2)_nBr$; ii. (a) BuLi/THF/-78 °C, (b) $CH_2=CHCHO$; iii. isomer separation; iv. 2% Ru(II)/benzene/rt; v. aq TFA/MeCN

Protonation of related lithiated anions derived from the bis(lactim) ethers (106) affords predominantly the *trans*-diastereoisomer when R^1 = iPr and R^2 = Me (S. Hünig, N. Klaunzer and H. Wenner, Chem. Ber., 1994, 127, 165).

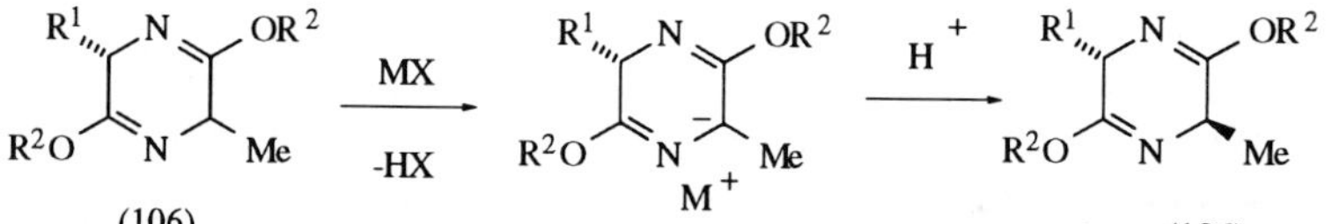

(ii) Tetrahydropyrazines

Asymmetric hydrogenation of the tetrahydropyrazine (107) in the presence of the catalyst [(R)-BINAP(COD)Rh]OTf afforded the corresponding piperazine in high yield (96%) and high ee (99%) (K. Rosen *et al.*, Tetrahedron Letters, 1995, 36, 6419).

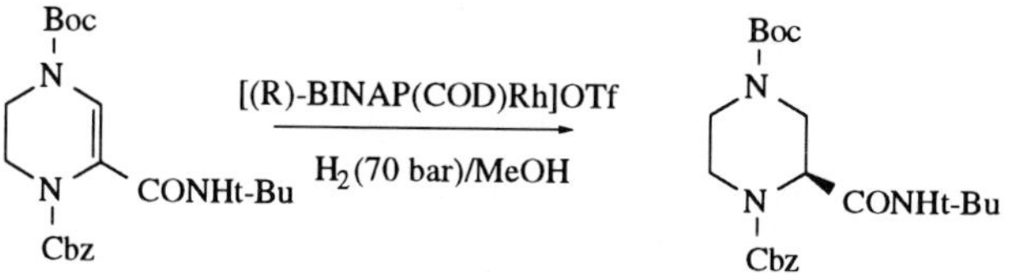

Photooxygenation of 1,4-diacetyl-5,6-diphenyl-1,2,3,4-tetrahydropyrazine (108) yields the expected labile but isolable 1,2-dioxetane derivative which on subsequent thermolysis at 20 °C undergoes ring cleavage with concomitant chemiluminescence. Epoxidation of (108) with m-CPBA is also accompanied by chemiluminescence (W. Adam, M. Ahrweiler and P. Vlček, J. Amer. chem. Soc., 1995, 117, 9690).

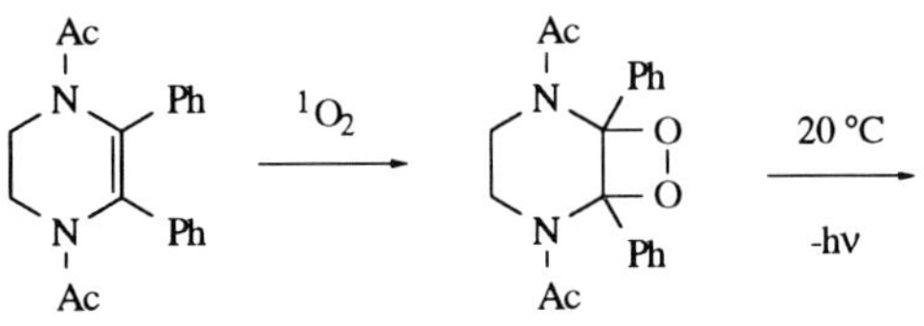

(108) 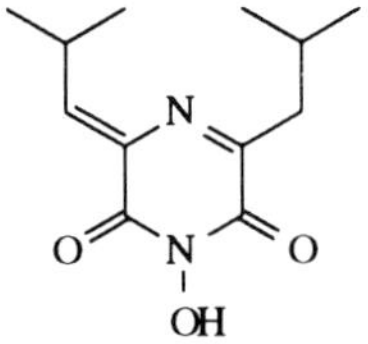 (109)

Flutimide (109), a novel natural product which inhibits influenza virus A, has been prepared in eight steps from L-leucine (S. B. Singh, Tetrahedron Letters, 1995, 36, 2009).

(iii) Hexahydropyrazines (Piperazines)

1,4-Disubstituted piperazines have attracted considerable attention on account of their diverse pharmacological activity. Syntheses of (a) the antihistamine cetrizine hydrochloride (110) (C. J. Opalka *et al.*, Synthesis, 1995, 766), (b) the potential antipsychotic agent (111) (N. Gil, P. Bosch and A. Guerrero, Tetrahedron, 1997, 53, 15115), (c) the growth hormone secretagogues (112) (K. J. Barakat *et al.*, Bioorg. med. Chem. Letters, 1998, 8, 1431), and (d) the potential CNS active compound (113) (R. Ollivier *et al.*, Heterocycles, 1997, 45, 723) have been reported.

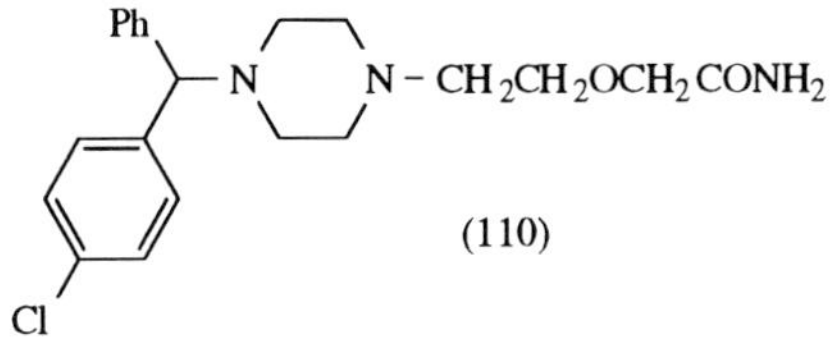

(110) 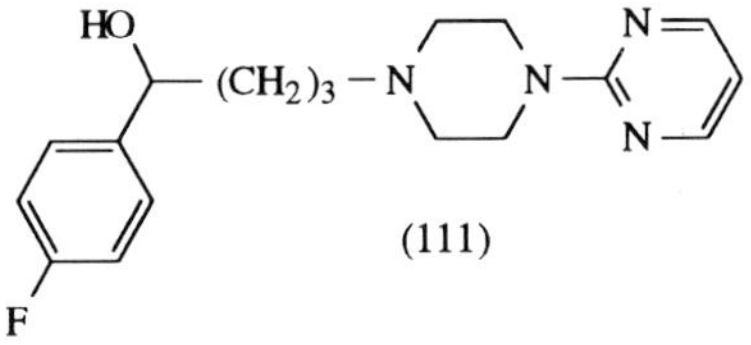 (111)

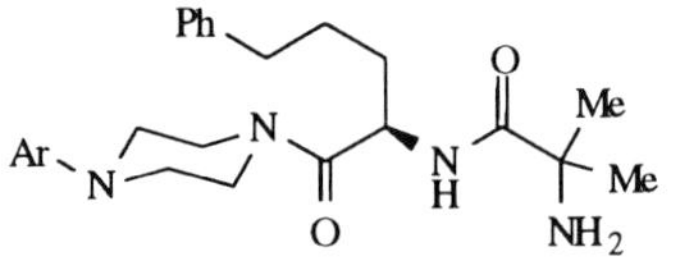

(112) 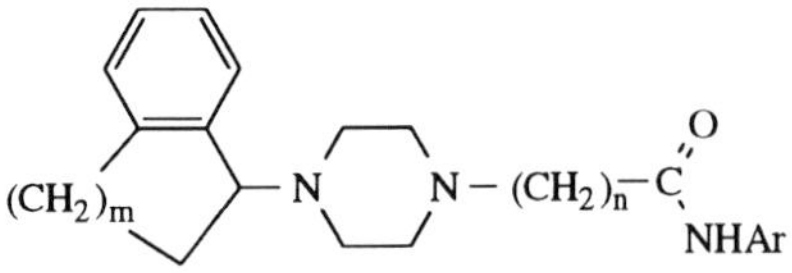 (113)

Simple *N*-alkyl or *N*-aryl piperazines can be readily prepared (20 - 90%) by treatment of 2-oxazolidinone derivatives (114) with HBr in glacial acetic acid followed by heating the resulting ring-opened salts in alcoholic solvents (G. S. Poindexter *et al.*, Tetrahedron Letters, 1994, 35, 7331).

HBr/HOAc, rt; EtOH, Δ

(114)

Improved procedures for the classical synthesis of simple 1-arylpiperazines by the condensation of bis(2-haloethyl)amines with substituted anilines have been reported including the use of (a) microwave radiation (H. G. Jaisinghani and B. M. Khadilkar, Tetrahedron Letters, 1997, 38, 6875) and (b) alumina solid supports (M. J. Welch *et al.*, Tetrahedron Letters, 1996, 37, 319). Microwave irradiation of mixtures of ethanolamine and anilines also provides 1-arylpiperazines in good yield (H. G. Jaisinghani, B. R. Chowbury and B. M. Khadilkar, Synth. Comm., 1998, 28, 1175).

H_2N-Ar, microwave irradiation

X = Br, Cl, OH, OTs

S_NAr displacement reactions between piperazine and activated fluoroaromatic compounds in either (a) solution phase (L. Neuville and J. Zhu, Tetrahedron Letters, 1997, 38, 4091), or (b) solid phase (S. M. Dankwardt, S. R. Newman and J. L. Krstenansky, Tetrahedron Letters, 1995, 36, 4923; Y. Yamamoto, K. Ajito and Y. Ohtsuka, Chem. Letters, 1998, 379) also affords 1-arylpiperazines conveniently. Alternatively, a variety of other haloarenes have been aminated by (a) unprotected piperazines (S.-H. Zhao *et al.*, Tetrahedron Letters, 1996, 37, 4463; M. Nishiyama, T. Yamamoto and Y. Koie, *ibid.*, 1998, 39, 617) and (b) *N*-Boc protected piperazine (F. Kerrigan, C. Martin and G. H. Thomas, *ibid.*, 1998, 39, 2219) in the presence of appropriate Pd-catalysts.

Treatment of piperazine, linked as a carbamate to a Wang resin, with cyanoacetic acid anhydride affords the corresponding cyanoacetamide which condenses readily with carbonyl compounds and iminoethers under mild Knoevenagel conditions to provide the novel monoacetylated piperazines (e.g. 115) after cleavage from the solid support (F. Zaragoza, *ibid.*, 1995, 36, 8677).

P O O N N H (NC O)$_2$O P O O N N O CN i, ii H_2 N + $CF_3CO_2^-$ N O CN Ar

(P) = Wang polymer support

i. ArCHO/DMF/piperidine; ii. TFA/CH_2Cl_2

(115)

2-(Trifluoromethyl)piperazine has been prepared in modest yield by the condensation of 3,3,3-trifluoro-2-oxopropanal with *N*-benzylethylenediamine, followed by *in situ* reduction with sodium cyanoborohydride and catalytic hydrogenolysis (L. W. Jenneskens *et al.*, Rec. Trav. chim., 1995, 114, 97).

NH_2 NH Bn + O CF_3 O H i. DMF ii. $NaBH_3CN$ H N CF_3 N Bn Pd.C cyclohexene H N CF_3 N H

(116)

The enantiomers of 1-allyl-*trans*-2,5-dimethylpiperazine (117), obtained by optical resolution of the racemate using (+)- or (-)-camphoric acid, have been transformed into the δ-opioid receptor agonist SNC 80 (118, R = OMe) and related compounds (K. C. Rice *et al.*, J. med. Chem., 1997, 40, 695).

O Et_2N Cl R + H N Me Me N i, ii O Et_2N H R N Me Me N

(-)-(117)

i. K_2CO_3/MeCN/Δ; ii. crystallise from aq. MeCN

(+)-(118)

A total synthesis of the marine alkaloid dragmacidin B (119) has been achieved in three steps from 1,4-dimethylpiperazine-2,5-dione (C. R. Whitlock and M. P. Cava, Tetrahedron Letters, 1994, 35, 371).

i. NBS/AIBN/CCl_4

ii. 6-bromoindole/DMF

iii. BH_3.THF

(119)

Several syntheses of piperazine-2-carboxylic acid, a conformationally restricted, nonproteinogenic amino acid, have been reported. The *N,N'*-orthogonally protected (S)-piperazine-2-carboxylic acid (120) has been prepared in four steps and in 40% overall yield from *N*-t-Boc-L-serine β-lactone (A. M. Warshawsky, M. V. Patel and T.-M. Chen, J. org. Chem., 1997, 62, 6439). Cyclisation of L- or D-serine with α-amino acids followed by $LiAlH_4$ reduction of the resulting piperazine-2,5-dione yields piperazines (121) which on *N*-protection, side-chain oxidation and deprotection furnish the piperazine-2-carboxylic acid derivatives (122) without loss of chirality at C(5) (M. Falorni *et al.*, Synthesis, 1994, 391).

(120)

R = iPr, iBu

(121)

(122)

i. $(Boc)_2O$/MeCN; ii. $NaIO_4/RuO_2$/aq. acetone/rt
iii. aq. HCl/EtOAc/rt

Piperazine-fused β-lactams, derived from the [2 + 2] cycloaddition of alkoxyketenes and chiral imines, are transformed into the anhydrides (123) on treatment with sodium hypochlorite with TEMPO as a catalyst. On reaction with nucleophiles such as amino acid esters, the anhydrides (123) ring open with concomitant decarboxylation to yield the dipeptides (124) (C. Palomo *et al.*, Tetrahedron Letters, 1997, 38, 4643).

NHBoc
R
TBDMSO
N
Ph
i, ii
BnO
H
NHBoc
R
O
N
OH
Ph
iii-vii
BnO
H
R
NBoc
O
N
Ph
viii

R = Me, iPr

Boc
N
R
Ph
N
H
O
H
N
O
OMe
R'
(124)
ix
O
H
R
NBoc
O
N
O
Ph
(123)

i. $BnOCH_2COCl/Et_3N/CH_2Cl_2$; ii. TBAF/THF/rt; iii. $MsCl/Et_3N/CH_2Cl_2$; iv. TFA/CH_2Cl_2
v. Et_3N/CH_2Cl_2; vi. $(Boc)_2O/CH_2Cl_2$; vii. H_2/Pd.C/MeOH; viii. TEMPO/NaOCl
ix. (L)-$H_2NCHR'CO_2Me/CH_2Cl_2$

Racemic piperazine-2-carboxamide is kinetically resolved by the enzyme L-leucine aminopeptidase to produce the (S)-acid (125) and the (R)-amide, each in high optical purity (N. Balasubramanian *et al.*, Synthetic Comm., 1995, 25, 2673).

H
N
$CONH_2$
N
Boc
±
i, ii
H
N
CO_2H
.2HCl
N
H
(125)
+
H
N
$CONH_2$
N
Boc

i. L-leucine aminopeptidase, pH 8.6; ii. Dowex 1X8 anion exchange resin

A one-pot, four-component Ugi condensation between an *N*-alkyl-ethylenediamine, chloroacetaldehyde, an isonitrile and a carboxylic acid affords the piperazine-2-amide derivatives (126) in moderate but acceptable yields (35-65%). This procedure is not however suitable for

R^1
NH
NH_2
+
Cl
O
H
+ R^3-N ≡ C
O
+ R^2 - C
OH
$NaHCO_3$
MeOH
R^1
N
N
R^2
O
O
H
N
R^3

R^1 = H, alkyl; R^2 = H, alkyl, aryl; R^3 = alkyl

(126)

asymmetric syntheses of piperazines such as (126) (K. Rossen, J. Sager and L. M. DiMichele, Tetrahedron Letters, 1997, 38, 3183).

Optically active 2,6-disubstituted piperazines (128) are obtained from the lactam (127) by a reaction sequence involving diastereoselective alkylation, reductive removal of the lactam carbonyl group, metallation and reaction with an appropriate electrophile, and finally removal of the *N*-protecting groups (J.-C. Quirion *et al.*, Synthesis, 1996, 833).

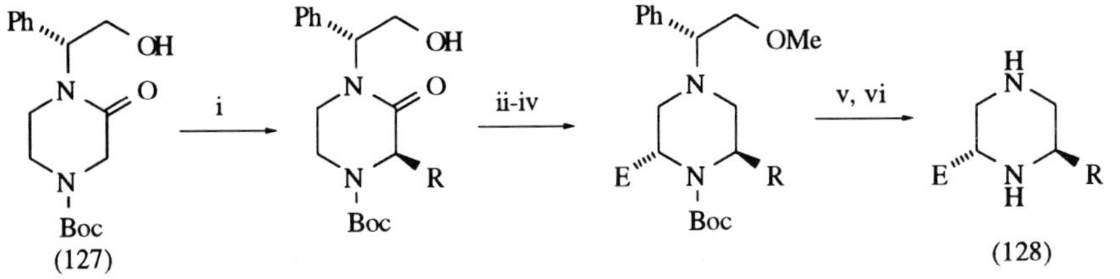

(127) (128)

i. tBuLi/THF/HMPA/R-X; ii. $BH_3.Me_2S$/THF; iii. NaH/THF/DMF/MeI; iv. sBuLi/THF/E-Y; v. $H_2/Pd(OH)_2$/MeOH; vi. aq. HCl/MeO

R = Me, $PhCH_2$
E = Me, CO_2H, RCHOH

(iv) Ketopiperazines

The novel nitrone (129), derived from 1-benzylpiperazin-2-one, undergoes regioselective [3+ 2] cycloadditions with alkynes and alkenes to yield Δ^4-isoxazolidines and isoxazolidines respectively, which are subsequently converted into 1-benzylpiperazin-2-one derivatives (130) and (131) by reductive cleavage of the N–O bond (R. C. Bernotas and G. Adams, Tetrahedron Letters, 1996, 37, 7339). Nitrone (129) has also been transformed into the constraines *N*-arylpiperazinone (132) by a sequence involving a [3+ 2] cycloaddition, reductive N–O bond cleavage and an intramolecular S_NAr reaction (*idem*, *ibid.*, 1996, 37, 7343).

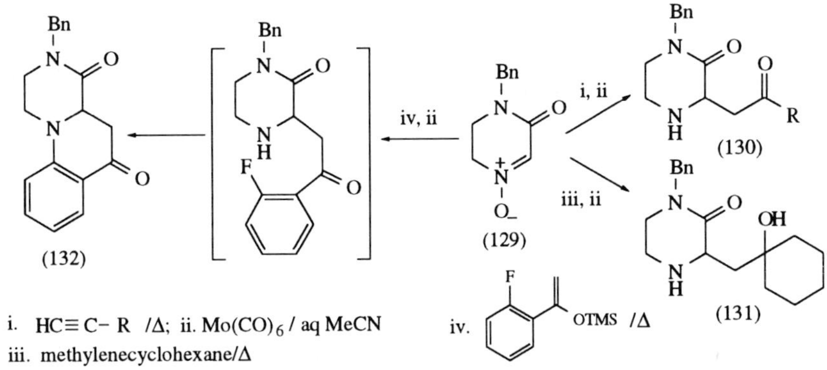

(132) (129) (130) (131)

i. HC≡C–R /Δ; ii. $Mo(CO)_6$ / aq MeCN
iii. methylenecyclohexane/Δ
iv. (2-fluorophenyl)C(=CH₂)OTMS /Δ

N-Arylpiperazinones (133) were readily obtained from the corresponding anilines by sucessive treatment with chloroacetyl chloride then ethanolamine, followed by intramolecular Mitsunobu cyclodehydration of the resulting amidoalcohols (S. A. Weissman *et al.*, Tetrahedron Letters, 1998, 39, 7459).

R^1 N OH O N H Ar

PBu_3/EtOAc

di-t-butyl azodicarboxylate or
1,1'-(azocarbonyl)dipiperidine

R^1 N O N Ar

(133)

Reductive alkylation of the ethylenediamine salt (134) with either an aldehyde or a ketone in the presence of sodium triacetoxyborohydride affords the unsymmetrically 1,4-disubstituted 2,3-diketopiperazines (135) in good yield (C. J. Dinsmore and J. M. Bergman, J. org. Chem., 1998, 63, 4131).

NH_2.HCl O OMe N Ph O

R^1 O R^2

$Na(AcO)_3BH$

R^1 R^2 N O N O Ph

(134) (135)

Treatment of 1,4-dialkylpiperazine-2,3-diones (136) with organolithium reagents provides a convenient method for the synthesis of symmetrically and unsymmetrically substituted α-diones depending on the nature of the *N*-alkyl group R (U. T. Mueller-Westerhoff and Ming Zhou, J. org. Chem., 1994, 59, 4988).

R N O N O R

i. R^1Li/THF/-78 °C
ii. R^2Li/THF/-78 °C

R N R^1 O^- Li^+ O^- Li^+ N R^2 R

H_3O^+

R^1 O O R^2

(136) R = Me, n-decyl

1,4-Disubstituted piperazine-2,5-diones (137) undergo free-radical bromination significantly faster when R = Me than when R = Ac (T. W. Badran *et al.*, Aust. J. Chem., 1995, 48, 1379). Under similar conditions, piperazine-2,5-diones (138a-c) selectively yield the monobrominated compounds (139a-c), whereas with (138d) the dibromo compound (140) is formed instead. Compound (140) can be debrominated on treatment with sodium iodide in acetonitrile to provide the α-methylene derivative (141) (C. L. L. Chai, D. B. Hay and A. R. King, Aust. J. Chem., 1996, 49, 605). The rate of bromination of the *trans*-isomer (142) is found to be about nine times faster than that of the *cis*-isomer (143); these differences have been rationalised on the basis of differences in the conformations adopted by each isomer (C. L. L. Chai, D. C. R. Hockless and A. R. King, Aust. J. Chem., 1996, 49, 1229).

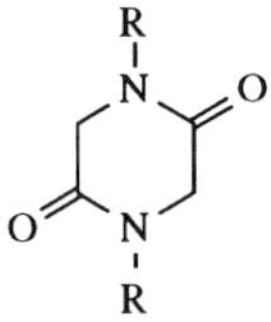

(137)

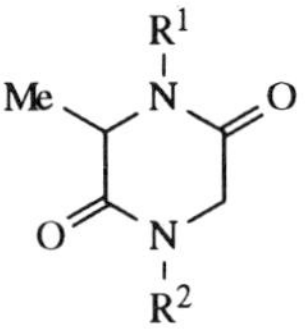

(138a) $R^1 = R^2 = Me$
(138b) $R^1 = R^2 = Ac$
(138c) $R^1 = Ac$, $R^2 = Me$
(138d) $R^1 = Me$, $R^2 = Ac$

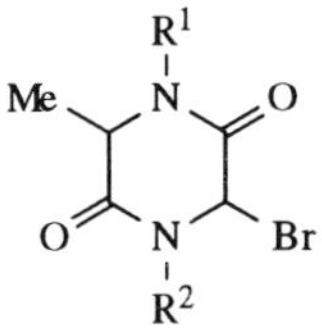

(139a) $R^1 = R^2 = Me$
(139b) $R^1 = R^2 = Ac$
(139c) $R^1 = Ac$, $R^2 = Me$

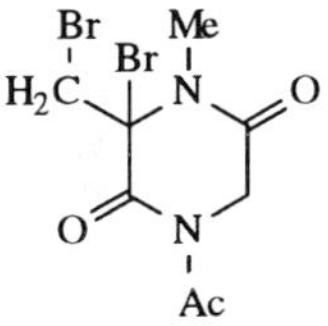

(140)

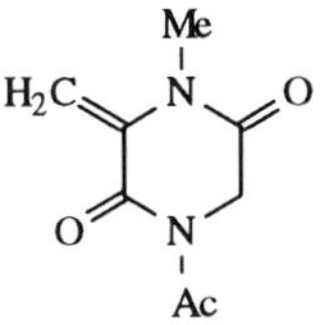

(141) 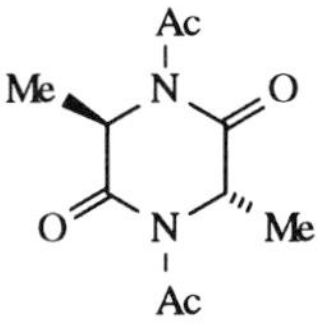 (142) 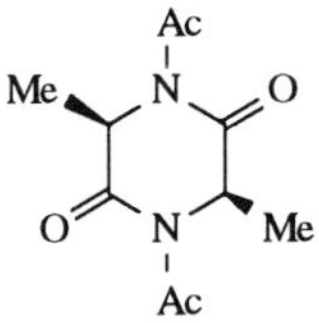 (143)

Brominated piperazine-2,5-diones (144) can readily transformed into α-methylene derivatives (145) via the Arbuzov reaction to give the required phosphonates, followed by the Wadsworth-Emmons reaction (C. L. L. Chai and A. R. King, J. chem. Res. S, 1997, 382). Radical addition reactions to (145, R = Me, $R^1 = R^2$ = Ac) are reported to take place in a stereoselective manner (*idem, ibid.*, Tetrahedron Letters, 1995, 36, 4295).

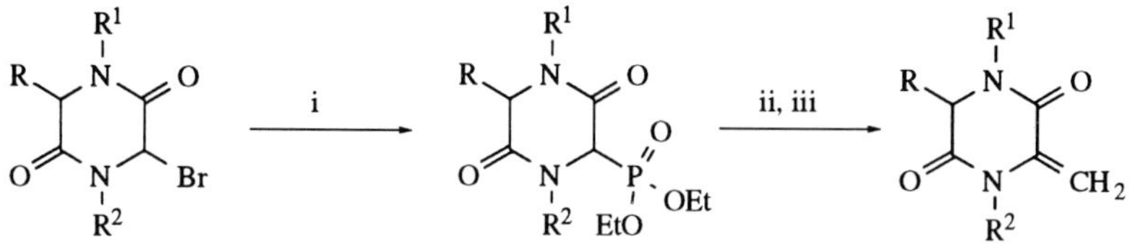

(144) (145)

i. $P(OEt)_3/CH_2Cl_2/\Delta$; ii. NaH; iii. paraformaldehyde

Based-catalysed condensation of (146) with benzaldehyde (147) yields (148) which can be elaborated to give (149). This compound serves as an ABC ring analogue of the antibiotic sanframycins (A. Kubo *et al.*, Chem. pharm. Bull., 1997, 45, 1120).

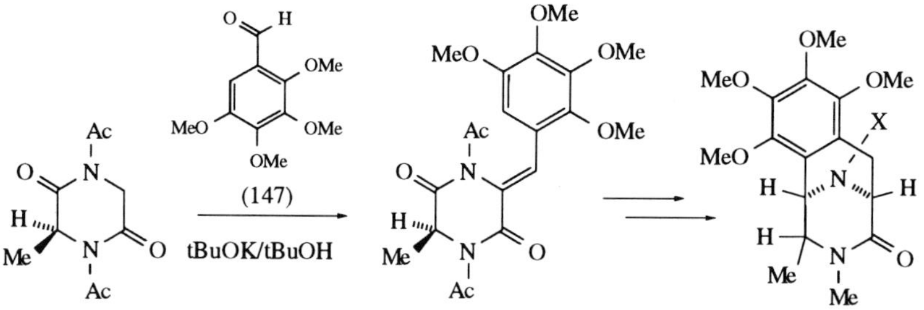

(+) - (146) (147) tBuOK/tBuOH (148) (149)

(3S)-*N,N'*-Bis(*p*-methoxybenzyl)-3-isopropylpiperazine (150) is found to have enhanced diastereocontrol as compared to Schölkopf's chiral auxiliary (105) (*vide supra*) during enolate alkylation (S. G. Davies *et al.*, Chem. Comm., 1998, 659).

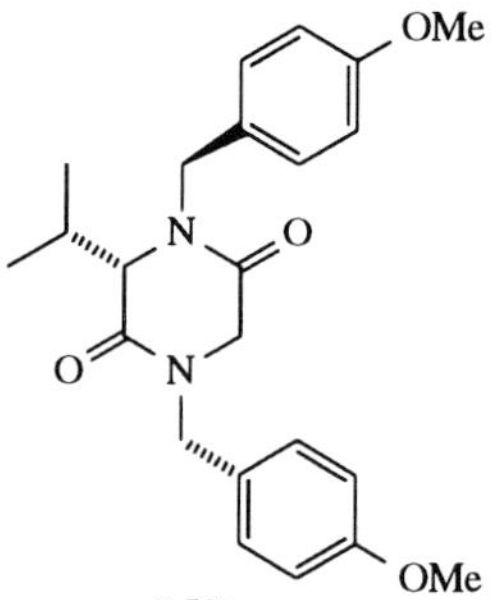

(150)

(v) Miscellaneous fused pyrazines

Syntheses of a variety of conformationally restricted piperazines have been reported including (a) the tricyclic piperazine (151) derived from (R)-cysteine A. Gonzalez, S. L. Vorob'eva and A. Linares, Tetrahedron Asymmetry, 1995, 5, 1357), (b) fused indenopiperazines (152) (M. Gelbcke *et al.*, Bull. Soc. chim. Belg., 1996, 105, 287), and (c) the 2,7-disubstituted octahydropyridopyrazines (153) (F. Compernolle *et al.*, Tetrahedron, 1994, 50, 1811).

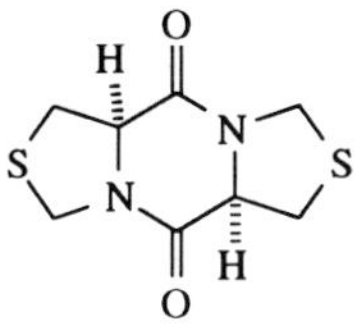

(151) 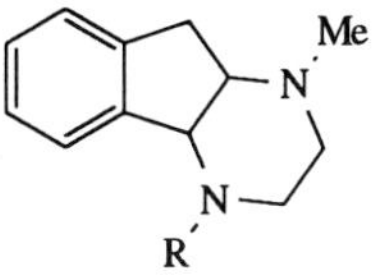 (152) 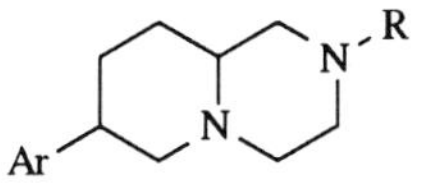 (153)

2. Quinoxalines, benzopyrazines

(a) Introduction

This second supplement is intended to cover advances in the chemistry of quinoxalines over the period 1994 until June 1998. These have largely centred on the synthesis and properties of biologically active quinoxaline derivatives.

Chloroquinoxaline sulfonamide (CQS) (1) has been shown to have considerable potential as an anticancer agent (A. B. Conley *et al.*, Cancer Chemother. Pharmacol., 1995, 37, 139; 1995, 37, 483;1997, 40, 415). The X-ray crystal structure of CQS (1) has been determined (M. Liu, J. R. Ruble and S. K. Aroba, Acta Crystall., 1994, C50, 2032). Highly substituted quinoxaline 1,4-dioxides such as (2) are selective cytotoxins of hypoxic cells in solid tumours (A. Monge *et al.*, J. heterocyclic Chem., 1995, 32, 1213; J. med. Chem., 1995, 38, 1786).

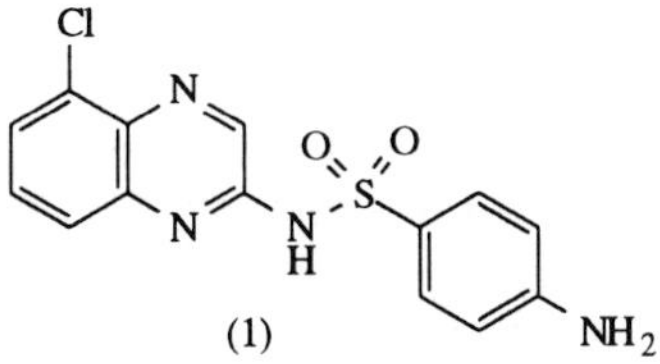

(1) 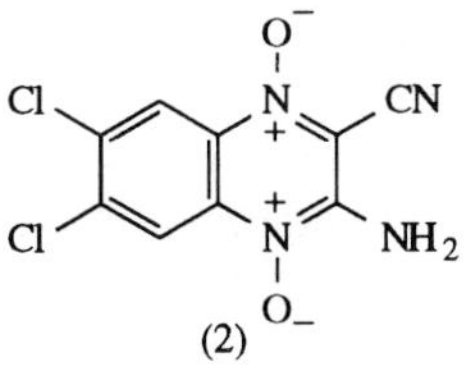 (2)

6,7-Dichloro-5-nitroquinoxalin-2,3-dione (3) has been found to be a potent antagonist at *N*-methyl-D-aspartate (NMDA)/glycine sites in the brain (R. M. Woodward *et al.*, Mol. Pharmacol., 1995, 47, 568; Euro. J. Pharmacol., 1995, 273, 187).

Aryl-substituted quinoxalines, e.g. (4), prepared by Pd-catalysed, cross-coupling reactions, exhibit useful herbicidal properties (review: T. P. Selby *et al.*, A. C. S. Symposium Series, 1995, 584, 171).

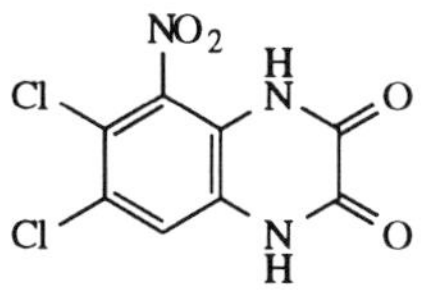

(3) 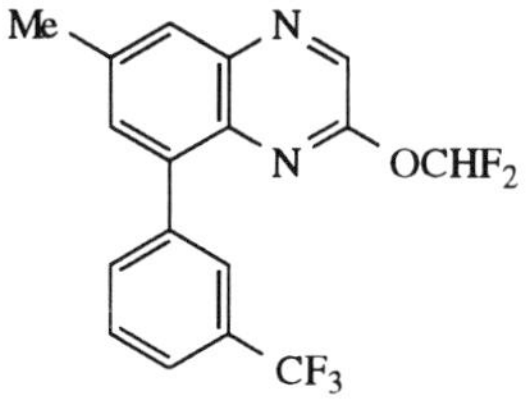 (4)

Recent reviews on quinoxaline chemistry include (a) a general review of the chemistry of 1,4-diazines which includes quinoxalines (N. Sato "Comprehensive Heterocyclic Chemistry II", Vol. 6, ed. A. R. Katritzky, C. W. Rees and E. F. V. Scriven, Pergamon, Oxford, 1997, pp. 233), and (b) recent advances in the chemistry of quinoxaline *N*-oxides and *N,N'*-dioxides (Y. Kurasawa, A. Takada and H. S. Kim, J. heterocyclic Chem., 1995, 32, 1085).

(b) Physical and spectroscopic properties

Ab initio SCF-MO calculations have shown that, in common with other diazines, the quinoxaline HOMO energy correlates well with pKa and first π-electron ionization energy (H. J. S. Machado and A. Hinchcliffe, Theochem - J. mol. Structure, 1995, 339, 255). Calorimetry has been used to determine (a) the standard enthalpies of formation of quinoxaline (M. A. V. R daSilva *et al.*, J. chem. Soc., Faraday Trans., 1995, 1907) and 2,3-dimethylquinoxaline (M. A. V. R daSilva *et al.*, Struct. Chem., 1996, 7, 329), and (b) the strength of the interactions between quinoxaline molecules in the solid state (R. Sabbah, D. Tabet and M. E. S. Eusebio, Thermochimica Acta, 1998, 315, 93).

Isomers of 2-substituted quinoxalines, which are often obtained from the condensation of 2-oxoaldehydes with unsymmetrical *o*-phenylenediamines, can be conveniently distinguishing using a ^{13}C nmr INEPT technique (R. Andersson, W. Tian and S. Grivas, J. chem. Res. S, 1993,

428). Detailed analyses of the mass, nmr and IR spectra of 2,3-diphenylquinoxaline have been reported (H. Z. Alkhathan and H. A. Al Lohedan, J. chem. Res. S, 1995, 10). The ^{1}H, ^{13}C and ^{15}N nmr chemical shifts have been measured for a series of 2,3-disubstituted quinoxalines with heterocyclic moieties in the 2-position and correlated with the preferred substituent conformations (E. Kleinpeter *et al.*, J. mol. Struct., 1998, 444, 199).

Studies of the enolization of phenacylquinoxalines by ^{1}H and ^{13}C nmr spectroscopic techniques indicate that, in contrast to pyrazines, the enaminone tautomer predominates for quinoxaline derivatives (R. A. M. O'Farrell and B. A. Murray, J. chem. Soc., Perkin Trans. 2, 1994, 2461).

The X-ray crystal structures of sodium and potassium salts of the 2,3-diphenylquinoxaline radical anion as 1,2-dimethoxyethane solvates indicate that the major structural changes occur in the vicinity of the nitrogen centres which bear the highest effective nuclear charge (H. Bock *et al.*, Helv., 1994, 77, 1505).

(c) General methods of synthesis

General strategies for the synthesis of quinoxalines have been discussed in detail elsewhere (K. J. McCullough, Rodd's Chemistry of Carbon 2nd Ed. Vol. IV, Part IJ, 1989, pp. 301; 1st Supplement, 1995, pp. 126; N. Sato "Comprehensive Heterocyclic Chemistry II", Vol. 6, ed. A. R. Katritzky, C. W. Rees and E. F. V. Scriven, Pergamon, Oxford, 1997, pp. 233).

The classical Hinsberg synthesis of 2,3-disubstituted quinoxalines (5) by the acid-catalysed condensation of an α-dicarbonyl compound with an *o*-phenylenediamine can be substantially accelerated by irradiation of the solvent-free reaction mixture with focused microwaves (D. Villemin and B. Martin, Synth. Commun., 1995, 25, 2319). The analogous condensation between acyloins and *o*-phenylenediamine can also be

NH_2 NH_2 — Method A or B → (5) N R^1 N R^2

Method A: O R^1 O R^2 / p-TSA / focused microwaves, 15-30 s

Method B: O R^1 HO R^2 / microwaves, 3-6 min

mediated by microwave irradiation in a commercial microwave oven though the yields are variable (20-90%) (F. Juncai *et al.*, Synth. Commun., 1998, 28, 193).

Several novel quinoxaline derivatives have been synthesised by condensation of *o*-phenylenediamines with (a) 1,2-diaroyl-1,2-dibromoethanes yielding the disubstituted quinoxalines (6) which exist preferentially in the enaminone form (B. Insuasty *et al.*, J. heterocyclic Chem., 1998, 35, 977); (b) polycyclic diones to give the corresponding fused quinoxalines (7) (R. N. Warrener *et al.*, Synlett, 1998, 590); (c) *cis*-hexene-2,5-dione unexpectedly affording 2-methylquinoxaline (80% yield) when the *o*-phenylenediamine is in excess (G, Adembri, A. M. Celli and A. Sega, J. heterocyclic Chem., 1997, 34, 541); and (d) tetraacetylethylene to provide the pentane-2,4-dione derivative (8) *via* an intermediate isoxazolo[2,3-*a*]quinoxaline (*idem, ibid.*).

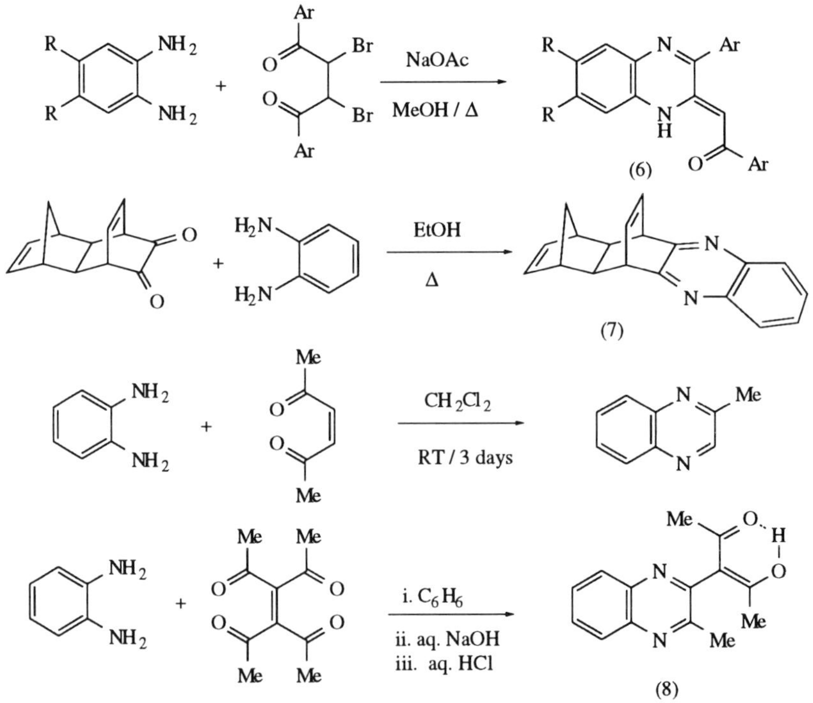

NaOAc, MeOH / Δ → (6)

EtOH, Δ → (7)

CH_2Cl_2, RT / 3 days

i. C_6H_6, ii. aq. NaOH, iii. aq. HCl → (8)

Although the reaction of *o*-phenylenediamine with aromatic aldehydes generally produces benzimidazoles, 2,3-diarylquinoxalines (9) can be obtained in significant quantities by heating the reaction mixture at

350 °C for 3-10 min. (C. Ochoa and J. Rodrígues, J. heterocyclic Chem., 1997, 34, 1053).

4 ArCHO
350 °C

(9)

The reaction of *o*-diisocyanoarenes with *trans*-bromo(methyl)bis-phosphinepalladium(II) reagents forms monomeric and oligomeric (3-substituted quinoxalin-2-yl)palladium(II) complexes (10, n = 1-6) which can be variously transformed into the corresponding quinoxaline derivatives (11) (Y. Ito *et al.*, Heterocycles, 1996, 42, 597).

$MePd(II)BrL_2$
$L = PMe_2Ph$
n = 1-6

RMgBr

(10)

(11)

The acid catalysed condensation of 1,2-diaminocyclohexane with either 2-iminoanilides of 2,3-dioxoacids (12, X = NPh, Y = O) or *vic*-2,3-dioxoanilides (12, X = Y = O) provides the tetrahydroquinoxalines (13) directly whereas with 2-iminothioanilides of 2,3-dioxoacids (12, X = NPh, Y = O), the acyclic compounds (14), which are initially isolated, require to

X = NHPh, O
Y = O

X = NHPh
Y = S

(12)

(13)

(14)

(15)

i. EtOH / AcOH / Δ ; ii. EtOH / Δ

be heated in ethanol at reflux temperature to effect cyclization to the thione derivatives (15) (B. Zaleska and B. Bialas, Pharmazie, 1996, 51, 428).

Tetrachloro-1,2-benzoquinone reacts with bis(benzal)ethylenediamine to yield a mixture of the quinoxaline derivatives (16) and (17) which were separated by a combination of crystallisation and chromatography (A. A. Aly *et al.*, Pharmazie, 1997, 52, 282).

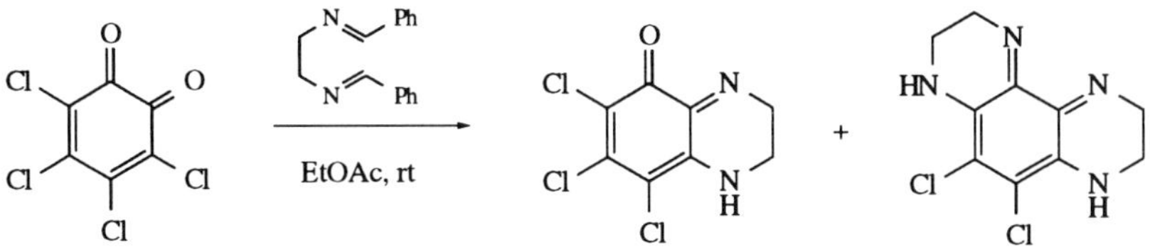

EtOAc, rt

(16) (17)

Flash vacuum pyrolysis (FVP) of 1-(2-azidophenyl)pyrazoles (18) at temperatures above 550 °C affords the corresponding 2,6-disubstituted quinoxalines in significant quantities (B. A. J. Clark, H. McNab and C. C. Sommerville, Chem. Commun., 1996, 1211). Quinoxalines (18) are also obtained in low yields from FVP of 1,5-benzodiazepines (D. Lloyd *et al.*, Chem. Commun., 1995, 2337; Tetrahedron, 1998, 54, 9667).

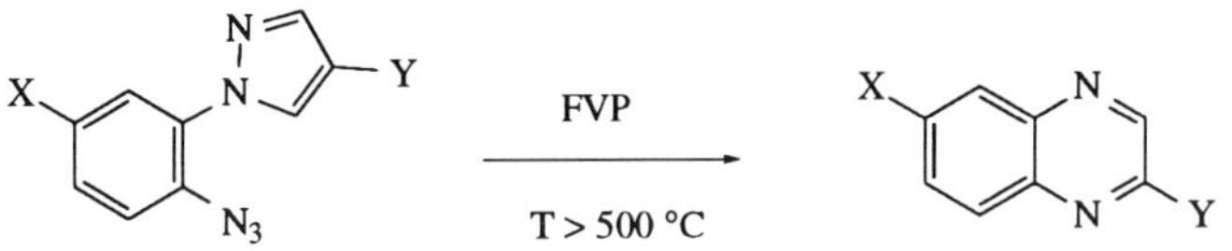

FVP

T > 500 °C

(18)

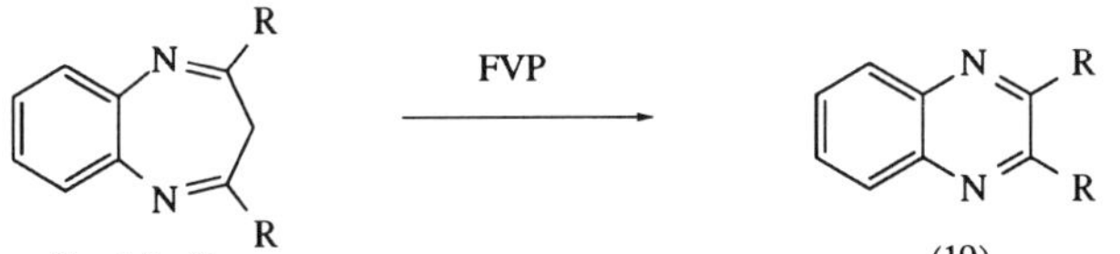

FVP

R = Me, Ph (19)

(d) Quinoxaline, its homologues and derivatives

(i) Quinoxaline, alkyl-, alkenyl-, alkynyl- and aryl-quinoxalines

Quinoxaline forms crystalline n→σ* charge transfer complexes with iodine (R. D. Bailey *et al.*, J. chem. Soc., Perkin Trans. 2, 1997, 2773).

Biotransformation of quinoxaline by (a) *Streptomyces badius* ATCC 39117 affords 3,4-dihydro-2(1*H*)-quinoxalinone and 2(1*H*)-quinoxalinone as the major metabolites (J. B. Sutherland *et al.*, Letters appl. Microbiol., 1996, 22, 199), and (b) *Streptomyces viridosporus* T7A ATCC 39115 yields 2(1*H*)-quinoxalinone (8%) and 1-methyl-3(1*H*)-quinoxalinone (12%) (J. B. Sutherland, J. P. Freeman and A. J. Williams, Appl. Microbiol. Biotechnol., 1998, 49, 445).

Quinoxaline readily undergoes NaOD-catalysed protium-deuterium exchange in supercritical D_2O to give quinoxaline-d_6 (42%, >85% deuteration) (T. Junk and W. J. Catallo, Tetrahedron Letters, 1996, 37, 3445). With 2-methyl- and 2,3-dimethylquinoxalines, the isotope exchange reaction is carried out at subcritical temperatures to minimise product decomposition (T. Junk, W. J. Catallo and J. Elguero, Tetrahedron Letters, 1997, 38, 6309).

Treatment of 2-methyl- and 2,3-dimethylquinoxalines with iridium powder in aqueous ethanol containing ammonium chloride at reflux temperature afforded the corresponding 1,2,3,4-tetrahydroquinoxalines in *ca.* 90% yield (C. J. Moody and M. R. Pitts, Synlett, 1998, 1029). 2-Methylquinoxaline is enantioselectively hydrogenated to (–)-(2S)-2-methyl-1,2,3,4-tetrahydroquinoxaline (20) with 90% ee in methanol in the presence of *fac-exo*-(R)-[$IrH_2\{C_6H_4C^*H(Me)N(CH_2CH_2PPh_2)_2\}$] as a catalyst precursor (C. Biachini *et al.*, Organometallics, 1998, 17, 3308).

cat / H_2 (5 atm)

MeOH / 100 °C

(20)

The reaction of 2-methylquinoxaline with two equivalents of ninhydrin results in the formation of the novel spiro compound (21 (G. Heinisch *et al.*, Heterocycles, 1996, 43, 1665).

Δ / 100 °C

xylene

2 molar equivalents

(21)

2-Methylquinoxaline derivatives react with ethyl bromopyruvate to produce the corresponding pyrrolo[1,2-*a*]quinoxalines (22) (A. Gueiffier *et al.*, J. heterocyclic Chem., 1995, 32, 1317).

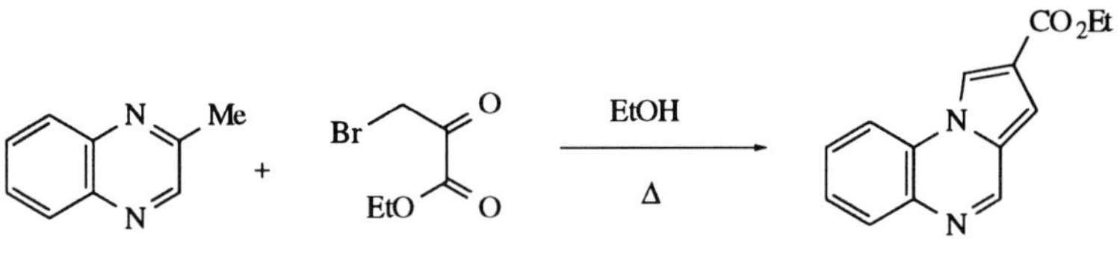

EtOH

Δ

(22)

Quinoxaline-2,3-quinodimethane, generated *in situ* by debromination of 2,3-bis(bromomethyl)quinoxaline using NaI in DMF at 130 °C, could be readily trapped by diethyl azodicarboxylate and dimethyl acetylenedicarboxylate yielding the 1:1 adducts (23) and (24) respectively; with less reactive dienophiles such as methyl vinyl ketone and methyl acrylate, no cycloadducts were formed (J. Stephanidou-Stephanidou *et al.*, Tetrahedron Letters, 1995, 36, 6777).

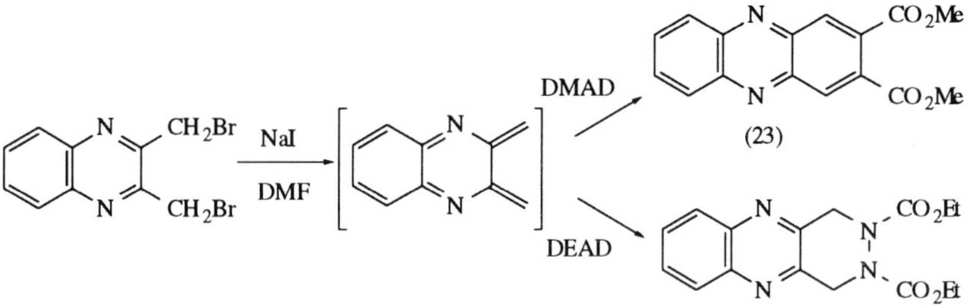

NaI

DMF

DMAD

DEAD

(23)

(24)

A series of quinoxalines (25) with perfluoroalkyl substituents have been prepared by condensation of the appropriate fluorinated benzil derivative with *o*-phenylenediamine (K. J. L. Paciorek *et al.*, J. fluor. Chem., 1996, 78, 27).

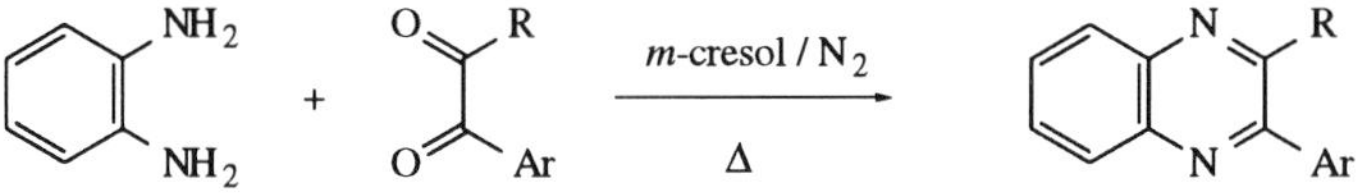

m-cresol / N_2

Δ

(25)

R = $C_6H_4C_8F_{17}$, C_7H_{15}

(ii) Haloquinoxalines

The kinetics of solvolysis of 2-chloroquinoxaline have been investigated (R. D. Patel, Internat. J. chem. Kin., 1994, 26, 403). Rates of reaction of 2-chloroquinoxaline with piperidine in DMSO over a wide range of amine concentration and temperature have been measured (J. Nasielski and C. Rypens, J. phys. org. Chem., 1994, 7, 545).

2-Chloroquinoxaline undergoes nucleophilic displacement with potassium 2-nitrophenoxide to yield (26) without the formation of cyclisation products (A. Cuenca, C. Bruno and A. Taddei, Tetrahedron, 1994, 50, 1927; A. Cuenca and C. Bruno, *ibid.*, 1996, 52, 11665).

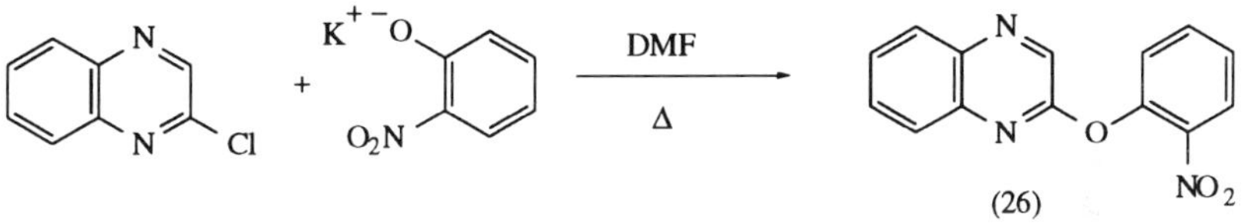

The 2,3-dichloroquinoxaline (27) reacts with the disodium salt of triethylene glycol to yield the crown ether (28), and with two molar equivalents of the sodium salts of ethylene glycol, diethylene glycol and triethylene glycol, the corresponding podands (29) were obtained. Treatment of the disodium salts of (29, n = 1,2) with (27) under high dilution conditions affords the crown ethers (30) (A. R. Ahmad, L. K. Mehta and J. Parrick, J. chem. Soc., Perkin Trans. 1, 1996, 2443).

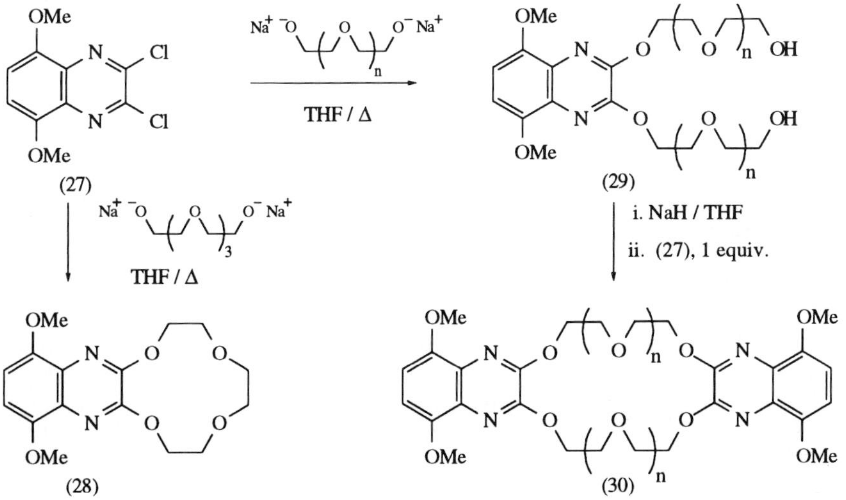

Quaternization of 6,7-difluoroquinoxaline using Meerwein's reagent yields the tetrafluoroborate salt (31) which undergoes selective nucleophilic substitution of the C(7) fluorine when treated with cyclic amines such as morpholine to give (32) (G. A. Mokrushina *et al.*, Russ. J. org. Chem., 1998, 34, 109). Reaction of quinoxalines (31) and (32) with 1,3-dicarbonyl compounds results in formation of the corresponding fluorinated furo[2,3-*b*]quinoxalines (33) and (34) respectively (V. N. Charushin *et al.*, Mendeleev Comm., 1998, 133).

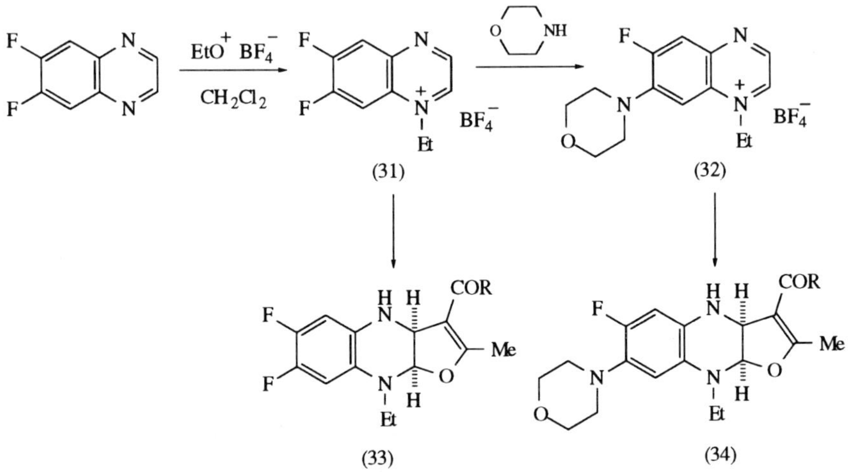

New 5,6,7,8-tetrafluoroquinoxalines (35) have been prepared by Hinsberg condensation of 3,4,5,6-tetrafluoro-1,2-phenylenediamine with the corresponding 1,2-dicarbonyl compounds (A. Heaton, M. Hill and F. Drakesmith, J. fluor. Chem., 1997, 81, 133).

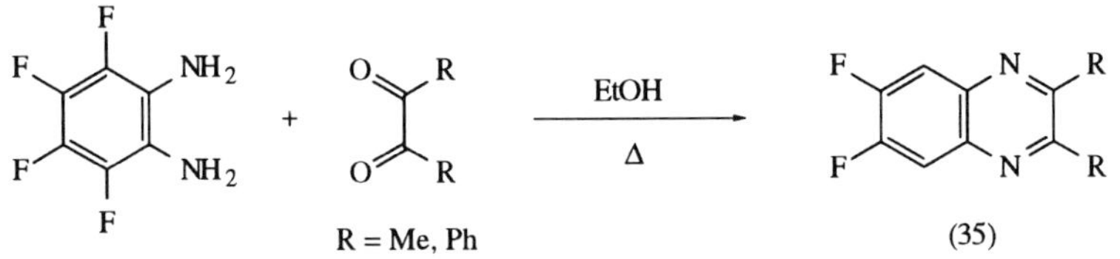

Hexafluoroquinoxaline is readily transformed into the 2,3-dibromo compound (36) on reaction with hydrogen bromide in a sealed tube at 75 °C. Hydrogenation of (36) over a Pd catalyst gives the 1,2,3,4-tetrahydro-quinoxaline (37) whereas the tetrafluoroquinoxaline (38) is obtained quantitatively from (36) using a Lindlar catalyst. Compound (36) will also undergo Pd-mediated cross-coupling reactions with pent-1-yne to yield (39) (R. D. Chambers *et al.*, J. chem. Soc., Perkin Trans. 1, 1998, 1705).

i. HBr / CH_3CN / 70 °C; ii. H_2 / Pd.C; iii. H_2 / Lindlar catalyst
iv. CuI / $(Ph_3P)_3PdCl_2$ / Et_3N then $C_3H_7C{\equiv}CH$

(iii) Aminoquinoxalines

Treatment of 2,3-dichloro-6-(trifluomethyl)quinoxaline (40) with ammonium carbonate in DMF affords the regioisomeric monoamino compounds (41) and (42) in the ratio (78:22). Subsequent reaction of (41) and (42) with 4-methylpyridine gives the pyrido[1',2':1,2]imidazolo[4,5-*b*]quinoxalines (43) and (44) respectively (S. Iwata, M. Sakajyo and K. Tanaka, J. heterocyclic Chem., 1994, 31, 1435).

A series 2,3-disubstituted 6-aminoquinoxalines (45) prepared by reaction of 2,3-dichloro-6-nitroquinoxaline with an appropriate nucleophile (amine or alcohol) followed by catalytic hydrogenation; compounds (45a) and (45c) were found to be useful high-sensitivity fluorescence derivatization agents for the HPLC analysis carboxylic acids (A. Katoh, M. Takahashi and J. Ohkanda, Chem. Letters, 1996, 369).

i. H–Ẍ

ii. H_2 / Pd.C / MeOH

(45a-d)

a: X = OMe b: X = NHPh

c: X = N(morpholino) d: X = N(piperidino)

Reaction of 6-amino-2,3-dimethylquinoxaline with acetaldehyde in ethanol provided the tetrahydropyrido[3,2-*f*]quinoxaline derivative (46) which on subsequent treatment with acid is converted into a separable mixture of (47) and (48) (S. Kondo, T. Shiozawa and Y. Terao, Chem. pharm. Bull., 1997, 45, 722).

(46) (47) (48)

i. CH_3CHO / EtOH / Δ; ii. H^+ / Δ

*(iv) Quinoxalin-2(1*H*)-ones and related compounds*

o-Phenylenediamine condenses smoothly with ethyl ethoxycarbonylformimidate in ethanol to yield the corresponding 3-aminoquinoxalin-2(1*H*)-ones (49); when $R^2 = NO_2$, only the 6-nitro isomer was obtained (A. McKillop *et al.*, Synthesis, 1997, 301).

EtOH, Δ

(49)

R^1, R^2 = H, H; Me, Me; Cl, Cl; H, Me; H, NO_2

Treatment of the pyridazinecarboxamide (50) with sodium hydride in DMF results in the formation of the quinoxalinyl pyrazole (51) *via* a cyclization-ring contraction process (G. Heinisch, B. Matuszczak and K. Mereiter, Heterocycles, 1994, 38, 2081). Several derivatives of (51) have been synthesised and investigated as potential anti-HIV and anti-malarial agents (B. Matuszczak and K. Mereiter, Heterocycles, 1997, 45, 2449; B. Matuszczak, T. Langer and K. Mereiter, J. heterocyclic Chem., 1998, 35, 113).

NaH / DMF, Δ

(50) (51)

Condensation of dialkyl (E)-2,3-dicyanobutendioates with *o*-phenylenediamine affords the α-methylene quinoxalinone derivatives (52) which are stabilised by intramolecular hydrogen bonding (Y. Yamada and H. Yasuda, J. heterocyclic Chem., 1998, 35, 1389)

EtOH, Δ

R = Me, Et, Pr, i-Pr, Bu, t-Bu

(52)

3-Methoxycarbonylmethylenequinoxalin-2-one (53) has been transformed into (a) furo[2,3-*b*]quinoxalinones (54) on reaction with isocyanates in the presence of sodium hydride in DMSO (T. Okawara *et al.*, Heterocycles, 1995, 41, 1951); (b) flavazoles (55) and triazole derivatives (56) (H. S. Kim *et al.*, J. heterocyclic Chem., 1996, 33, 1855); and (c) pyridazino[3,4-b]quinoxalines (57) (H. S. Kim *et al.*, J. heterocyclic Chem., 1997, 34, 39).

(53) → i, ii → (54) R = alkyl, aryl

(53) → iii - v → (55) Ar = p-$NO_2C_6H_4$

(53) → iii, v, vi → (56)

(53) → v, vii → (57)

i. NaH / DMSO / Δ; ii.R-N=C=O; iii. $ArN_2^+Cl^-$; iv. $POCl_3$ / pyridine; v. $NH_2NH_2H_2O$ / EtOH; vi. $NaNO_2$ / AcOH / H_2O; viii. NH_2NH_2.2HCl / 10%aq. HCl

Several studies of the tautomeric equilibrium between the hydrazone imine and diazenyl enamine forms of quinoxalinones (58) in DMSO have been reported. Although tautomeric form A predominates when $R^1 = CO_2Et$ or $R^1 = H$ and R^2 = alkyl, the equilibrium favours tautomer B when $R^2 = NO_2$ or CO_2Et (Y. Kurasawa *et al.*, J. heterocyclic Chem., 1994, 31, 527; 1994, 31, 1661; 1995, 32, 445). The proportion of tautomer B increases with increasing acid concentration until it is the exclusive form in trifluoroacetic acid (Y. Kurasawa *et al.*, J. heterocyclic Chem., 1995, 32, 531; 1996, 33, 421; 1997, 34, 305).

(58)

tautomer A ⇌ tautomer B

Natural abundance ^{17}O and ^{14}N nmr spectroscopic studies of quinoxalin-2(1*H*),3(4*H*)-diones suggest that they exist predominantly in the non-planar keto form analogous to that observed previously in the solid state (I. P. Gerothanassis *et al.*, Tetrahedron Letters, 1996, 37, 3191; I. P. Gerothanassis and G. Varvounis, J. heterocyclic Chem., 1996, 33, 643).

6,7-Dichloro-5-nitroquinoxaline-2,3-dione (3) is known to be a potent antagonist at NMDA/glycine sites in the brain (R. M. Woodward *et al.*, Mol. Pharmacol., 1995, 47, 568; Euro. J. Pharmacol., 1995, 273, 187). A series of di-, tri- and tetra-substituted quinoxaline-2,3-diones have been synthesised by conventional chemistry as analogues of (3) (J. F. W. Keanna *et al.*, J. med. Chem., 1995, 38, 4367). 5-Aminoquin-oxaline-2,3-diones (59) have also been investigated as potential α-amino-3-hydroxy-5-methylisoxazole-4-propionic acid (AMPA) receptor antagonists (Y. P. Auberson *et al.*, Bioorg. med. Chem. Letters, 1998, 8, 65; 1998, 8, 71; P. Acklin *et al.*, *ibid.*, 1998, 8, 493).

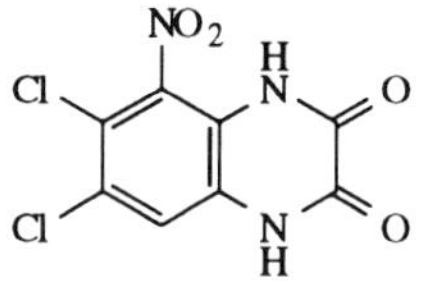

(3)

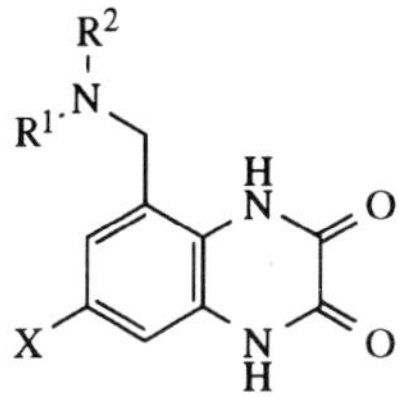

(59) $X = NO_2$, Br

Attempted nitration of 6,7-dialkoxyquinoxaline-2,3-diones (60) using fuming nitric acid in acetic acid resulted in the formation of the 5-acetoxy derivatives (61) in *ca.* 50% yield instead of the expected 5-nitro compounds. The acetoxy compounds are readily hydrolysed in aqueous acid to the corresponding 5-hydroxy compounds which are versatile intermediates for the synthesis of a variety of related compounds (Z.-L. Zhou, E. Weber and J. F. W. Keana, Tetrahedron Letters, 1995, 36, 7583).

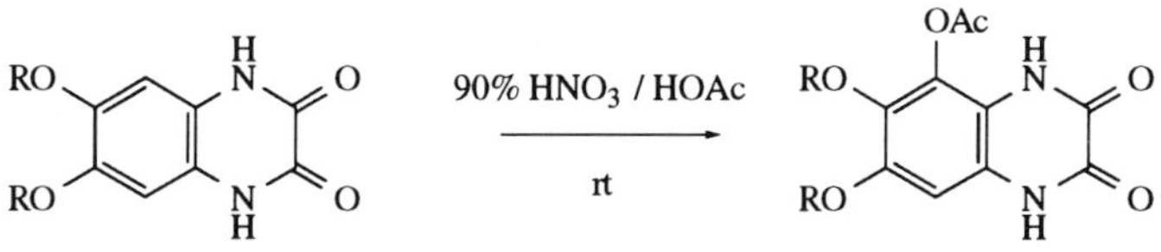

(60) R = alkyl $\xrightarrow[\text{rt}]{90\%\ HNO_3\ /\ HOAc}$ (61)

(v) Quinoxaline N-*oxides*

α-Iminooximes (62), derived from the condensation of (E)-benzil monooxide with a suitable aniline, undergo oxidative cyclization to the corresponding 2,3-diphenylquinoxaline 1-oxides (63) (A. J. Maroulis, K. C. Domzaridou and C. P. Hadjiantoniou-Maroulis, Synthesis, 1998, 1769).

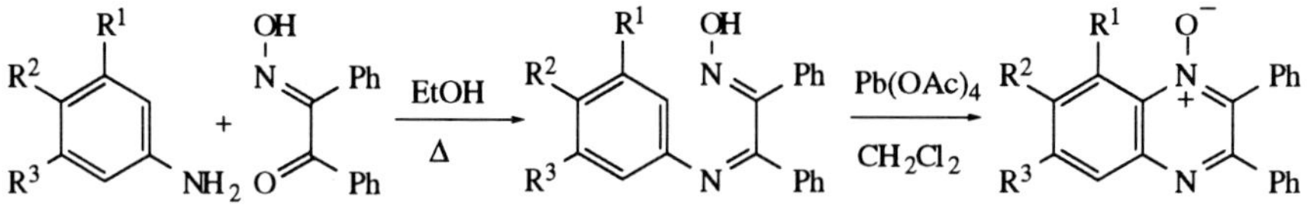

(62)

(63)

6-Chloro-2-(1-methylhydrazino)quinoxaline 1-oxide (64) reacts with acetylenedicarboxylates to yield the dihydropyrido[3,4-*b*]quinoxalines (65) as the major products; the pyrazolylquinoxaline oxides (66) are also produced in *ca.* 10% yield (Y. Kurasawa *et al.*, J. heterocyclic Chem., 1996, 33, 757).

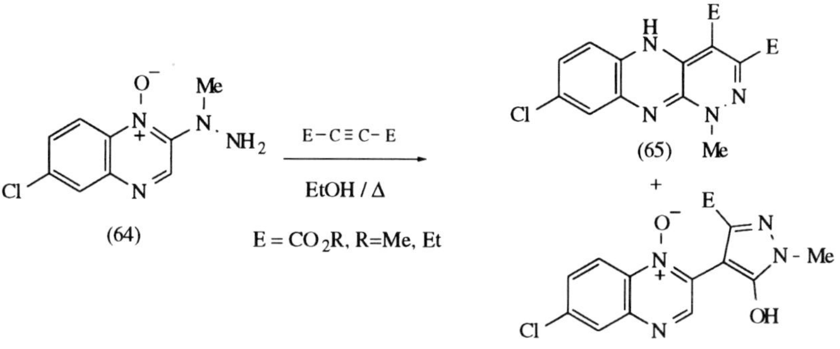

(64)

E−C≡C−E

EtOH / Δ

E = CO_2R, R=Me, Et

(65)

(66)

The outcome of the reaction between 6-chloro-2-(1-methylhydrazino)quinoxaline 4-oxide (67) and acetic anhydride is sensitive to the reaction conditions: (a) in the absence of solvent, intramolecular cyclization takes place producing the oxadiazino[5,6-*b*]quinoxaline (68), (b) in the presence of pyridine or acetic acid, the diacetyl derivative of (69) is obtained, and (c) in dioxane or chloroform, the monoacetyl compound (70) is produced and this can be converted into (71) using $POCl_3$ (H. S. Kim *et al.*, J. heterocyclic Chem., 1998, 35, 445). In contrast, treatment of the corresponding *N*-methylthiocarbamoyl derivative of (71) with acetic anhydride by under a variety of conditions affords the cyclized product (72) (Y. Kurasawa *et al.*, J. heterocyclic Chem., 1996, 33, 1859; H. S. Kim *et al.*, J. heterocyclic Chem., 1997, 34, 1539).

The cyclocondensation between benzofuroxan and 1,3-dicarbonyl compounds to produce 2,3-disubstituted quinoxaline 1,4-dioxides (usually called the Beirut reaction) can be efficiently catalysed by silica gel or molecular sieves at 90-110 °C (T. Takabatake, Y. Hasegawa and M. Hasegawa, J. heterocyclic Chem., 1993, 30, 1477; T. Takabatake, T. Miyazawa and M. Hasegawa, J. heterocyclic Chem., 1996, 33, 1057; M. Hasegawa *et al.*, Yakugaku Zasshi, 1996, 116, 491).

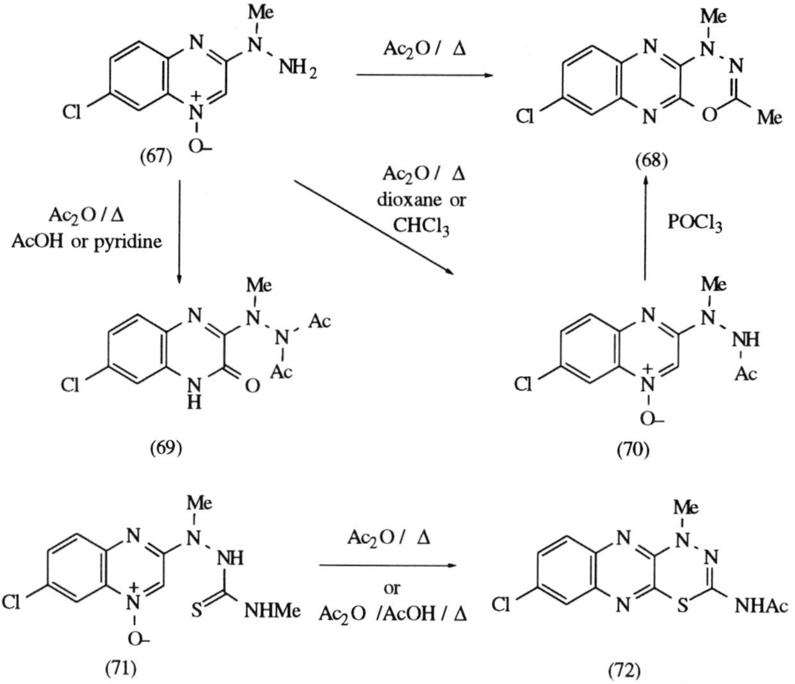

(67) Ac_2O / Δ → (68)

(67) Ac_2O / Δ, AcOH or pyridine → (69)

(67) Ac_2O / Δ, dioxane or $CHCl_3$ → (70)

(70) $POCl_3$ → (68)

(71) Ac_2O / Δ or Ac_2O /AcOH / Δ → (72)

Highly substituted quinoxaline 1,4-dioxides (73), which have considerable potential as selective cytotoxins of hypoxic cells in solid tumours, are readily prepared by the Beirut reaction (A. Monge *et al.*, J. heterocyclic Chem., 1995, 32, 1213; J. med. Chem., 1995, 38, 1786; M. M. El-Abadelah et al., Heterocycles, 1995, 41, 2203).

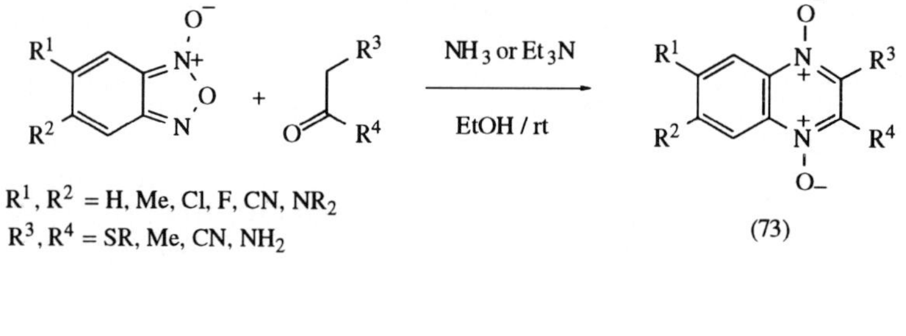

NH_3 or Et_3N, EtOH / rt → (73)

R^1, R^2 = H, Me, Cl, F, CN, NR_2

R^3, R^4 = SR, Me, CN, NH_2

5,6-Difluorobenzofuroxan reacts readily with enamines derived from cyclopentanone and cyclohexanone to provide the corresponding annelated quinoxaline 1,4-dioxides (74). The 7-fluoro substituent in (74) can be selectively displaced by secondary amines such as morpholine to give compounds (75) (S. K. Kotovskaya *et al.*, Russ. J. org. Chem., 1998, 34, 369).

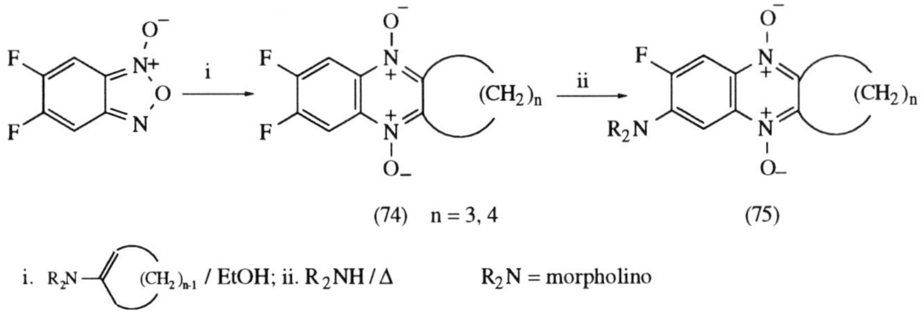

(74) n = 3, 4 (75)

i. R_2N–(enamine, $(CH_2)_{n-1}$) / EtOH; ii. R_2NH / Δ R_2N = morpholino

The reaction of the 3-aminoquinoxaline-2-carbonitrile 1,4-dioxides with 2-chloroethyl isocyanate in dry toluene at 100-110 °C affords the oxadiazolo[2,3-*a*]quinoxaline 5-oxides (76) (A. Monge *et al.*, J. heterocyclic Chem., 1996, 33, 1671).

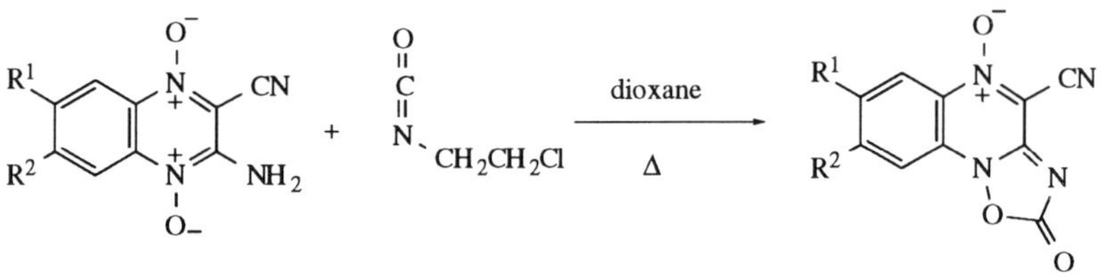

dioxane, Δ

R^1, R^2 = H, Me, Cl (76)

3-Aminoquinoxaline-2-carbonitrile 1,4-dioxide (77) can be selectively mono- or di-deoxygenated using either ethanolic hydrochloric acid or methanolic sodium dithionite respectively; 3-aminoquinoxaline-2-carbonitrile (78) is a versatile intermediate for subsequent synthetic manipulation (A. Monge *et al.*, J. heterocyclic Chem., 1994, 31, 1135). Treatment of (77) with excess *N,N*-dimethylaminoethylamine results in the formation of the cyclic amidines (79, 45%) and (80, 10%) (A. Monge, J. A. Palop and J. C. del Castillo, J. heterocyclic Chem., 1994, 31, 33).

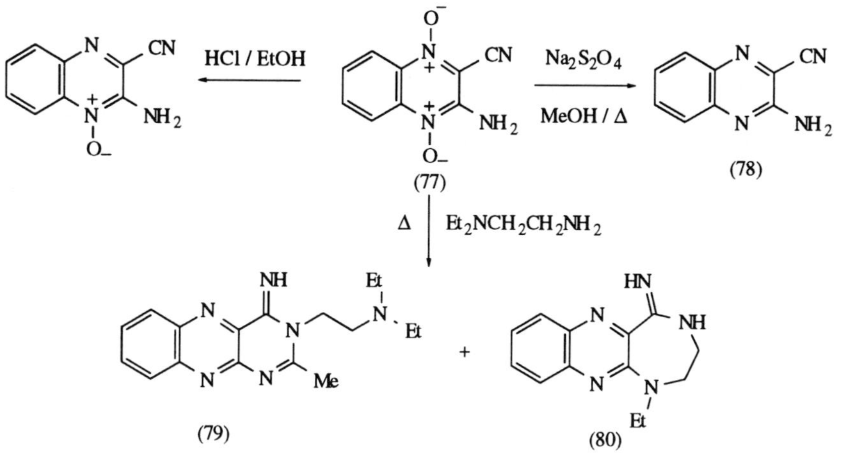

Reaction conditions have been reported for the selective preparation of the mono-*N*-oxides (81) and (82) derived from Carbadox, a growth stimulant widely used in farm animals (D. J. R. Massy and A. McKillop, Synthesis, 1996, 1477).

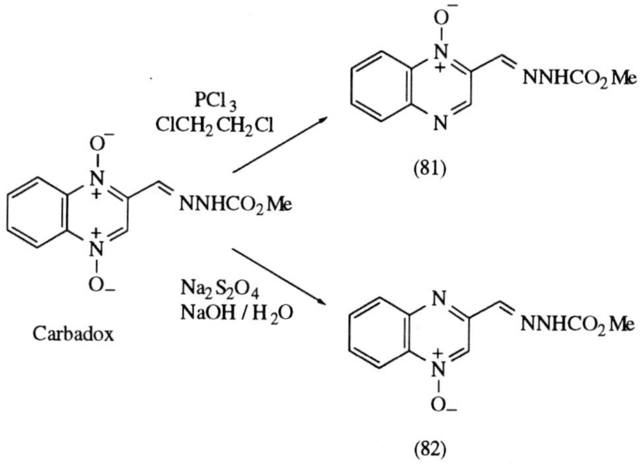

Heats of formation and N–O bond energies have been determined for a series of quinoxaline 1,4-dioxides (W. E. Acree *et al.*, J. org. Chem., 1997, 62, 3722). 2,3-Dimethylquinoxaline 1,4-dioxide forms stable 2:1 donor-acceptor complexes in both the solution phase and the solid state with the electron-deficient olefin, TCNE (S. C. Blackstock *et al.*, Tetrahedron Letters, 1997, 38, 7665).

(vi) Miscellaneous Quinoxaline Derivatives

Condensation of 2,3-dicyanoquinoxaline with silicon tetrachloride in the presence of urea, quinoline and tri-n-butylamine affords the dichlorosilicon complexes of the tetra-2,3-quinoxalinoporphyrazines (83) (S. V. Kudrevich and J. E. van Lier, Can. J. Chem., 1996, 74, 1718).

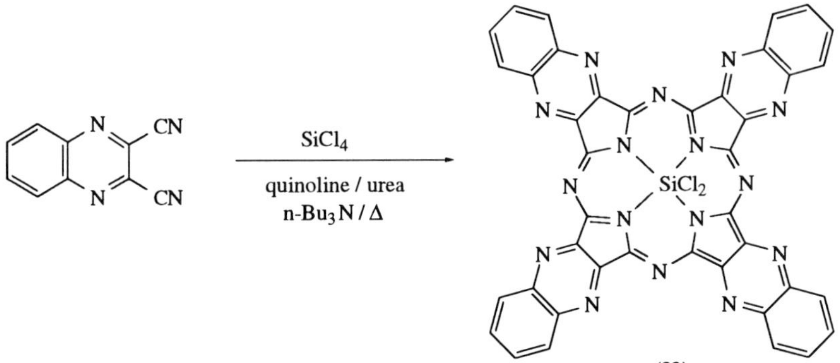

$SiCl_4$, quinoline / urea, $n\text{-}Bu_3N$ / Δ

(83)

Ethyl 3-methylquinoxaline-2-carboxylate has been transformed by convention chemical methods into a variety of oxadiazoyl-, triazolyl- and pyridoquinoxaline derivatives (84)-(86) respectively (A. M. K. El-Dean and A. A. Geies, Heterocyclic Comm., 1998, 4, 367).

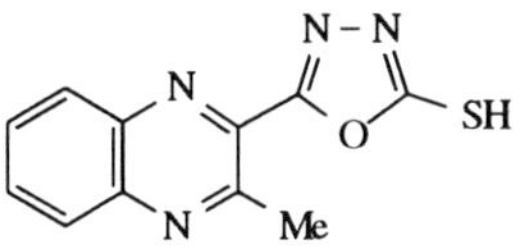

(84) 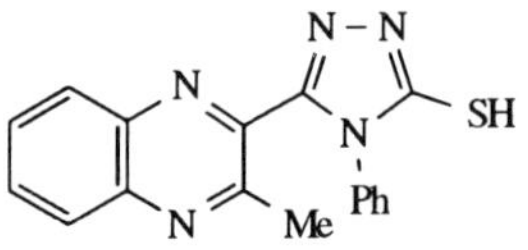 (85) 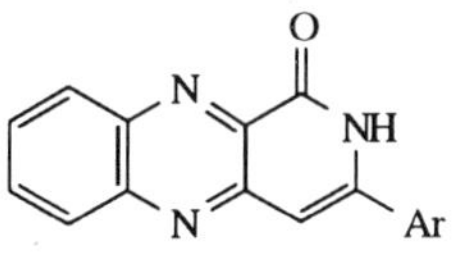 (86)

Ethyl 2-aminoquinoxaline-2-carboxylate reacts with (a) formamide to give pyrimido[4,5-*b*]quinoxalin-4(3*H*)-one (87), and (b) phenylisothio-cyanate to give an intermediate phenylthioureido derivative which cyclizes on treatment with alcoholic KOH to the thione (89) (O. S. Moustafa, E. A. Bakhite and M. Z. A. Badr, Afinidad, 1998, 55, 285).

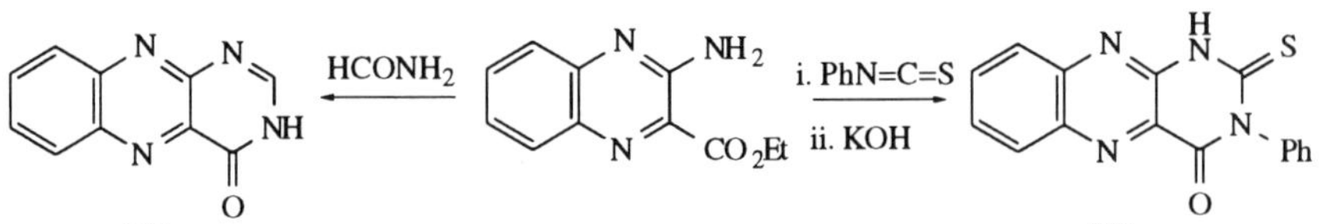

$HCONH_2$; i. PhN=C=S, ii. KOH

(87) (88)

An improved synthesis of 6,7-dichloroquinoxaline-5,8-dione (89) from 4-aminophenol in eight steps (27% overall yield) has been reported. Compound (89) is a useful synthetic intermediate for the development of new anticancer drugs because it can be readily elaborated to a variety of condensed systems such as (90)-(92) (S. W. Park *et al.*, Heterocycles, 1996, 43, 2495; H.-W. Yoo, M.-E. Suh and S. W. Park, J. med. Chem., 1998, 41, 4716).

O Cl N Cl N O (89)

i. NaN_3 / DMF
ii. $NaBH_4$
iii. R^1COCOR^2

R^1 N O N R^2 N O N (92)

N NH_2 EtOH / Δ

NH_2 NH_2 EtOH / Δ

O N N N N O (90)

OH Cl N N N N (91)

The tetrasubstituted quinoxalines (93), synthesised in six steps from 3,5-dimethoxyaniline, were oxidised to the diones (94) on treatment with nitric acid in acetic anhydride and subsequently aminated using sodium azide to yield 7-amino-5,8-diones (95), potential analogues of the anti-tumour antibiotic, streptnigrin (K. V. Rao and C. P. Rock, J. heterocyclic Chem., 1996, 33, 447).

MeO N Cl N Ph OMe (93) —i→ MeO O N Cl N Ph O (94) —ii→ MeO O N Cl H_2N N Ph O (95)

i. Ac_2O / HNO_3 / H_2SO_4 ; ii. NaN_3

Oxidation of the 2,3-disubstituted 5,6-dimethoxyquinoxalines (96, X = OMe, SEt) with cerium(IV) ammonium nitrate (CAN) affords the corresponding quinoxaline quinones (97). The 2,3-dichloro compound (96, X = Cl) was converted into the tetrachloroquinoxaline quinone (98) on treatment with a mixture of concentrated hydrochloric and nitric acids (J. Parrick *et al.*, J. chem. Research-S, 1998, 224).

i X = OMe, SEt ii X = Cl

(97) (96) (98)

i.CAN/ MeOH/ H_2O; ii. conc. HCl /conc. HNO_3

(vii) Reduced Quinoxalines

The reductive metallation of 6-methyl-2,3-diphenylquinoxaline by sodium in THF results in the generation of a monomeric anion which readily undergoes alkylation with iodomethane or benzyl chloride, acylation with ethyl chloroformate and annelation with 1,3-dichloropropane or 1,4-dichlorobutane to produce a variety of dihydroquinoxaline derivatives (99)-(102) (N. Öcal, Z. Turgut and S. Kaban, J. heterocyclic Chem., 1998, 35, 1349).

MeI $Cl-(CH_2)_n-Cl$ $ClCO_2Me$ $PhCH_2Cl$

(99) (102) (101) 100)

The *N*-methyl quinoxalinone derivatives (103) and (104) were unambiguously prepared in three steps from 2-nitroanilines. Coupling of acid (100a) with crown ethers provided a series of fluoroionophores (103b, c). On heating compound (104) in acetic anhydride, the dimeric red pigment (102) was obtained (A. R. Ahmad, L. K. Mehta and J. Parrick, Tetrahedron, 1995, 51, 12899).

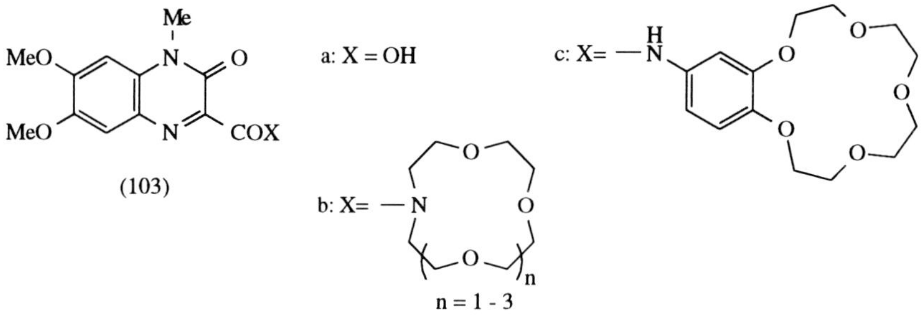

(103)

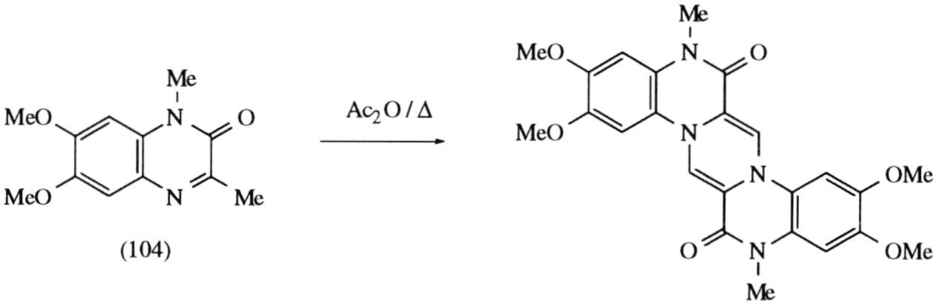

a: X = OH

b: X= —N ... n = 1 - 3

c: X= —NH—

(104) $\xrightarrow{Ac_2O / \Delta}$ (105)

Partial hydrogenation of the nitro group of the oxamates (106) yielded the corresponding 1-hydroxy-1,4-dihydroquinoxaline-2,3-diones (107) which were found to be significantly less potent antagonists of NMDA receptor glycine sites than the analogous 1,4-dihydroquinoxaline-2,3-diones such as (3) mentioned above (J. F. W. Keanna *et al.*, Bioorg. med. Chem. Letters, 1996, 6, 439). Conversely, 1,4-dihydroquinoxaline-2,3-dione analogues with amino acid residues at N(1) (108) exhibit strong affinities for AMPA receptors (G. Sun *et al.*, J. med. Chem., 1996, 39, 4430).

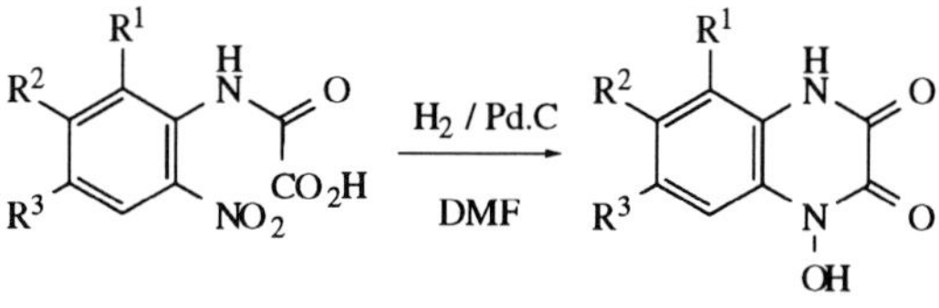

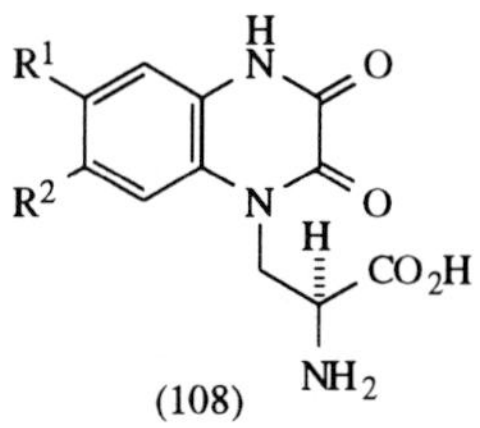

(106) $\xrightarrow[DMF]{H_2 / Pd.C}$ (107)

(108)

Treatment of the 3,4-dihydroquinoxalin-2(1*H*)-ones (109) with excess fuming nitric acid in TFA results in oxidative nitration to provide the 5-nitroquinoxaline-2,3-diones (110) in good yield (J. F. W. Keanna *et al.*, J. med. Chem., 1995, 38, 5839; S. X. Cai *et al.*, J. med. Chem., 1997, 40, 730).

conc. HNO_3 / TFA

(109) R^1, R^2 = Me, F, Cl, Br (110)

Catalytic hydrogenation of 5,6-disubstituted quinoxaline derivatives (111) in methanol over PtO_2 at 50 psi afforded the corresponding 1,2,3,4-tetrahydroquinoxalines (112) (S. A. Munk *et al.*, Bioorg. med. Chem. Letters, 1994, 4, 459).

H_2 / PtO_2 / MeOH / 50 psi

(111) R = H, Me, Br (112)

Thermolysis of 2,2-dialkyl-1,3-dibenzyl-2,3-dihydrobenzimidazoles at 200 °C results in formation of the isomeric tetrahydroquinoxalines (113) (G. M. Reddy, P. 1. Prasunamba and P. S. N. Reddy *et al.*, Tetrahedron Letters, 1996, 37, 3355).

Δ / 200 °C

(113)

R^1, R^2 = Me, Me; Me, Et; Me, Pr;- $(CH_2)_4$-; - $(CH_2)_5$-

The condensation of 2-amino-*N,N*-dimethylanilines with alloxan affords the spiro-compounds (114) which on subsequent treatment with acidic hydrogen peroxide undergo rearrangement to the red-coloured quinoxaline carboxyureides (115) (P. J. Zeegers and M. J. Thompson, Heterocycles, 1996, 43, 1873).

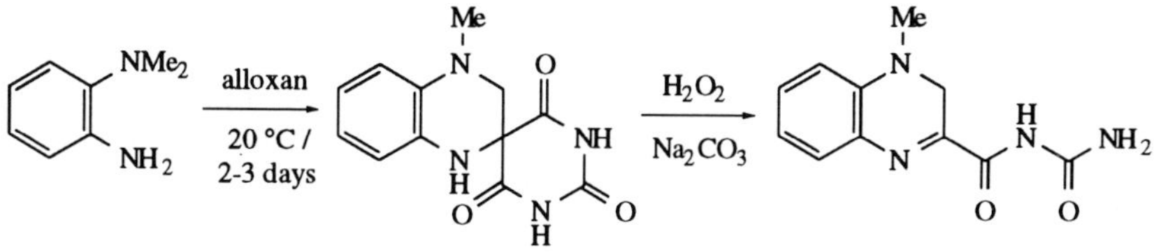

(114) (115)

1,2,3,4-Tetrahydroquinoxaline (116) is readily transformed into the bis(lactam) (117) which is subsequently reduced using $LiAlH_4$ to the cyclophane (118) (R. H. Khan, R. C. Rastogi and A. C. Ghosh, Ind. J. Chem., 1995, 34B, 646).

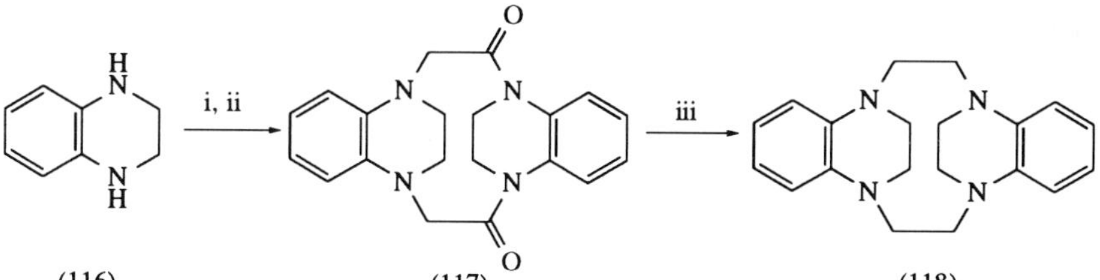

(116) (117) (118)

i. $ClCH_2COCl$ (2 equiv); ii. (116) (1 equiv)/ K_2CO_3/DMF/Δ; iii. $LiAlH_4$/THF/ Δ

A series of 2,3-disubstituted hexahydroquinoxaline stereoisomers (119) were synthesised by the condensation of the appropriate 1,2-diaminocyclohexane and 1,2-dicarbonyl compound. Oxidation of (119) using SeO_2 affords the corresponding 5,6,7,8-tetrahydroquinoxaline (120) (M. L. Jimeno *et al.*, Anal. Quimica, 1994, 90, 423). The *cis*-isomer of 2,3-dimethylhexahydroquinoxalines (119, R = Me) has been found to be an effective DNA strand-breaking agent, particularly in the presence of Cu^{2+} ion (T. Yamaguchi *et al.*, Biol. Pharm. Bull., 1996, 19, 1261).

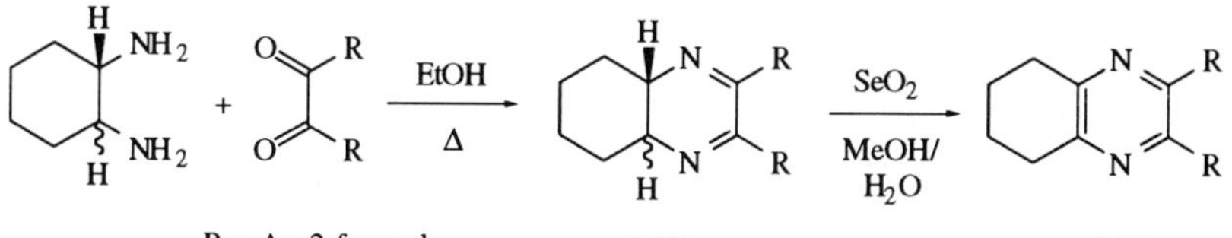

R = Ar, 2-furanyl (119) (120)

3. Phenazines, dibenzopyrazines

(a) Introduction

Although several new phenazine natural products have been isolated and their biological properties extensively investigated, relatively few developments in the fundamental chemistry of phenazines have been reported. No significant new reviews on phenazine chemistry have been noted since the 1st Supplement (K. J. McCullough, Rodd's Chemistry of Carbon 2nd Ed. Vol IV, Part IJ, 1st Supplement, 1995, pp. 93).

The isolation of several antibiotics (1)-(3) derived from phenazine-1-carboxylic and 1,6-dicarboxylic acids have been reported (Table 1). Although most compounds show only weak antibacterial activity, pelagiomicin A (3a) has also been found to be an antitumour agent (Table 1, ref. 4). The C_2-symmetrical phenazine-1,6-diol and the corresponding 5,10-di-*N*-oxide have been isolated from some actinomycetes (I. Tanabe, M. Kuriyama and H. Nonomura, J. Ferm. Bioeng., 1995, 79, 384).

Structurally novel metabolites, which contain two phenazine substructural units, have been isolated and identified: (a) the phenazostatins A and B (4) and (5), isolated from *Streptomyces* sp. 833 as neuronal cell protecting substances (I.-D. Yoo *et al.*, Tetrahedron Letters, 1996, 37, 8529), and (b) diphenazithionin (6), an inhibitor of lipid peroxidation from *Streptomyces griseus* ISP 5236 (Y. Okami *et al.*, Tetrahedron Letters, 1996, 37, 9227).

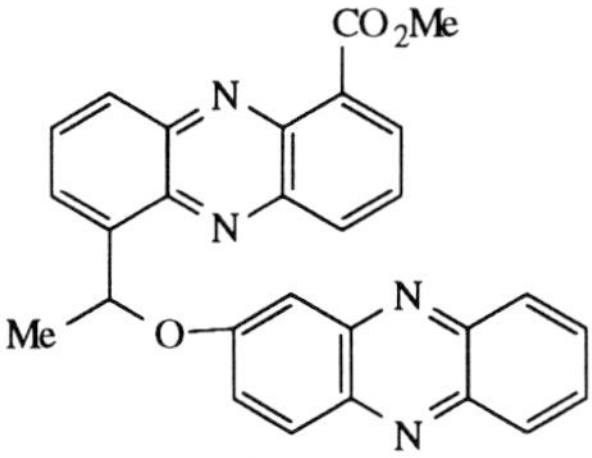

(4)

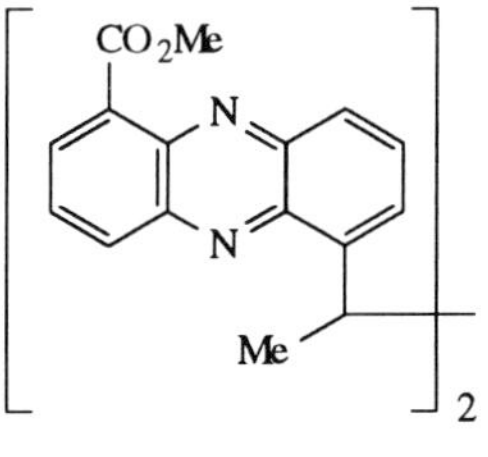

(5)

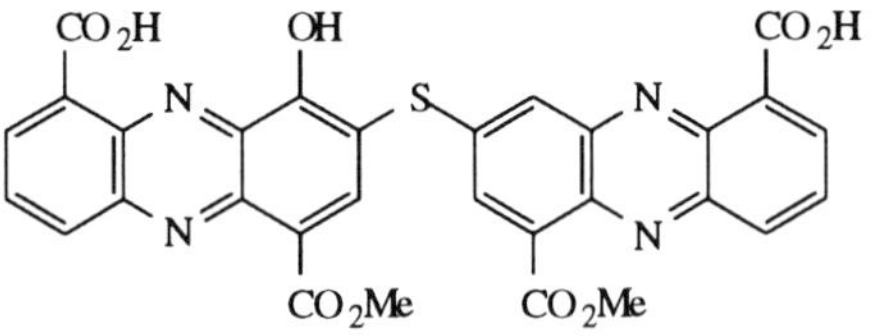

(6)

TABLE 1
PHENAZINE ANTIBIOTICS

Structure	Substituents	Source
(1)	a: $R^1 = R^2 = H$	*Streptomyces* Y-9031725[1]; unidentified *Streptomyces*[2]
	b: R^1 = OH, R^2 = H	unidentified *Streptomyces*[3]
	c: R^1 = OH, R^2 = S-CH$_2$-CH(NHCOMe)-CO_2H	unidentified *Streptomyces*[3]
(2)		unidentified *Streptomyces*[2]
(3)	a: R = CH_2O-CO-CH(NH_2)-C(OH)(Me)Me	*Pelagiobacter viriablis*[4]; culture LL-14I352[5]
	b: R = CH_2O-CO-CH(NH_2)-CH(Me)Me	*Pelagiobacter viriablis*[4]
	c: R = CH_2O-CO-CH_2NH_2	*Pelagiobacter viriablis*[4]
	d: R = CH_2OH	culture LL-14I352[5]

1. S. Chatterjee, E. K. S. Vijayakumar, C. M. M. Franco, R. Maurya, J. Blumach and B. N. Ganguli, J. Antibiot., 1995, 48, 1353.
2. K. Pusecker, H. Laatsch, E. Helmke and H. Weyland, J. Antibiot., 1997, 50, 479.
3. M. L. Gilpin, M. Fulston, D. Payne, R. Cramp and I. Hood, J. Antibiot., 1995, 48, 1081.
4. M. Imamura, M. Nishimura, Takadera, K. Adachi, M. Sakai and H. Sano, J. Antibiot., 1997, 50, 8.
5. M. P. Singh, A. T. Menendez, P. J. Petersen, W.-D. Ding, W. M. Maiese and M. Greenstein, J. Antibiot., 1997, 50, 785.

(b) Physical and spectroscopic properties

Electrochemical and esr spectroscopic studies show that 2,7-disubstituted phenazines undergo stepwise, two-electron reduction in weakly acidic media (1% or 2% TFA in acetonitrile/$NaClO_4$ solution) under N_2. This involves initial reduction of singly protonated phenazine molecules, followed by proton transfer to form dihydrophenazine radical cations. These are subsequently reduced to produce dihydrophenazines (T. Machida and H. Sayo, Chem. pharm. Bull., 1996, 44, 1448).

i. $+e^-$, $+H^+$

ii. $+e^-$

X = MeO., Eto, H, Cl, Br, CO_2Et

Phenazine di-*N*-oxide forms stable, violet-coloured, 1:1 donor-acceptor complexes with TCNE in solution and also in the solid state (S. C. Blackstock *et al.*, Tetrahedron Letters, 1997, 38, 7665).

The absorption spectrum of the phenazine di-*N*-oxide radical cation, generated electrochemically, has been measured. The reactions of this reactive species with aromatic hydrocarbons have also been investigated (E. M. Koldasheva *et al.*, Russ. chem. Bull., 1996, 45, 1889).

A series of 3,7-di-*t*-butylphenazin-5(10*H*)-yl radicals (7) have been characterised by esr and ENDOR spectroscopy. X-ray crystallographic analysis indicates that crystalline compounds (7a) and (7c) form radical pairs in the solid state (F. A. Neugebauer *et al.*, Ann.-Recueil, 1997, 473).

(7)

a: R = Me
b: R = Et
c: R = Pr
d: R = Bzl
e: R = 4-$C_6H_4NO_2$
f: R = 4-$C_6H_4NH_2$

(c) General methods of synthesis

Condensation of 3,3,6,6-tetrachlorocyclohexane-1,2-dione (8) with *o*-phenylenediamines, followed by treatment with pyridine provides an efficient, one-pot procedure for the preparation of a series of new 1,4-dichlorophenazine derivatives (9) (A. Guirado *et al.*, Tetrahedron, 1997, 53, 6183).

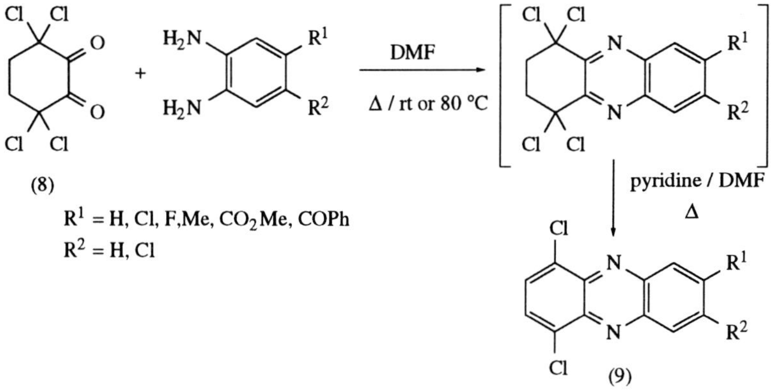

Dimeric nitrosochlorides (10), derived from unsaturated hydrocarbons such as 1-methylcyclohexene, react with *o*-phenylenediamine to yield the corresponding hexahydrophenazines (11). Under similar conditions, cyclohexene nitrosochloride affords 1,2,3,4-tetrahydrophenazine (P. A. Petukhov and A. V. Tkachev, Tetrahedron, 1997, 53, 9761)

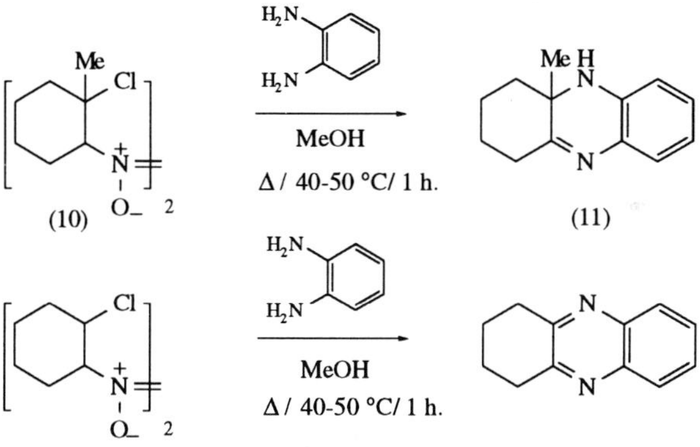

Condensation of quinoxaline-1,4-dioxide (12) with aromatic aldehydes, followed by base-catalysed cyclization of the resulting cinnamoyl derivatives by either a sequential or a one-pot procedure affords the 2-arylphenazine-5,10-dioxides (13) in good yield. In addition, the highly substituted phenazine di-*N*-oxides (14) can be conveniently prepared in a one-pot reaction by treatment of a mixture of the appropriate *o*-nitroanilines and either catechol, or resorcenol, or dihydroquinone with sodium hypochlorite (A. A. Fadda *et al.*, Monat. Chem., 1995, 126, 1217).

ArCHO

NaOH / MeOH

(12)

NaOH / MeOH

(13)

$R = H, Me$

NaOCl

NaOH / MeOH

$R^1 = OH, R^2 = R^3 = H$
$R^2 = OH, R^1 = R^3 = H$
$R^3 = OH, R^1 = R^2 = H$

(14) $R^1 = OH, R^2 = H$
$R^1 = H, R^2 = OH,$
$R^1 = R^2 = H$

(d) Phenazine, its homologues and derivatives

(i) Phenazine and its derivatives

Phenazine is transformed into 5,10-dihydrophenazine by *Pseudomonas cepacia* IFO 15124 in growing cultures at low oxygen concentrations (H. Kawashima and S. Ogawa, Biosc. Biotech. Biochem., 1996, 60, 1186).

Cathodic reduction of phenazine in acetonitrile containing 1% TFA and 0.1M $NaClO_4$ solution under air affords 2-cyanomethylphenazine (15) in 40% yield; 2,7-diethoxyphenazine behaves in a similar fashion (T. Machida and Y. Yamaoka, Chem. pharm. Bull., 1998, 46, 207).

CH_3CN

1%TFA / 0.1M $NaClO_4$

electrolysis

(15)

Although phenazine does not readily undergo Friedel-Crafts alkylation, reaction of the related 5-acetyl-5,10-dihydrophenazine with t-butyl chloride in the presence of $AlCl_3$ yields the 3,7-di-t-butyl compound (16), which on subsequent oxidative deacylation provides 2,8-di-t-butyl-phenazine (17) in 51% yield (F. A. Neugebauer *et al.*, Ann.-Recueil, 1997, 473).

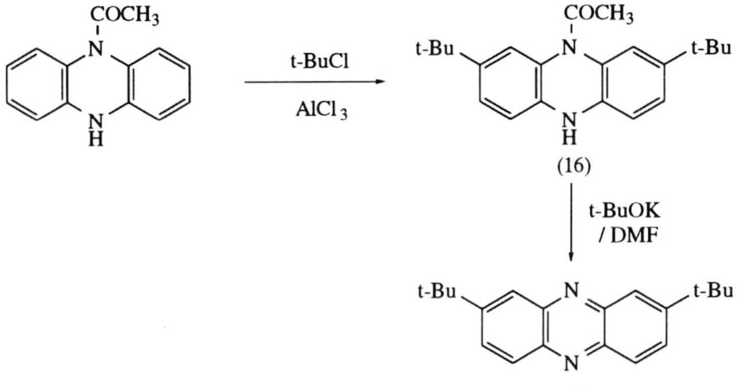

(16)

(17)

Amination of *N*-alkylphenazinium salts using amino alcohols affords selectively the 2-substituted phenazines (18), which are key intermediates in the synthesis of phenazine-tethered oligonucleotides (J. Chattapadhyaya *et al.*, Tetrahedron, 1997, 53, 10409; *ibid.*, 1998, 54, 5667; *ibid.*, 1998, 54, 8183). 2-Aminophenazinium salts (19), also readily prepared by nucleophilic substitution of the corresponding *N*-methyl-phenazinium salts, can be further aminated under mild conditions by treatment with acetic anhydride, followed by a suitable primary or secondary amine to produce the 2,8-disubstituted derivatives (20) (V. N. Sil'nikov, N. P. Luk'yanchuk and G. V. Shishkin, Russ. chem. Bull., 1996, 45, 1955).

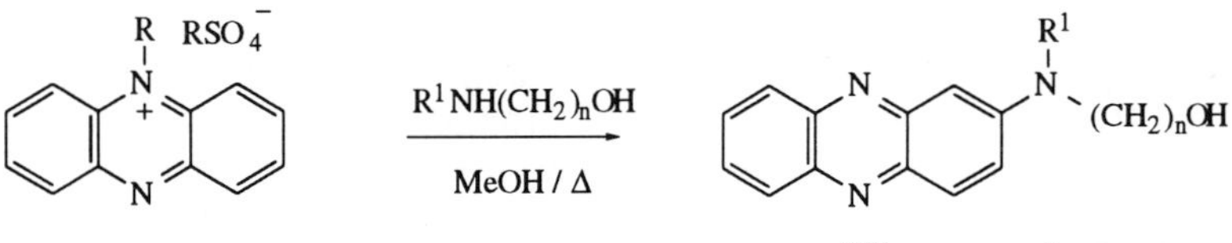

R = Me, Et

(18) R^1 = H, Me

n = 2, 4, 6

R^1NH_2

(19)

R^1 = Me, $CH_2CH_2CO_2Me$
R^2, R^3 = H, Me; morpholino

i. Ac_2O / $BF_3.Et_2O$
ii. R^2R^3NH / MeOH

(20)

Phenazine-1-carbohydrazide reacts with 2,4,6-tri- and 2,3,4,6-tetra-substituted pyrilium salts to form the corresponding pyridinium salts (21) which exhibit antifungal activity (A. Nicolae, T. Cristea and C. T. Supuran, Rev. Roum. Chim., 1997, 42, 301).

EtOH
Δ

(21) R^1 = alkyl; R^2 =H, Me
R^3 = Me, Ph; R^4= alkyl, Ph

(ii) Reduced Phenazines

Reactions of trialkylsilyloxyphenazines (22) with either triflates, derived from allylic alcohols, or allylic bromides under high pressure (15 kbar, 24 h) affords, after desilylation, the *N*-alkylated phenazinones (23a-c) and lavanducyanin (24), which are related to the phenazine pigment, pyocyanin (T. Kitahara *et al.*, Synlett, 1995, 186). The antileukemic natural product, phenazinomycin (25), and its enantiomer were also conveniently synthesised by the latter high-pressure procedure (method C) (Y. Kinoshita and T. Kitahara, Tetrahedron Letters, 1997, 38, 4993).

OSiR$_3$ + X- R^1 →(i, ii)

R = Me, Et, Bu (22)

X = OH, Br
R^1 = allyl, prenyl, geranyl

(23) a: R^1 = allyl
b: R^1 = prenyl
c: R^1 = geranyl

i. Method A (X = OH): Tf_2O, 2,6-di-t-Bu-4-MeC_5H_2N, CH_2Cl_2
Method B (X = OH): CF_3SO_3Ag, CH_2Cl_2
Method C (X = Br): CH_2Cl_2,high pressure (15 kbar)
ii. $(n\text{-}Bu)_4NF$ or$(Me_2N)_3S(Me_3Si)F_2$

(24)

(25)

Reaction conditions have been reported for the preparation of a series of *N*-mono- and *N,N'*-diacylated 5,10-dihydrophenazines (26) and (27) by the reaction of either dihydrophenazine or the 5,10-bis(trimethylsilyl) derivative with acid chlorides (M. Mikulla and R. Müllhaupt, Chem. Ber., 1994, 127, 1723).

RCOCl, THF / rt ← → RCOCl (excess), pyridine / Δ

R = Ph, OMe, OEt, OPh, OAr (26)

X = H, $SiMe_3$

(27)

The Beirut reaction between 5,6-difluorobenzofuroxan and 1-aminocyclohexene enamine derived affords the tetrahydrophenazine 5,10-dioxides (28). Displacement of one fluorine can be effected by treatment of (28) with a secondary amine such as morpholine to give (29) (S. K. Kotovskaya *et al.*, Russ. J. org. Chem., 1998, 34, 369).

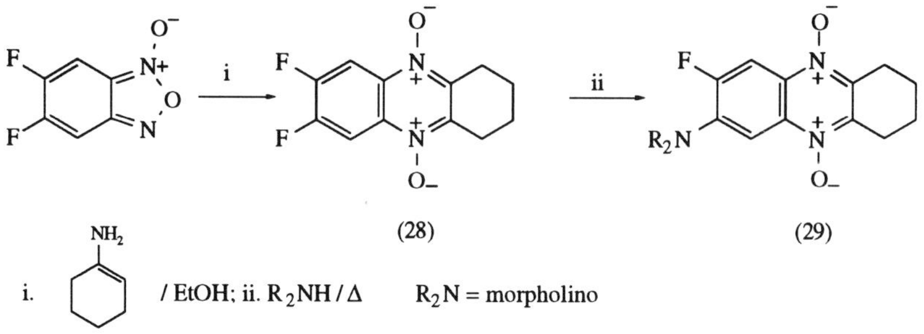

(28)

(29)

i. [1-aminocyclohexene] / EtOH; ii. R_2NH / Δ R_2N = morpholino

(iii) Miscellaneous Phenazine Derivatives

Irradiation of benzo[*b*]phenazine in solution results in the formation of the centrosymmetrical [4+4] cyclodimer (30) as the sole isolable product in dichloromethane, and as the major product along with alternative cyclodimer (31) (ratio 4:1) in isopropanol. Sensitised photo-oxygenation of benzo[*b*]phenazine in methanol yields the crystalline endoperoxide (32) (E. Fasani, M. Mella and A. Albini, Heterocycles, 1995, 40, 577).

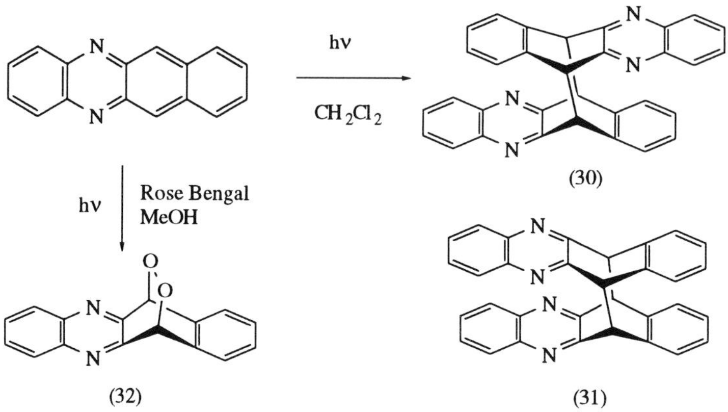

hν / CH_2Cl_2

hν / Rose Bengal / MeOH

(30)

(32)

(31)

5,7-Dihydrobenzo[*a*]phenazin-5-ones (33), which are of interest as fluorescent probes, have been synthesised by the condensation of o-phenylenediamines with 2-hydroxynaphthoquinones in the presence of acetic acid (G. M. Rehberg and J. L. Rutherford, J. heterocyclic Chem., 1995, 32, 1643).

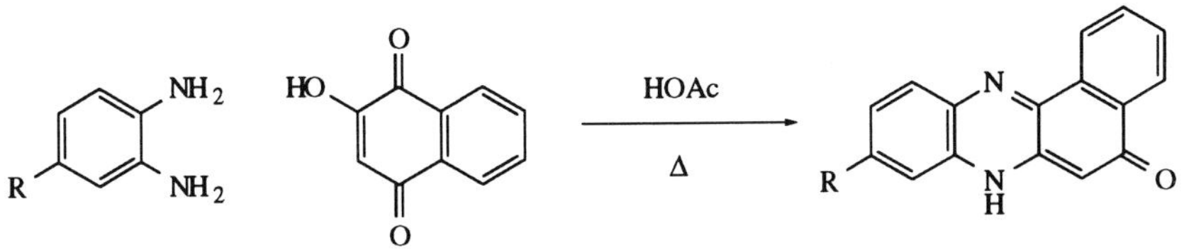

R = H, OMe, Cl, NO_2

HOAc

Δ

(33)

The syntheses of several structurally novel compounds containing the phenazine moiety have been reported, including (a) the bridged tetrahydrophenazine (34) (R. Gleiter, T. Doerner and H. Irngartinger, Liebigs Ann., 1996, 381) (b) the hexamethylhexahydrobenzo[*b*]phenazine (35) M. Sainsbury *et al.*, J. chem. Soc., Perkin Trans. 1, 1995, 3117), (c) the phenazine-fused [60]fullarene (36) (J. Mattay *et al.*, J. org. Chem., 1997, 62, 2752), (d) the betaine (37) derived from the condensation of mitomycin A and *N*-methyl-1,2-phenylenediamine (S. Wang and H. Kohn, J. org. Chem., 1996, 61, 9202).

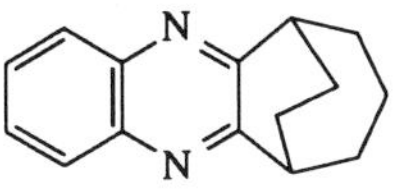

(34)

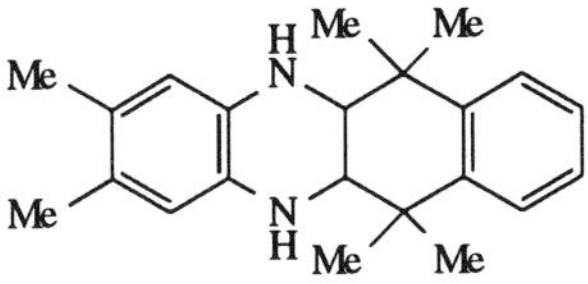

(35)

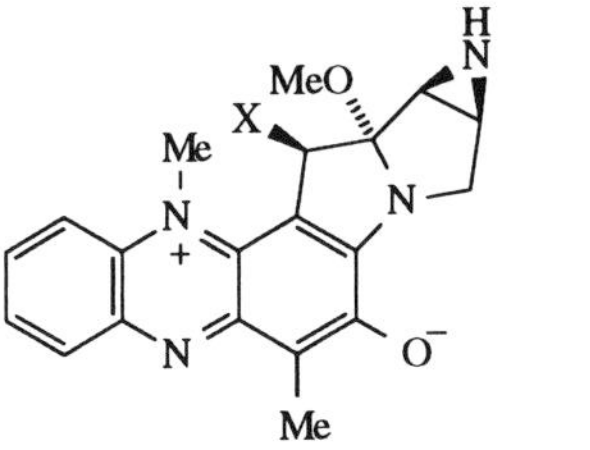

(36) X = CH_2OCONH_2

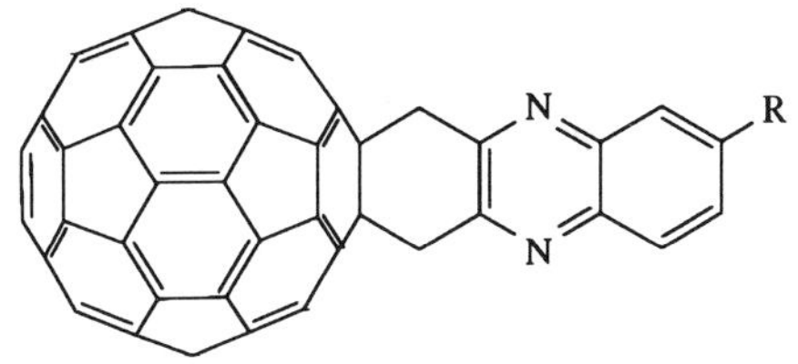

(37) R = H, OMe

Second Supplements to the 2nd Edition of Rodd's Chemistry of Carbon Compounds, Vol.IV IJ, edited by M. Sainsbury

Chapter 45. PHENAZINE, OXAZINE, THIAZINE AND SULFUR DYES

R. BOLTON

This Chapter deals with the chemistry of a range of dyes. Earlier reviews [N. Hughes, *Rodd's CCC, 2nd Ed., Vol IV J,* 45, 403 (1989); G Hallas and A.D. Towns, *Suppl. 2nd Ed. Vol. IV I/J,* 45, 193 (1995); H. Zollinger, *Color Chemistry: Synthesis, Properties and Applications of Organic Dyes and Pigments.* (VCH:Weinheim 1987, 367] dealt comprehensively with methods for the syntheses of these compounds, many of which are simple and inexpensive. Few improvements have come in this area, though some synthetic innovations of industrial importance have been made [T.Yomo *et al., Eur. J. Biochem.*, 1989, 179, 293; A.G. Miller and R.J. Balchunis, EP 339,869 (1990)]. The more recent reports of these compounds focus attention more sharply upon their applications than their fundamental synthesis and reactions (*Prospects of the Development of Chemistry and Technology of Dyeing and Synthesis of Dyes: Interuniversity Collection of Scientific Papers.* Ed., B.N. Mel'nikov, (Ivanovskii Khim-Tekhnol. Inst., 1989). This review will necessarily reflect such interests.

The synthesis of perfluoromethylated Nile Red derivatives (*e.g.* **1,** 11%) by heating *bis* (trifluoroacetyl) peroxide in the presence of alkylated Nile Red compounds, typifies the more recent chemistry [M. Mitani *et al.*, JP 07 82,495 (1995)].

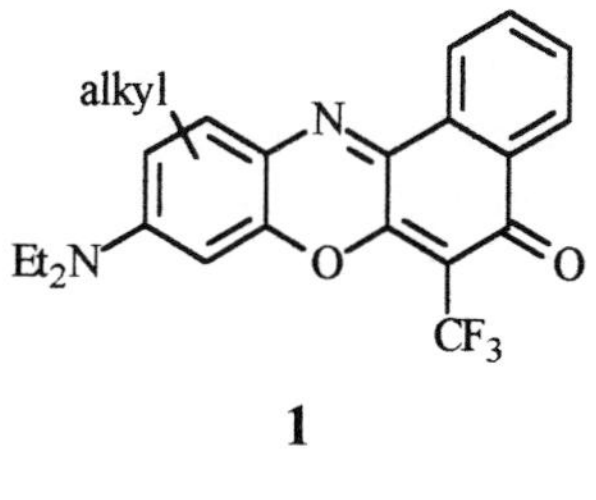

1

The tendency for these cationic dyes to influence free-radical processes is used in tumour photosensitization (*e.g.* L. Cincotta and J.W. Foley, *Proc. SPIE-Int. Soc. Opt. Eng.*, 1989, 997, 145). This then suggests the synthesis of radiolabelled dyes capable of attacking the tumour in a number of ways. In the synthesis of ^{125}I-, ^{127}I-, or ^{211}At-labelled Methylene Blue, both nucleophilic substitution (halogen exchange using inorganic halides and molten crown ether as solvent and catalyst, and halogeno-dediazoniation) and electrophilic halogenation (using chloramine-T as oxidant and radioactive inorganic halide as source) were tried. Halogen exchange gives products with the highest extent of radiolabelling. Yields were poor, with only low specific activities, in both the other processes (I. Brown *et al., J. Radioanal. Nucl. Chem.*, 1986, 107, 337).

A review of the influence of the Vilsmeier-Heck reaction upon this area of the dye industry covers some appropriate and interesting chemistry (G. Seybold, *J. prakt. Chem. / Chem. Ztg.*, 1996, 338, 392).

The reduction of a number of sparingly soluble colouring agents, including sulfur dyes as well as indigo, is achieved by the complexes of Fe(II) with triethanolamine (T. Bechtold *et al., J. Electrochem. Soc.*, 1996, 143, 2411). This system is advocated over many others.

1. Dye-mediated electron-transference

(a) Application in polymerisation

Phenazinium salts (earlier, phenazonium) are exemplified by Phenosafranine (**2**). Here the electron-donating amino-groups at C-3 and C-7 provide opportunities for charge-delocalisation and hence the development of colour. A second consequence of this delocalisation is the easy formation of a radical cation through electron transfer. Much of the present use of these dyes rests upon their catalytic effect upon electron-transfer processes, especially the oxidation or reduction of other species in the solution.

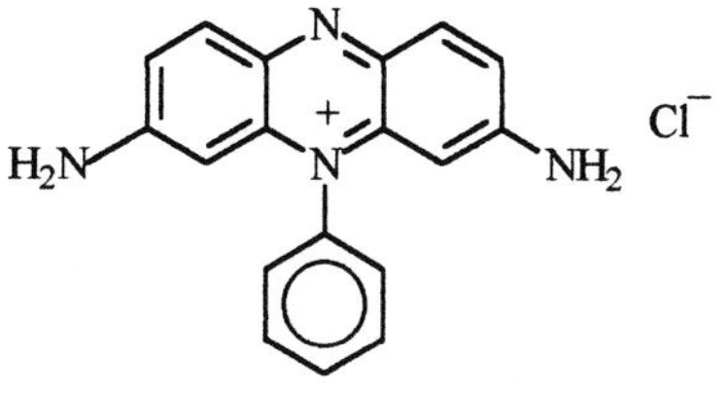

2

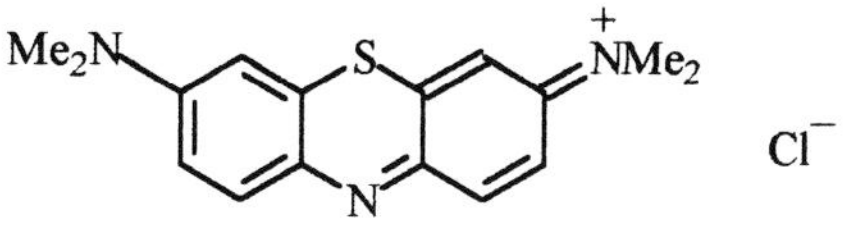

3

This is exemplified in lithography, where the phenothiazinium dye Methylene Blue (C.I. Basic Blue 9; **3**), upon photosensitisation, initiates the radical-induced polymerisation of a suitable alkene [S. Matsumoto *et al.*, JP 09,134,007 (1996). K. Kishi, JP 09,235,310 (1996)]. Interestingly, **3** is also reported to *inhibit* vinyl polymerisation (T.Shimizu, O. Morya and K. Shigematsu, JP 08 81,397 (1996)).

(b) Application to biological systems

In photochemical cancer therapy, **3** has been advanced as a solid state photosensitizer (X. He, X. Feng and H. Shen, *Fenxi Kexue Xuebao*, 1995, 11, 1), as it has in photoelectrochemical solar cells for hydrogen generation (I. Bayer, I. Eroglu and L. Turker, *Hydrogen Energy Prog. XI, Proc. World Hydrogen Energy Conf., 11th.* 1996, 1, 919). However, it was not the best sensitizer in a study of the photodegradation of Arochlor 1254 by simulated sunlight (Y. Lin, G. Gupta and J. Baker *Bull. Environ. Contam. Toxicol.*, 1996, 56, 566).

Methylene Blue (**3**) activates large Ca^{2+} - activated K^{+} channels, a process which may be associated with its action as a free-radical inhibitor of guanylyl cyclase (J.D. Stockand and S.C. Sansom, *Biochim. Biophys. Acta,* 1996, 1285, 123).

Again, the photo-generation of reactive radical species from some commercially available phenothiazinium dyes substantially improved their antibacterial action against five common pathogens, both Gram positive and Gram negative (M. Wainwright *et al.*, *FEMS Immunol. Med. Micro-biol.*, 1997, 19, 75). D. Philips (*Progr.React.Kinetics*, 1997, 22, 175) has discussed the relative merits of aluminium phthalocyanine and **4** in a detailed and thorough review of photodynamic therapy.

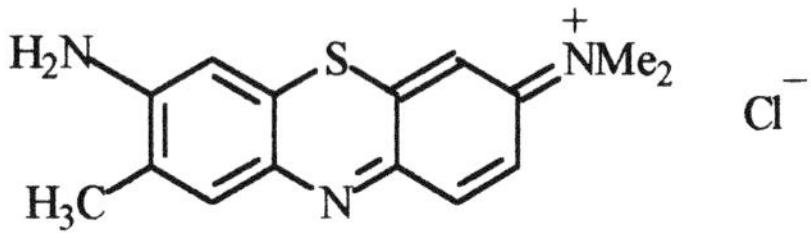

H_2N S $\overset{+}{N}Me_2$ Cl^-

H_3C N

4

The cariogenic bacterium *Streptococcus mutans* is killed by laser i.r. radiation in the presence of Toluidine Blue O (**4**). This arises from membrane damage induced by lipid peroxidation by reactive oxygen species. D_2O increases the kill, while 0.1*M*-methionine or 0.5*M*-NaN_3 each gives 98% protection against lethal photo-sensitisation (T. Burns, M. Wilson, and G.J. Pearson, *Proc. SPIE-Int. Soc. Opt. Eng.*, 1996, 2625, 288). **4** is also used to sensitise *P. gingivalis* to He/Ne laser light (M. Bhatti *et al.*, *Photochem. Photobiol.*, 1997, 65, 1026). *Helicobacter pylori* was similarly killed by laser photoirradiation in the necessary presence of dyes; **3** and **4** were effective, though Crystal Violet and thionine (**5**) were not. Tritiated-**4** was used to measure the uptake by *H. pylori* (C.E. Millson *et al.*, *Proc. SPIE-Int. Soc. Opt. Eng.*, 1995, 2371, 72; Millson *et al.*, *J. Med. Microbiol.*, 1996, 44, 245; Millson *et al.*, *Lasers Med. Sci.*, 1997, 12, 145). In general, these antibacterial properties of **3** and **4** (C.E. Millson *et al.*, *J. Photochem. Photobiol.*, 1996, 32B, 59; D Philips, *loc.cit.*) rely upon the preferential absorption of the dye by the pathogen cells and the consequent selective generation of radical species such as singlet oxygen within the cell, with lethal consequences. *Staphylococcus aureus* is also killed by laser treatment in the presence of **4** (M. Wilson and C. Yianni, *J. Med. Microbiol.*, 1995, 42, 62).

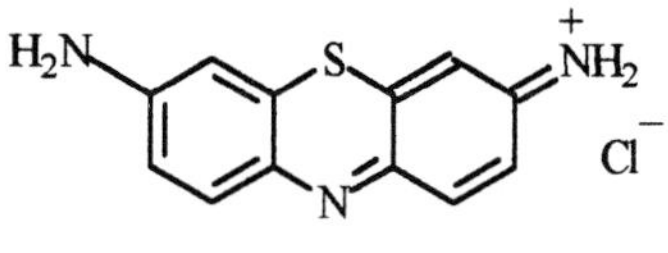

5 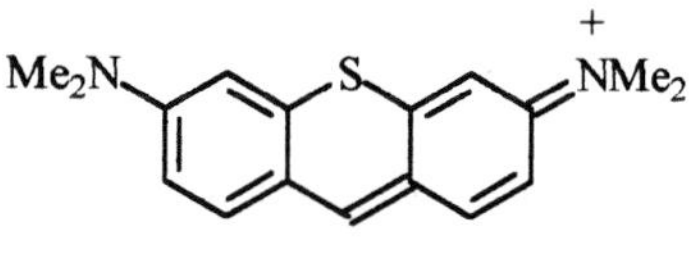 **6**

Viral contaminants in body fluids may be inactivated similarly by using **3** (L. Wolf, W. Brattan and J. Foley, US 5,527,704 (1996); *Chem.Abstr.*, 1996, 125, 123839u), or by the absorption of **3**, **5**, or thiopyronine (**6**) on Qβ bacteriophage. The concentration of the dye used is critical, since an excess causes aggregation and less effective sensitising (S. Jockusch *et al., Proc. Natl. Acad. Sci. U.S.A.* 1996, 93, 7446). Extracellular enveloped virus in blood and blood components may also be inactivated [S J Wagner, US 5,545,516 (1996)], and similar chemistry is the basis of the detection of biomolecules by **3** (F. Schubert, U. Moeller and A. Knaf, DE 4,403,780). A device based upon the use of **4** to remove leukocytes and viral inactivity reagents from the blood has also been described [F. Castino *et al.*, WO 95 18665 (1995)].

Methylene Blue is also claimed [R. Ramasamy and S. Schaefer, WO 96 19,225 (1996)] to afford cardioprotection after ischaemia. Such dyes also control nitrogen oxide concentrations to moderate skeletal muscle contractions [J. Stamler and L. Kobzik, US 5,545,614 (1996)]. Methylene Blue also influences the neurotoxic mechanisms of ifosfamide encephalopathy in cancer therapy (A. Kuepfer, C. Aeschlimann and T. Cerny, *Eur. J. Clin. Pharmacol.*, 1996, 50, 249). Kuepfer and Cerny claim [WO 95 13,079 (1995)] that dyes such as **3** may be bound to a biomolecule where, when irradiated with light, they generate singlet oxygen which is trapped by an alkene to give a dioxetane. Hydrolysis of an appropriate substituent in the dioxetane causes its decomposition and the generation of light. The reported detection limit is 10^{-20} moles of analyte. The cytotoxic properties of **3** are enhanced in the presence of the iron-chelating agent deferoxamine (Y.S. Lee, S.K. Han, and R.D.Wurster, *Arch. Pharmacol. Res.*, 1995, 18, 159). The effect is blocked by the presence of stoichiometric amounts of Fe(III), hydroxyl radicals, antioxidants, and calcium release blockers, which suggests that hydroxyl radical and Ca^{++} may mediate this potentiation. It also implies the application of **3,** together with deferoxamine, in the treatment of human brain tumours.

The sensitising effect of dyes such as the phenothiazinium salts may be reversed, at least in some instances, by the presence of antibiotics (G. Lazarova and H. Tashiro, *Microbios*, 1995, 82, 332, 187) While the mechanism of the process is not certain, the polyenic structure of effective compounds such as amphotericin B may afford a sacrificial route for removing the various oxygen-based radicals which are thought to cause membrane damage when cells are irradiated in the dual presence of oxygen and these dyes.

2. Analytical applications

The general method of specific analysis of organic compounds, especially those which are biologically important, has been copiously reported. In essence, both an appropriate dye and an enzyme are trapped on an electrode. In some cases these adjuncts are absorbed onto the electrode itself (*e.g.* carbon); in others, they are held by a second phase (*e.g.* silica gel) in contact with the electrode. Electrochemical studies of the properties and behaviour of such electrodes abound (*e.g.* B. Persson *et al., Electroanal.*, 1995, 7, 935; B. Grundig *et al., J. Electroanal. Chem.*, 1995, 395, 143; Q.J. Chi and S.J. Dong, *J. Mol. Cat.*, 1996, 105, 193; G. Nagy, I. Kapui, and L. Gorton, *Anal. Chim. Acta,* 1995, 305, 65; L. Ding and E.K. Wang, *Bio-electrochem. Bioenergetics,* 1997, 43, 173; *ibid., Chin. Sci. Bull.*, 1997, 42, 203; L.T. Kubota *et al., Electrochim. Acta,* 1996, 41, 1465; M.Tian and S.J. Dong, *Electroanal.*, 1996, 8, 454). New dehydrogenase-based electrodes have been made through electro-polymerisation of anilines or phenylenediamines onto Au, Pt or carbon electrodes (A. Curulli *et al., Biosens. Bioelectron.*, 1997, 12, 1043; P.N. Bartlett, P.R. Birkin and E.N.K. Wallace, *J. Chem. Soc., Faraday Trans.*, 1997, 93, 1951), and NADH-based systems based upon electrodeposited films derived from 3,4-dihydroxybenzaldehyde have also been reported (F. Pariente *et al., Anal. Chem.*, 1997, 69, 4065).

Cationic dyes have also been used to study cavities in a variety of materials. Thus, recent publications have reported the electro-chemistry and photoelectrochemistry of phenothiazine dyes included into β-cyclodextrin (C. Ratna Raj and R Ramaraj, *J. Electroanal. Chem.*, 1996, 405, 141), and into zeolites (X. Li and V. Ramamurthy, *J. Amer. Chem. Soc.*, 1996, 118, 10666), together with the effect of the monoazacrown system upon benzothiazolium dyes (I.K. Lednev, R.E. Hester and J.N. Moore, *J. Chem. Soc., Faraday Trans.*, 1997, 93, 1551).

(a) Biosensors

Glycerol dehydrogenase (GDH) and lipase, immobilised in a carbon electrode also containing Meldola Blue (**7**), Nile Blue (**8**) or Toluidine Blue O (**4**), are used in the amperometric determination of glycerol and of triglycerides. Responses of 2-9 nA/m*M* glycerol are in part determined by the amount of GDH immobilised (V. Laurinavicius *et al., Anal. Chem. Acta,* 1996, 330, 159).

A reagentless biosensor for lactic acid uses a screen-printed carbon electrode containing **7**, which markedly improved the stability of the electrode system (M. Khayyami *et al., Electroanal.*, 1997, 9, 523), lactate dehydrogenase (LDH), NAD^+ and cellulose acetate (S.D. Sprules *et al., Anal. Chim. Acta*, 1995, 304, 17 and 1996, 329, 215; *Electroanalysis*, 1996, 8, 539). An analogous system based upon the electrochemical detection of NADH produced by the reaction of lactate with NAD^+ in the presence of LDH and a carbon electrode dosed with **4** has also been reported. The catalytic current increased linearly when $4 \times 10^{-5}M$ < [NADH] < $1.5 \times 10^{-3}M$ (H.X. Ju, L. Dong and H.Y. Chen., *Talanta,* 1996, 43, 1177). LDH and **4**, co-immobilised on a graphite electrode and sheathed in a dialysis membrane, were used to measure L-lactate in the range $1 \times 10^{-6}M$ < [L-lactate] < $60 \times 10^{-6}M$ with a lower limit of $4 \times 10^{-7}M$ (M.J.F. Villamil, A.J.M. Ordieres and P.T. Blanco, *Anal. Chem. Acta*, 1997, 345, 37). Other such sensors have been described (H.C. Yoon and H.S. Kim, *Anal. Chim. Acta.*, 1996, 336, 57). Correspondingly, a system based upon the **4**-sensitised reaction of NAD^+ in the presence of D-lactate dehydrogenase at an electrode surface gave excellent selectivity towards D-lactate even in the presence of the enantiomer, and response in the range of 0.05 m*M* - 5 m*M* and a detection limit of 0.03 m*M* (H. Shu *et al.*,

Biotech. Bioeng., 1995, 46, 270).

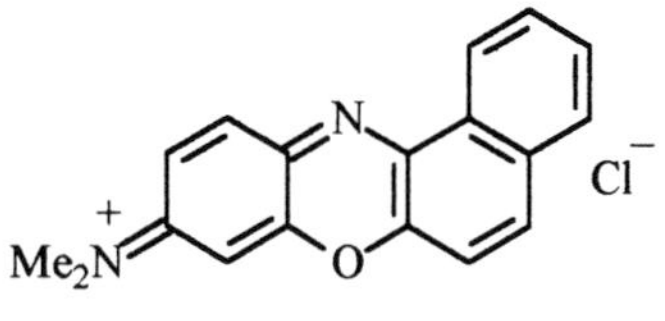

7

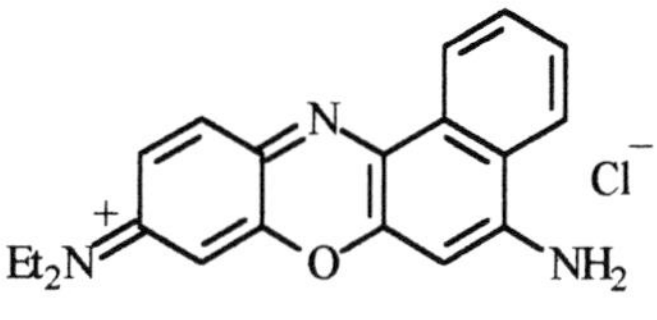

8

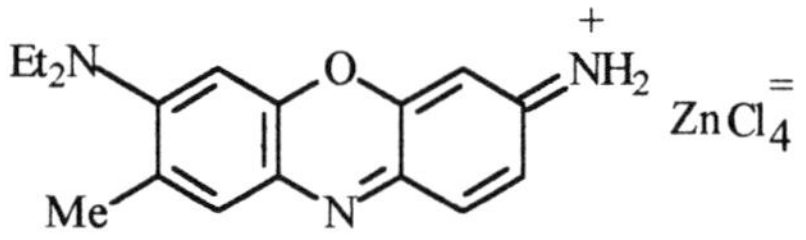

9

The combination of ascorbic acid and a phenothiazinium dye is also reported (L.E. Leon, *Talanta*, 1996, 43, 1275) as a means for the amperometric analysis of L-ascorbic acid at levels of 5-90 μg/mL (lower limit: 1.9 μg/mL) based on the photochemical reduction of **3**. A similar process using **4** measures ascorbic acid in the region 5-50 μg/mL (A. Safavi and L. Fotouhi, *Talanta,* 1994, 41, 1225).

Glucose may be determined by a similar system using glucose-oxidase (M. Khayyami *et al., Talanta*, 1996, 43, 957; H.Y. Liu *et al., Fresenius J. Anal. Chem.*, 1997, 357, 812); glucose dehydrogenase allows specific amperometric glucose determination measurements by the oxidation of NADH (L.I. Boguslavsky, *Biosens. Bioelectron.*, 1995, 10, 693). Fructose may be similarly measured with Meldola Blue (**7**) and fructose-5-dehydrogenase in the region 0.1m*M* < [Fructose] < 0.8m*M* (C.A.B. Garcia *et al., J. Electroanal. Chem.*, 1996, 418, 147).

Toluidine Blue (**4**), Meldola Blue (**7**), Nile Blue (**8**) and Brilliant Cresyl Blue (**9**) have each been advocated in enzyme immobilised electrodes for amperometric measurement of ethanol concentrations; alcohol dehydrogenase and **4** are the best combination, giving linear responses at 10μ*M* < [EtOH] < 100 μ*M* and a detection limit of 5μ*M* (M.J. Lobo, A.J. Miranda and P. Tunon, *Electroanal.*, 1996, 8, 591; see also J. Wang and N. Naser, *Electroanal.*, 1995, 7, 362). S.G. Mueller *et al. (Talanta*, 1996, 43,

779) propose the use of **7** in such systems. Sprules (S.D. Sprules *et al., Anal. Chim. Acta,* 1996, 329, 215) describes a screen-printed electrode pair in which an Ag/AgCl reference electrode is set on PVC close to a working carbon electrode modified with **7** and coated with alcohol dehydrogenase and NAD^+. Trehalose markedly improved the life of the electrode, which gave a linear response to 3.5 x $10^{-2}M$ EtOH. Unfortunately, butan-1-ol and both propanols gave positive responses.

Salicylate may analogously be determined with a carbon electrode coated with silica-gel containing Meldola Blue (**7**) and salicylate hydroxylase (L.T. Kubota *et al., Anal. Letters,* 1996, 29, 893). However, the oscillatory response suggests kinetic complexity which makes the process ineffective for analytical application (Kubota *et al., Chem. Phys. Letters,* 1997, 264, 662).

In a range of dyes, the best composite electrode for the determination of saccharin used **4**, which formed an association complex with saccharin in which the electrode showed a linear response at $8.1 \times 10^{-5}M$ < [saccharin] < $1.4 \times 10^{-2}M$, with a detection limit of $6.3 \times 10^{-5}M$ (O. Fatibello and C. Aniceto, *Anal. Letters*, 1997, 30, 1653).

Haemoglobin may also be estimated amperometrically, using a carbon fibre electrode activated by electrochemical oxidation, and then bonding **4** to the resulting $-CO_2H$ functionality by acylation (H.X. Ju, L. Dong and H.Y. Chen, *Anal. Letters,* 1996, 29, 587).

The electrochemical oxidation of NADH may be brought about by the use of Neutral Red (**10**) or of **4** covalently attached to polymeric electrodes. These may act either as sensors for reducable materials such as EtOH, glucose and vitamin K, or as systems which may bring about the enantiomerically selective reduction of aldehydes and ketones (Y. Okamoto, T. Kaku and R. Shundo, *Pure Appl. Chem.,* 1996, 68, 1417).

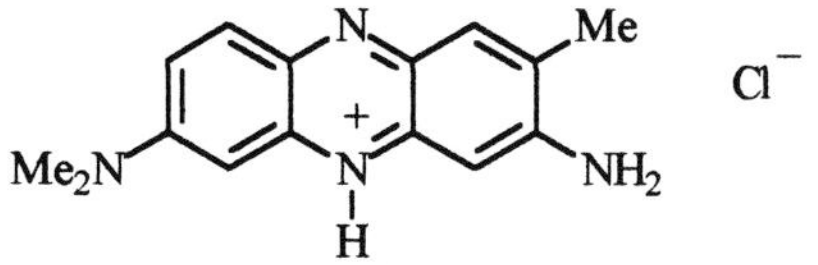

10

Hydrogen peroxide can be analysed amperometrically to levels of 0.1 µ*M*, using (*i*) a Methylene Blue-mediated electrode in which peroxidase is immobilised in a composite membrane (H.Y. Liu *et al., Fresenius J. Anal. Chem.*, 1997, 357, 302), or (*ii*) Methylene Green (**11**) and peroxidase immobilised in a fibroin/PVA membrane (X.-X. Wu *et al., Chin. J. Chem.*, 1996, 14, 359) or in a montmorillonite – modified BSA – glutaraldehyde matrix (C. Lei and J. Deng, *Anal. Chem.*, 1996, 68, 3344, who claim a limit of 0.4 µ*M*), or (*iii*) New Methylene Blue N (C.I. Basic Blue 24; **12**) - peroxidase in regenerated silk fibroin matrix (J. Qian *et al., Anal. Biochem.*, 1996, 236, 208; limit, 0.1 µ*M*). A similar system, using **12** on a Nafion gel electrode system, showed a claimed detection limit of 0.5µ*M* H_2O_2 (H. Liu *et al., J. Electroanal. Chem.*, 1996, 417, 59; *Anal. Chim. Acta*, 1996, 329, 97), whereas **12** on a montmorillonite matrix with modified horse-radish peroxidase immobilised upon it showed a linear response in the region 5.0µ*M* < $[H_2O_2]$ < 2.0m*M* and a reported detection limit at 0.6µ*M* (C. Lei *et al., Anal. Chim. Acta*, 1996, 332, 73).

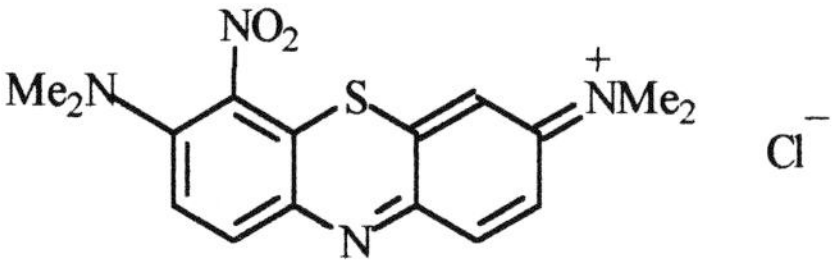

11

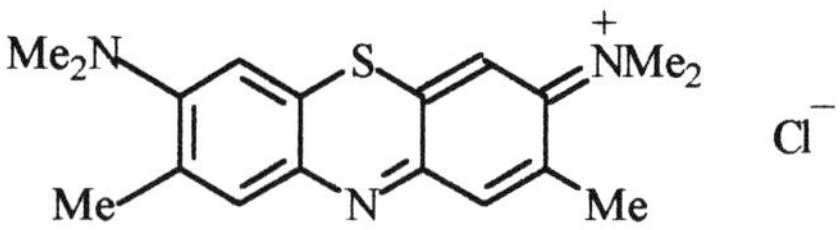

12

(b) Kinetic/catalytic application in analysis

A simple laboratory demonstration in which Methylene Blue (**3**) catalyses the atmospheric oxidation of alkaline solutions of glucose has been widely used to stimulate student discussion and understanding of kinetic processes; it has again been recently assessed and advocated as a teaching aid (W.R. Vandaveer, M. Mosher and G.L. Gilbert *J. Chem. Educ.*, 1997, 74, 402). This also reflects the wide range of applications by these dyes in synthetic and analytical chemistry. The oxidation of an amine in the presence of **3** affords a similar undergraduate experiment on photochemistry (D. Burget, C. Carre and P, Jacques, *Actual. Chim.*, 1994, 16). The second-order rate constant for the aerial oxidation of ascorbic acid mediated by **3** at pH 9 is 4.21 $l.mol^{-1}.s^{-1}$ (25°C) (P. Sevcik and B. Dunford, *Int. J. Chem. Kinet.*, 1995, 27, 925). Riboflavin is an even better catalyst for the light-induced oxidation of ascorbic acid in aqueous media; at 1.2 ppm., the relative photosensitising abilities are riboflavin (k_{rel}, 21) > **3** (15) > protoporphyrin IX (1). The same authors (M.Y. Jung, S.K. Kim and S.Y. Kim, *Food Chem.*, 1995, 53, 397) report second-order rate constants for the reaction of ascorbic acid and singlet oxygen as 6.63×10^8 $l.mol^{-1}s^{-1}$ (pH 7.5), 5.77×10^8 (6.0) and 5.27×10^8 (4.5), and observed the effect of added aminoacids upon these values.

Dyes may assist photochemical reactions, and this can offer a route to their analysis. For example, the rate of photo-oxidation of ascorbic acid is linearly linked to the phenothiazinium ion concentration in the region $5 \times 10^{-7}\ M < [\text{Dye}] < 1 \times 10^{-4}\ M$. When this reaction occurs in the presence of lucigenin there is a substantial enhancement of the luminescence of the reaction products. This in turn affords a sensitive means of measuring concentrations of the dyes Thionine (**5**), Azur C (**13**), Azur B (**14**), Thionine Blue (**15**; C.I. Basic Blue 25), Toluidine Blue O (**4**), Methylene Blue (**3**) and New Methylene Blue (**16**) at environmentally important levels (T. Perez-Ruiz *et al.*, *Anal. Quim. (Barcelona)*, 1996, 15, 326). Interestingly, a photographic developing system has been proposed for black-and-white film in which a number of amines, diamines, and azo-compounds, including **5**, catalyse or mediate the reduction of the light-sensitised silver halide granule by ascorbic acid [E. Okutsu and S. Hirano, JP 08,166,658 (1996)].

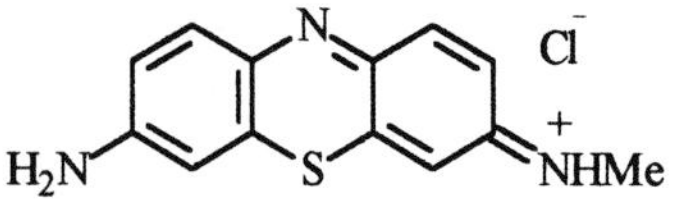

13 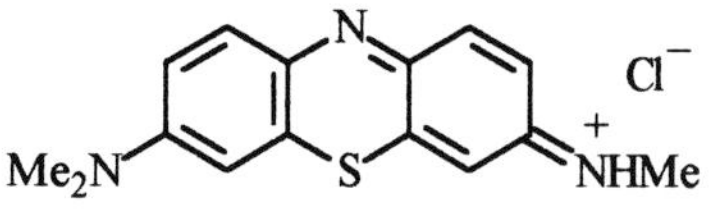 **14**

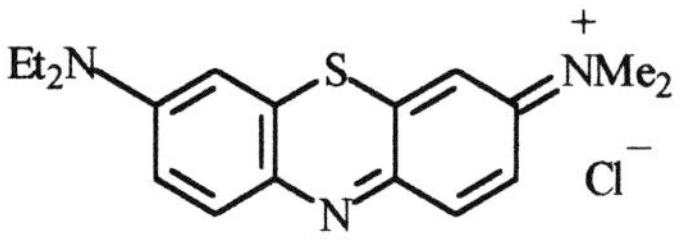

15 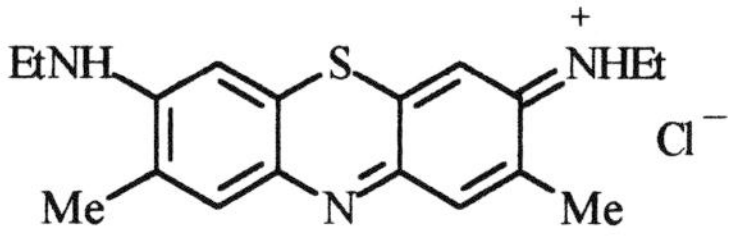 **16**

Similarly, the reductive decolorisation of Methylene Blue (**3**) by divalent sulfur is catalysed by trace amounts of selenium. Measurements of the fading time may be matched against the small concentrations of this element (R.Jiang, J.Gan, and Z.Han, *Haihuyan Yu Huagong*, 1996, 25, 18; *Chem. Abstr.*, 1997, 126, 342610u). Methods of measuring the time for fading, and a computer program to convert these into selenium concentrations, are reported and discussed (J. Hudak, V. Varga and E. Grafne Harsanyi, *Magy. Kem. Foly.*, 1995, 101, 395; *Chem. Abstr.*, 1994, 123, 328668m). This process for the determination of Se(IV) is made more sensitive by the presence of cetyltrimethylammonium bromide. The parameters of the process have been critically examined; the method is best at levels of 5-10 μM Se(IV) (B. Arikan, M. Tuncay and R. Apak, *Anal. Chim. Acta.*, 1996, 335, 155). 1-Carboxy-7-(dimethylamino)-3,4-dihydroxyphenoxazin-5-ium chloride (C.I. Mordant Blue 10, gallocyanine; **17**) has been used similarly (A.A. Ensafi, *Indian J. Chem* 1997, 36A, 105) in the measurement of Se(IV). The linear response is in the range 2.5-500 ng/mL, and the detection limit is 1 ng/mL.

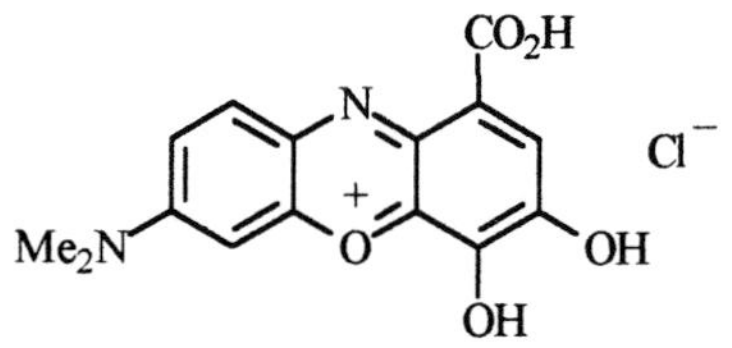

17

Tellurium(IV) at trace levels (0.096-1.250 μg/mL; detection limit, 80 ng/mL) can be measured from its catalytic effect on the reduction of Toluidine Blue by sulfide ion at pH 4. The reaction is followed from the decrease in the dye absorbance at 628 nm (M. Shamsipur and M.F. Mousavi, *Iran. J. Chem. Chem. Eng., Int. Eng. Ed*, 1995, 14, 1. *Chem. Abstr.*, 1996, 124, 192605z). A second, similar method claims (Mousavi and M.R. Almasian, *Anal. Letters*, 1996, 29, 1851) a linear range at 0.08μg/mL < [Te(IV)] < 3.8 μg/mL and a lower detection limit of 0.06 μg/mL.

The further effect of small quantities of higher oxidation-state metals on the rate of oxidation of **17** by bromate has allowed the determination of similarly small levels of Ce(IV) (linear range, 0.005-10 μg/mL; detection limit, 2 ng/mL. Ensafi and M. Hashem, *J.Sci. Islamic Repub. Iran*, 1996, 7, 239; *Chem. Abstr.*, 1997, 126, 206871p). Vanadium, as V(V), may be measured down to 2 pg/200μL (A.A. Ensafi and A. Kazemzadeh, *Microchem. J.*, 1996, 53, 139). and its catalytic effect upon the oxidation of Methylene Green (**11;** C.I. Basic Green 5) by $KBrO_3$ is the basis of a spectrophotometric analytical method which is claimed to give a linear response to 0.20μg/25 mL and a detection limit of 3.4 x 10^{-10}g/mL (S. Ding and Z. Zhang, *Guangpu Shiyanshi*, 1996, 13, 47; *Chem. Abstr.* 1996, 125, 346412u).

Osmium (as OsO_4) may be analogously determined to levels of 1-2 pg/mL over the range 0.01-100 ng/mL (A.A. Ensafi, *Anal. Lett.*, 1996, 29, 1177; Ensafi and E.S. Sollary, *Indian J. Chem.*, 1995, 34A, 1005). The latter method is reported to be free from most interferences, and especially from the effect of even large concentrations of platinum metals.

These properties of phenazine dyes also allows them to be used as indicators in redox titrations. In the presence of EDTA they can be used to give sharp endpoints in the $Fe^{2+} \rightarrow Fe^{3+}$ titration; complexation by EDTA or a similar species is necessary because many of these cations are formed by the Fe^{3+} oxidation of their leucobases, which suggests that the iron(II) - iron(III) couple and the leucobase - cation couple have similar E_o values so that complexation is necessary to lower the standing concentration of the metal ion sufficiently that oxidation of the leucobase only occurs after the formation of Fe(III) is essentially complete. Thus, Phenosafranine, Methylene Violet, Amethyst Violet, Safranine T, Wool Fast Blue BL, and Aposafranine have each been used as redox indicators in the estimation of Sb(III), hydroquinone ascorbic acid and hydrazine with chloramine-T (N.V. Rao and C.K. Sastri, *J. Ind. Chem. Soc.*, 1987, 64, 131) and thionite-bleached indicator papers containing Methylene Blue (**3**), Resazurin (**18**), Nile Blue (**8**), Janus Green (**19**), Neutral Red (**10**) or benzylviologen have been reported [P. Smigan, M. Greksak and P. Rusnak, CS 248,958 (1988)] as indicators of levels of airborne pollutants.

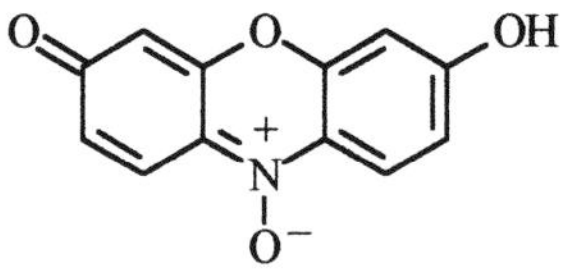

18

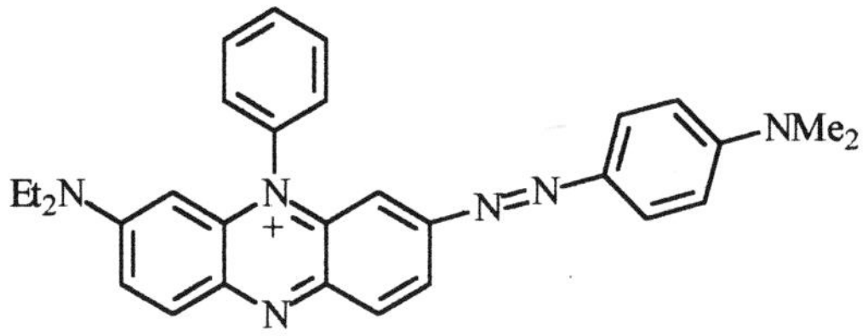

19, Janus Green B

(c) Spectroscopic applications to analysis

Phenothazin-5-ium cation has been similarly used in the spectrophotometric determination of thallium as Tl(III), although the analysis is sensitive to the presence of Au(III), Hg(II), Sb(III) and many other oxidising agents (Y. He *et al., Lihua Jianyan, Huaxue Fence* 1995, 31(5), 295). Thallium has also been determined by selectively extracting the ion-pair between Janus Blue cation and $TlCl_4^-$ into an organic solvent (Kh.A. Kerimova and I.K. Guseinov, SU 1,272,252. *Otkrytiya, Izobret.* 1987, 186, 180. *Chem. Abstr.*, 1987, 106, P148618k). The similar extraction of $[\text{safranine T}]^+$ (**20**) $TlBr_4^-$ from aqueous sulfuric acid (<2*M*) into isoamyl acetate allows measurements in the range 0.05 < [Tl] < 2.5 μg/mL (D.A. Mikaelyan, V.Zh. Artsruni and A.G. Khachatvyan, *Arm. Khim. Zh.*, 1988, 41, 672; 1989, 42, 568).

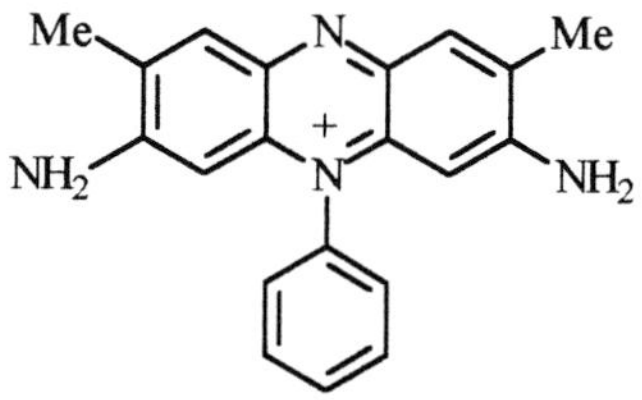

20

Sn(IV) in H_2SO_4-$H_2C_2O_4$ gives with **16** or with Toluidine Blue O (**4**) a highly sensitive absorptive complex which appears at -0.60v w.r.t. SCE and may be determined using absorption polarography. This allows the measurement of tin to levels of 0.8 ng/mL (Y. Zhang, G. Yue, and T. Zhu, *Fenzi Huaxue* 1995, 23, 1237; *ibid., Anal. Sci.*, 1996, 12, 295).

A range of phenothiazinium dyes has been studied in the extractive spectrophotometric determination of Ag, Hg and Cu. An initial associative complex between bromothymol blue and the metal ion, in $(CH_2Cl)_2$, after treatment with aqueous CN^-, then forms a ternary associated complex which is extracted into dichloroethane (B. Zuo *et al., Fenxi Ceshi Xuebao* 1996, 15, 17. *Chem. Abstr.*, 1997, 126, 180511u). The Ag^+-catalysed oxidation of Brilliant Cresyl Blue (**9**) in the presence of 1,10-phenanthroline affords a method for the determination of low levels of the metal; 0.3 ng/mL $< [Ag^+] <$ 1500 ng/mL is the reported linear range, and 0.1 ng/mL the detection limit (A. A Ensafi and S. Abbasi, *Anal. Letters,* 1997, 30, 327).

Mercury may be estimated through $[HgI_4]^=$ when the ion-pair with Cd(II)-phenanthroline is extracted into benzene, selectively removed with EDTA, and the iodine content of the aqueous phase is oxidised to iodate and then (I^-) converted into iodine. This is extracted into benzene, equilibrated with iodate in the presence of HCl and Rhodamine 6G as the $[ICl_2]^-$ ion-pair, and finally determined at 535 nm. An effective range of 0.08-2.5 ppb of mercury(II) is claimed. (P. Padmala *et al., Talanta*, 1994, 41, 255; *cf.* S. Mathew *et al., Anal. Letters,* 1992, 25, 1941).

Bioactive substances may be determined, generally at the levels 0.4-5.6 μg/mL, through their effect upon the reaction between chloramine-T and gallocyanine (**17**) (C.S.P. Sastry, K. Rama Srinivas, and K.M.M. Krishna Prasad, *Talanta* 1996, 43, 1625). Celestine Blue (C.I. Mordant Blue 14; **21**) allows a spectrophotometric means of analysing aspartame in artificial sweeteners by oxidising with an excess of N-bromosuccinimide and determining the amount of unreacted reagent through the decrease in absorbance by **21**.

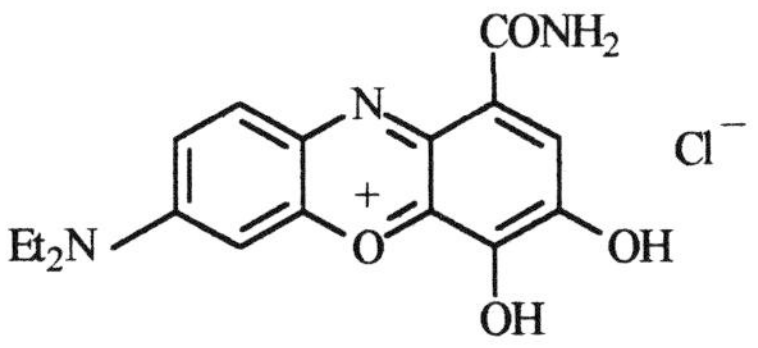

$CONH_2$
Cl^-
Et_2N
OH
OH

21

The reaction of BrO_3^- in aq. H_2SO_4 with nitrite ion is assisted by Janus Green (**19**). This has been applied to the determination of amounts of this anion, with a claimed sensitivity of 2.69×10^{-4} μg

N/mL. as NO_2^- (Z. Zhang, *Fenxi Huaxue*, 1989, 17, 122; *Chem. Abstr.*, 1989, 111, 12237y).

A similar application of selective analysis through extraction of an ion-pair has been applied to the estimation of Ta (as TaF_6^-). Ion-pairs from diazine cations such as Janus Green B, Janus Green V, and Janus Green D have been extracted from aqueous acid into benzene-acetone mixtures. The limits of measurement using these three dyes are 0.051, 0.020 and 0.010 μg Ta/mL respectively (I.K. Guseinov, A.M. Pashadzhanov and Yu.A. Azimov, *Azerb. Khim. Zh.*, 1986, 84. *Chem. Abstr.*, 1987, 107, 167828r).

Nile Blue (**8**) has been reported as a suitable derivatising agent for carboxylic acids. Benzoic acid was one of a number of models which, as the acyl chlorides, provided Nile Blue derivatives. These all showed a ten-fold decrease in molar absorptivity and quantum yield, and a 40 nm increase in Stokes shift. Conventional fluorescence detection after HPLC columning showed a limit of detection of 88 fmol. With a visible diode laser fluorescence detector this could be lowered to 2 fmol on the column (S.V. Rahavendron and H.T. Karnes, *J. Pharm. Biomed. Anal.*, 1996, 15, 83).

Sevron Blue 5G also allows estimation of the sweeteners saccharin, acesulfame, and cyclamate in foodstuffs. The ion-association complex formed at pH 7 is extracted into chloroform and measured at 655 nm; the process is claimed to be accurate to within 1% (C.S.P. Sastry *et al.*, *Analyst,* 1995, 120, 1793).

Eosinophils may be stained using the oxazine dyes C.I. Basic Blue 122 followed by C.I. Basic Blue 141 and a brief rinse with acetate buffer at pH 3.45 (L. Kass, *Biotech. Histochem.*, 1995, 70, 19). Basic Blue 148 has been also used to screen for T-helper cells (*ibid., idem,* 99) but the same author proposes C.I. Basic Blue 141 as a diagnostic tool for identifying and differentiating T helper cells from T cytotoxic/suppressor cells (*ibid., idem,* 1993, 68, 247).

Similarly, a general ribonuclase assay relies upon the shift in the visible absorption maximum when **3** intercalates with high molecular weight RNA (T. Greiner-Stoeffle, M. Grunow and U. Hahn, *Anal. Biochem.*, 1996, 240, 24). The detection of double-stranded RNA-protein interactions is also reported by **3**-mediated photocrosslinking (Z.-R. Liu *et al., RNA,* 1996, 2, 611, *Chem. Abstr.*, 1996, 125, 136926p). With **4** the decrease in absorption at the maximum (628 nm) is linear at low concentrations of dye (*e.g.* 2 x $10^{-5}M$), and under these conditions is also proportional to the amount of DNA present up to 4 μg/mL (A.A. Killeen, *Microchem. J.*, 1995, 52, 333). An analogous interaction is reported (X.-Z. Feng, X.-W. Hei and S.-X. Shen, *Gaodeng Xuexao Huaxue* 1996, 17, 878; *Chem. Abstr.*, 1996, 125, 260954k) with organic solvents.

Spectroelectrochemical studies have been carried out on a variety of phenothiazinium dyes, such as **3,** when immobilised on Nafion film (S.A. John and R. Ramaraj, *Langmuir*, 1996, 12, 5689).

(d) Analysis using luminescence/fluorescence

A chemiluminescent method, using lucigenin to respond to the oxidation products of ascorbic acid in the presence of **4**, offers a sensitivity of $1 \times 10^{-9}M$ < [ascorbic acid] < $3 \times 10^{-4}M$ in the analysis of this component of many foodstuffs and juices (T. Perez-Ruiz, C. Martinez Lozano, and A Sanz, *Anal. Chim. Acta* 1995, 308, 299).

Thionine Blue (**15**) gives a highly fluorescent leucobase upon photo-oxidation. This permits analysis of ascorbic acid in solution at levels of $8 \times 10^{-7}M$ to $5 \times 10^{-5}M$ (Perez-Ruiz *et al., Analyst,* 1997, 122, 115).

The fluorescence of Methylene Blue is quenched by nucleic acids. This is the basis of a new fluorimetric method for the determination of DNA and RNA (W.-Y. Li *et al., Anal. Letters,* 1997, 30, 527). The response towards calf thymus DNA is linear to levels of 40 μg/mL from a detection limit of 28 ng/mL); towards yeast RNA a linear response is found up to 55 μg/mL from a detection limit of 82 ng/mL. The quenching of the fluorescence of thionine (**5**) by H_2S appears to offer a means for the analysis of this compound, but lack of reversibility and long response times make this method ineffective (O. Kohls *et al.*, *Proc. SPIE-Int. Soc. Opt. Eng.*, 1996, 2836, 311). However, Mn in trace amounts (<0.15 μg/26 mL) catalyses the KIO_4-oxidation of **5** in the presence of NTA at pH 4 resulting in quenching of the fluorescence. This may be used to limits of 10^{-10}g/mL Mn (L. Chen and D. Hua, *Yejin Fenxi*, 1995, 15, 40; *Chem. Abstr.*, 1995, 123, 17277b).

The fluorescence of Cresyl Violet is quenched by sodium dodecyl sulfate. This quenching is eliminated by gliadins and allows a method of estimating these food components at levels above 2.5 x 10^{-7}g/mL (B. Gala, A. Gomezhens, D. Perezbendito, *Analyst*, 1996, 121, 1133).

3. Application in synthesis

Along with other sensitisers such as fullerenes and phthalocyanine systems, Methylene Blue (**3**) has been used as a solid photo-sensitiser in the quest for an oxygen-iodine laser, which may generate singlet O_2 ($^1\Delta_g$) (J. Kodymova *et al.*, *Proc. SPIE-Int. Soc. Opt. Eng.* 1996, 2767, 245).

The intermediate radical cation formed during the oxidation of **3** may abstract hydrogen to form resonance-stabilised radicals from suitable precursors. These may link together, or undergo attack. This is exemplified in the reaction of Methylene Blue-induced oxidation (O_2) of 2-arylindoles (Scheme 1) (K.-Q. Ling, *Youji Huaxue* 1996, 16, 518; *Chem. Abstr.*, 1997, 126, 218372n. *ibid.*, *Synth. Commun.*, 1995, 25, 3831).

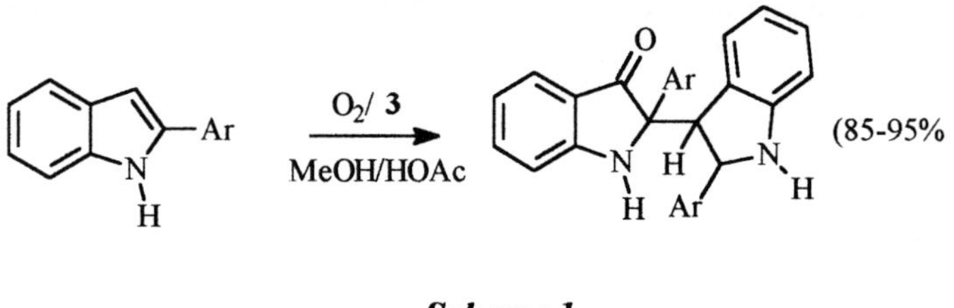

O_2/ **3**, MeOH/HOAc (85-95%

Scheme 1

When Ar = p-$C_6H_4.NO_2$, the initially formed product **22** loses methanol to give the isolated product **23**

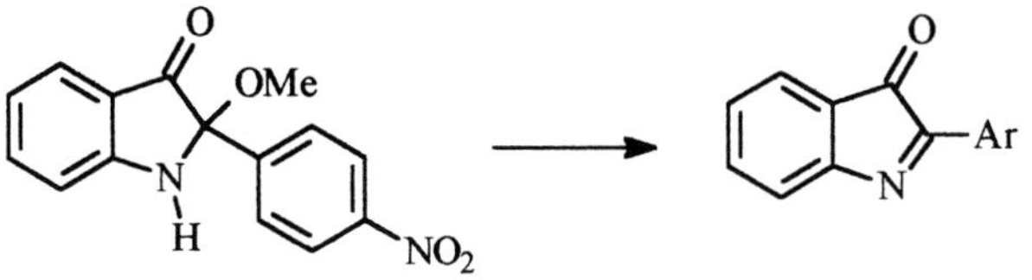

22 **23**

The 2-biphenylyl analogue **24**, however, forms 2-(2′-biphenylyl)-4H-3,1-benzoxazin-4-one, **25**.

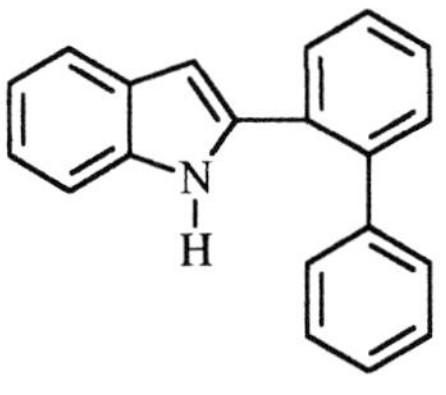

24 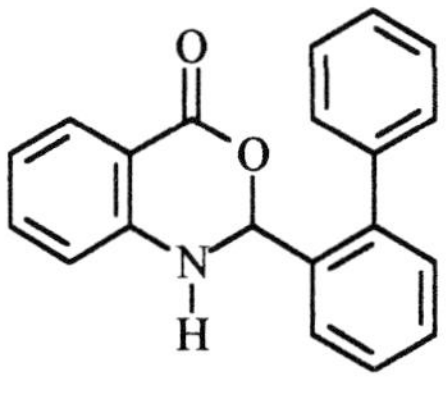 **25**

Methylene Blue (**3**) has also been used preparatively as a sensitiser in the introduction of the peroxo ligand when certain cationic complexes $[M(CO)(MeCN)(PPh_3)_2]^+$ are irradiated in the presence of oxygen and **3** to give $[M(CO)(MeCN)(PPh_3)_2(O_2)]^+$ (M = Ir, Rh; M. Selke *et al.*, *Inorg.Chem.*, 1995, 34, 5715; *ibid.*, *idem*, 1996, 35, 4519). Phenosafranine (**2**), thiopyronine (**6**), or **3** are used as electron-transfer agents when benzophenone triplets react with amines or alcohols to give, respectively, α-amino or ketyl nucleophilic radicals. The associated rate constants are near those expected from diffusion-controlled processes, but substituent effects are still clearly evident (S. Jockusch *et al.*, *J. Photochem. Photobiol.*,, 1996, 96A, 129).

4. Oxazine dyes

Oxazine dyes have been used as redox indicators in the titration of Fe(II) against quinones in 8-9*M*-phosphoric acid (K.V. Raju, M.R. Yadav and G.D. Sudhaker, *Proc. Natl. Acad. Sci., India,* 1994, 64A, 457).

5. Spiro-oxazines and spiropyrans

Spiro-oxazines undergo ring-opening to form merocyanines. Colour is developed under the influence of either heat (E.Pottier *et al., Helv. Chim. Acta,* 1990, 73, 305; C.Bohme *et al., J. Photochem. Photobiol.,* 1992, 66, 79) or light (C. Lenoble and R.S. Becker, *J. Phys. Chem.,* 1986, 90, 62), though the considerable interest in these materials is based mainly upon their photochromic behaviour. The reversibility of the system is affected by concomitant free-radical processes which consume the merocyanine system. Studies of the photodegradation of a model compound, 1,3-Dihydro-1,3,3,4,5-pentamethylspiro{2H-indole-2,3′-[3H]naphth[2,1-*b*][1,4]oxazine} (**26**) suggest that the photochemical formation of activated oxygen species does not impinge upon the degradation, which is a free-radical process. The effect of DABCO (1,4-diaza-[2,2,2]bicyclooctane) is consistent with the occurrence of electron-transfer processes, and so is the observed solvent effect (D. Eloy, C. Gay and P. Jardon, *J. Chim. Phys. Phys.-Chim. Biol.,* 1997, 94. 683).

The degradation of 5-X-1,3-dihydro-1,3,3-trimethylspiro{2H-indole-2,3′-[3H]pyrido[3,2-*f*][1,4]benzoxazine} (**27a**;, Y = N) and of 5-X-1,3-dihydro-1,3,3-trimethylspiro{2H-indole-2,3′[3H]naphth [2,1-b][1.4]oxazine} (**27b**, Y = CH) in polyurethane films under xenon light was shown to involve exclusively the photoxidation of the original dye, without the participation of any reaction intermediate (G. Baillet, G. Giusti, and R Guglielmetti, *Bull. Chem. Soc. Jpn.,* 1995, 68, 1220).

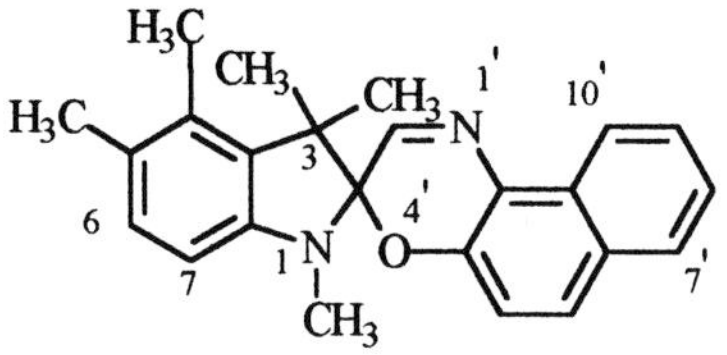

26 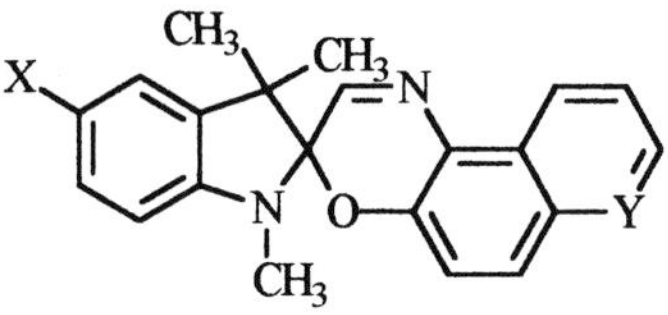 **27**

Among the identified reaction products were **28-31** which have obvious structural links with the merocyanine photolysis product.

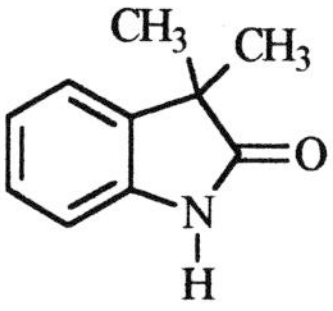

28 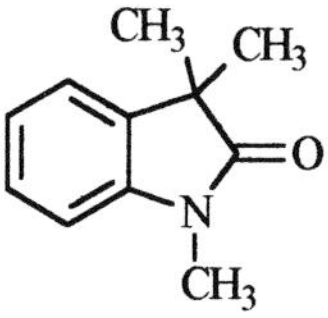 **29**

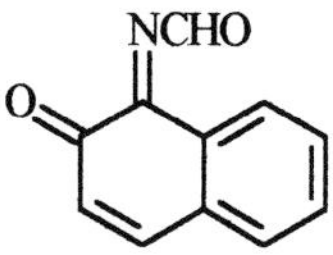

30 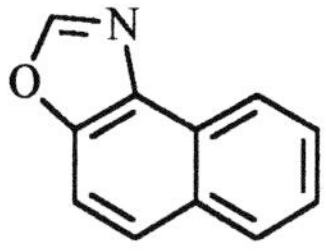 **31**

Some new spirodihydro[indole-3,2′-(1′,2′,3′,4′-tetrahydroquinoline)]-2,4′-diones have been obtained from indole-2,3-dione by a standard reaction sequence (Scheme 2) (M.S. Al Thebeiti, *Heterocycles*, 1998, 48, 145):

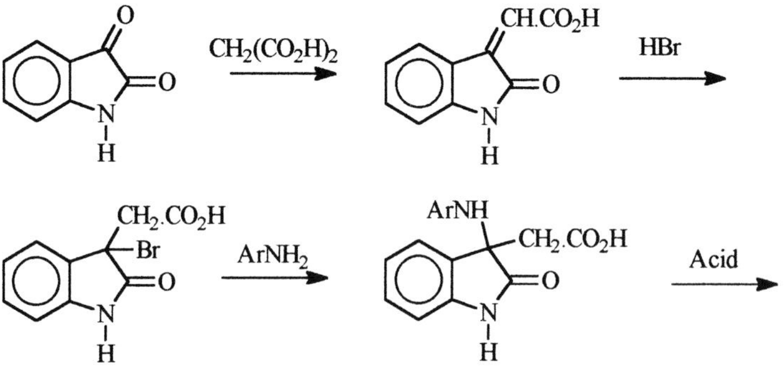

$CH_2(CO_2H)_2$ → HBr → $ArNH_2$ → Acid →

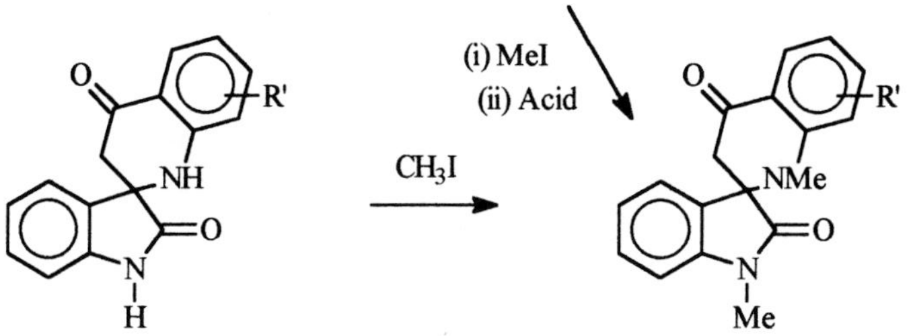

(i) MeI
(ii) Acid

CH_3I

Scheme 2

Spirodihydro[indole-2,4′-piperidines] may be prepared from 2-bromoaniline using either the Heck reaction or a free-radical process, though the yields are better in the latter case (J. Cossy, C. Poitevin and D.G. Pardo, *Synlett.*, 1998, 251)

The detailed mechanism of the interconversion of spiro-oxazines or -pyrans to merocyanines was supported when the zwitterionic merocyanine species were trapped by trimethylsilyl cyanide in dichloromethane solution. In the case of **27b** (X = H, Y = CH) the reacting species is either a zwitterion (**32**) or a similar, and highly dipolar, quinonoid system (**33**) which undergoes 1,6-Michael attack by Me_3SiCN to provide adducts (**34**) which may be isolated easily and in nearly quantitative yield (V. Malatesta *et al.*, *J. Amer. Chem. Soc.*, 1997, 119, 3451). X-ray diffraction analysis confirmed the proposed structure and the stereoselectivity of the addition.

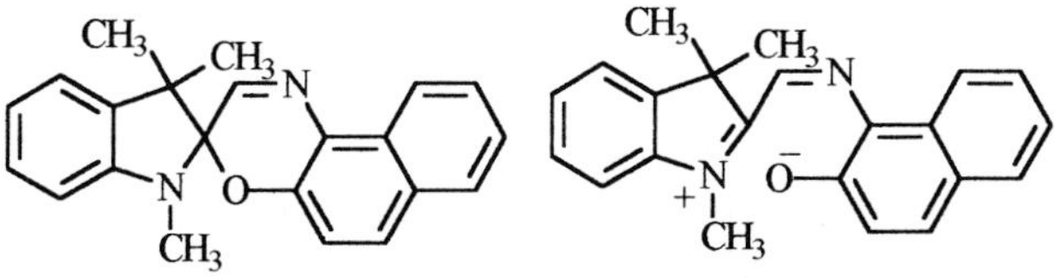

27b **32**

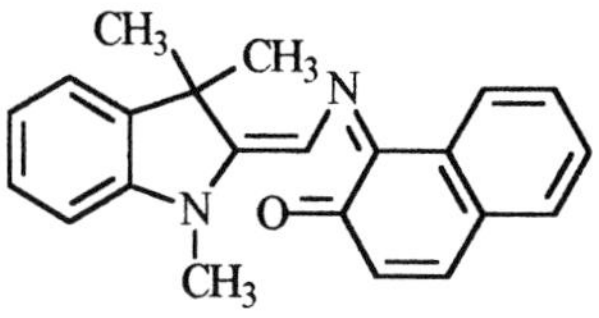

33 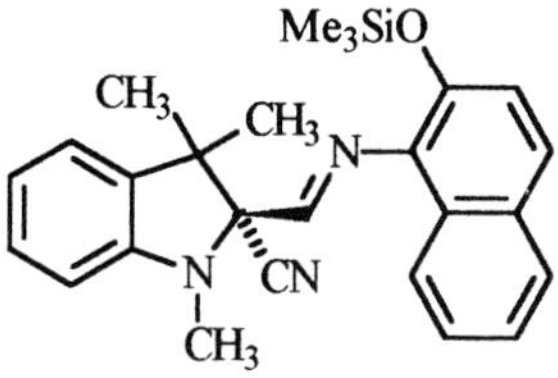 34

A study of 1,3-dihydro-1,2′,3,3,5-pentamethylspiro{2H-indole-2, 3′-[3H]naphth[2,1-*b*][1,4]oxazine} (X.Y. Zhang *et al., Sci. in China,* 1995, 38B, 257; *Chem. Abstr.,* 1996, 124, 31970z. *ibid., Res. Chem. Intermed.,* 1995, 21, 17; *Chem. Abstr.,* 1995, 123, 21847e) by nanosecond laser flash photolysis (248 nm) identified two species. One was short-lived, with a lifetime of 12 μs in MeCN, but only 0.8 μs in cyclohexane; it showed absorptions at about 460 and 640 nm. The second was long-lived, and was therefore thought to be the planar photomerocyanine; it showed absorption at about 620 nm. The substantial solvent effect upon the lifetime of the former species led the authors to propose a charge-transfer complex of unusually long life the structure of which appeared to lie between those of a zwitterion and a quinonoidal intermediate.

The 1,2′,3,3-Me_4-9′-OMe analogue gave a short-lived coloured product, but no long-lived species behaving analogously to the second component above (X.Y. Zhang *et al., Sci. in China,* 1994, 37B, 915; *Chem. Abstr.,* 1995, 122, 55519u). The putative charge-transfer complex showed similarly different lifetimes (13 μs in MeCN; 0.5 μs in cyclohexane) and bands at about 460 and 650 nm. Spiro-1′,3′-dihydro-[1-benzopyran-2,2′-indole], under these conditions, gave the same intermediate whether this spiro-compound or the derived stable *trans*-merocyanine was irradiated. The authors accordingly proposed that the intermediate was the *cis*-merocyanine, and that the triplet state of the *trans*-merocyanine is the transient with a lifetime of <10μs.

These spiro-compounds are often dispersed in a polymer matrix prior to irradiation, which has an effect upon the lifetime and physical properties of the irradiation products. The polarity of the matrix is thought to have a crucial role here. Less often, the spiro-oxetane system is combined with a polymerisable group so that the photogenic fragment becomes part of the polymer backbone. An example of this is 1-{4′-[2″-methacryloxy)ethoxycarbonyl]-benzyl-1,3-dihydro-3,3-dimethylspiro{2H-indole-2,3′-(3H)naphth-[2,1-*b*][1,4]oxazine} (**35**), dispersed in a PMMA matrix. Where **35** is doped throughout the polymer matrix the thermal decolouration rate falls off rapidly with increasing concentration of spirooxetane, but where the compound becomes incorporated in the polymer backbone the rate of thermal decolouration remains independent of concentration of spiro-oxetane (A.T. Hu, W.H. Wang and H.J. Lee, *J.Macromol. Sci.*, 1996, A33, 803).

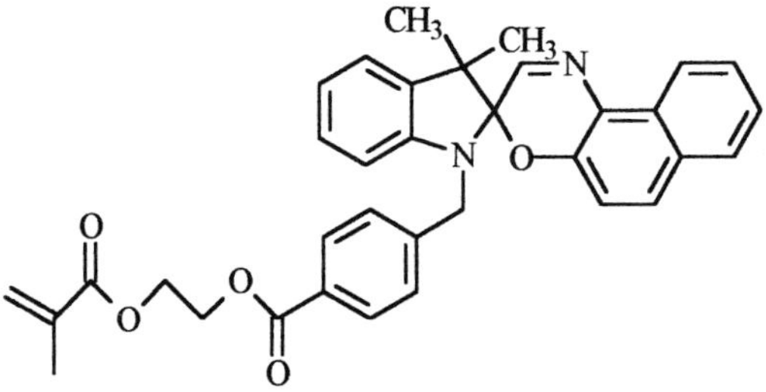

35

Spiro-oxetane and spiropyran structures have been doped into matrices derived from dialkoxysilane derivatives. Both the nature of the photochromic response and its kinetics are sensitive to the bulk properties of the matrix, and spirooxetanes dispersed in matrices derived from mixtures of triethoxysilane and methyl-diethoxysilane show direct, efficient, reversible, and remarkably fast photochromic responses (B. Schaudel *et al.*, *J. Mat. Chem.*, 1997, 7, 61).

6. Triphenodioxazine dyes

These materials, in which the advantages of the oxazine system are extended, have been often made through addition-elimination reactions of benzoquinones, themselves readily available through chlorinative oxidation of substituted anilines. The following sequence, Scheme 3, is typical. 2-Ethyl-3,4,6-trichlorobenzoquinone is first treated with N^2-(2-aminoethyl)-2,5-diaminobenzenesulfonic acid to provide **36**. Reaction with *N*-(2,4-dichloro-1,3,5-triazinyl)-3-aminobenzenesulfonic acid gave **37**. Further nucleophilic attack by $NH_2.CH_2.CH_2.OH$ gave **38**, which dyes cotton fibres blue but has no effect upon polyamide fibres [G. A Thomson and D.A.S. Philips, PCT Int.Appl., WO 95 14060 (1995)].

NH_2 Et $\xrightarrow{Cl_2,\ H_2SO_4}$ O Cl Et Cl Cl O

$NH.CH_2.CH_2NH_2$ SO_3H NH_2

Cl $H_2N.CH_2.CH_2.NH$ O N HO_3S SO_3H N O $NH.CH_2.CH_2.NH_2$ Et

36

SO_3H N H N Cl N N Cl

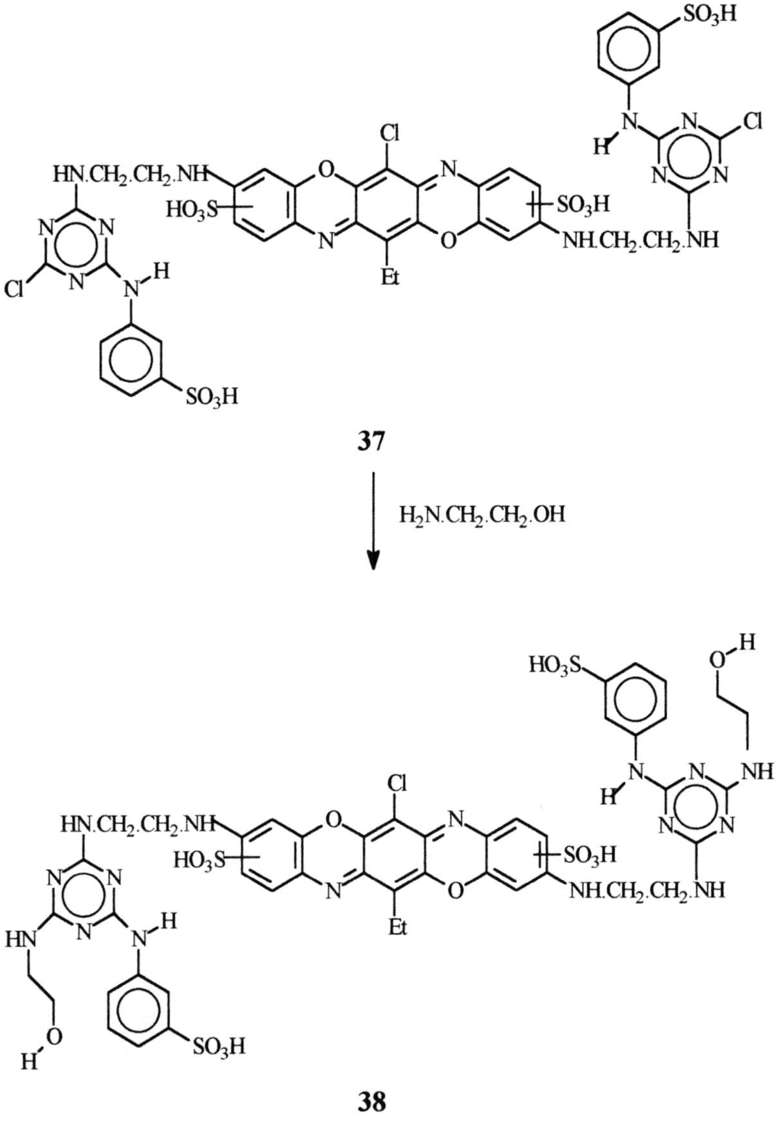

37

$H_2N.CH_2.CH_2.OH$

38

Scheme 3

Analogous structures **39** are derived from 9-R-3-aminocarbazoles (R = C_1-C_8), the condensation of which with chloranil is achieved by using Na_2CO_3 suspended in *o*-$C_6H_4Cl_2$ with $ArSO_2Cl$ to facilitate the thermal (130-165°) process. Although the subject of a patent [M Ikeda and M Oonishi, JP 07,145,327 (1995)], the process is generally similar to early methods of making slightly less complex analogues.

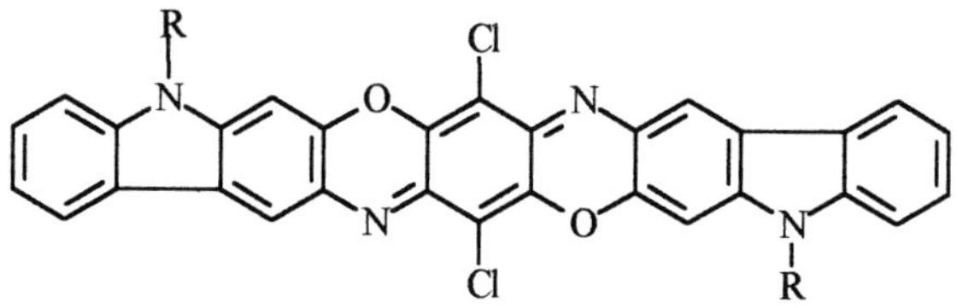

39

7. Sulfur dyes

A number of minor reviews have been published on aspects of these dyes (G S Shankarling, R Paul and J Thampi, *Colourage*, 1997, 44, 71; *Chem. Abstr.*, 1997, 127, 67431. *ibid.*, *idem*, 37; *Chem. Abstr.*, 1997, 127, 163085m. D. M. Lewis, *Sen'i Gakkai-shi*, 1997, 53, 213; *Chem. Abstr.*, 1997, 127, 163086n. T. van Thien, *Chem. Appl. Leuco Dyes*, 1997, 67; *Chem. Abstr.*, 1997, 127, 163089r). A Polish review of cationic dye-stuffs focuses attention upon the dyeing of polyacrylonitrile fibres (L. Maminska and J Wroblewski, *Barwniki, Srodki Pomocnicze*, 1996, 40, 135; *Chem. Abstr.*, 1997, 126, 61397b). J.H. Moro and J.R. Aspland (*Text. Chem. Color.*, 1997, 29, 20; *Chem.Abstr.*, 1997, 127, 52205f) have reported the successful application of transmittance spectro-photometric techniques to studies of sulfur dyes by using optically clear colloidal solutions. This should help the analysis and so the identification of these dyes, many of which are somewhat ill-defined mixtures of reaction products.

Sulfur dyes are essential components for heat stable, IR scannable inks which give water-fast images [R. Botros and S.P. Chavan, EP 764,700 (1997)].

Olive-coloured dyes are made by heating 2,4-dichloro-6-nitrophenol, 2,6-dichloro-4-nitrophenol, and 2,4,6-trichlorophenol with Na_2S at 90-170° [H. Hoetzel *et al.*, DD 227,717 (1987)], while brown-black dyes result when phenols, chlorophenols, and 2,4-dichlorophenol are heated with "Na_2S_4", NaOH and $CuSO_4$ at 230-240° for 5 hours [Hoetzel *et al.*, DD 251,775 (1987)]. The yields of sulfur dyes derived from 2,4-diaminotoluene and 4-aminophenol or their congenors are improved with ethylene glycol as solvent at 170-185° (N.V. Bragina *et al.*, SU 1,407,939 *Otkrytiya, Izobret.* 1988, 115; *Chem. Abstr.*, 1989, 110, P40500j).

Second Supplements to the 2nd Edition of Rodd's Chemistry of Carbon Compounds, Vol.IV IJ, edited by M. Sainsbury

Chapter 46

QUINAZOLINE ALKALOIDS

SIEGFRIED JOHNE

1. Bicyclic quinazolines

This review covers the literature period from August 1993 - December 1998, as well as some papers published before this period, but which were not considered in former reviews. Results of earlier investigations are mentioned briefly, where necessary. Reports on the synthesis of quinazolines are considered only where they have potential value for the preparation of quinazoline alkaloids. Developments in the field of quinazoline alkaloids are being continuously updated in Natural Product Reports of the Royal Society of Chemistry. A comprehensive review on the distribution of quinazoline alkaloids in the plant and animal kingdoms is available (A. L. D'yakonov and M. V. Telezhenetskaya, Chem. nat. Comp., 1997, **33**, 221).

Simple quinazolines are represented by a quinazoline nucleus (see 1), which may contain substituents in positions 1, 2, 3, 4 and 7. These substituents may be methyl, ethyl, hydroxyl, hydroxymethyl, benzyl, carbonyl, carboxyl, amide, bromo, or acetyl groups. More complex groupings are often found in position 3.

Quinazolin-4(3*H*)-one (2) and quinazoline-2,4-dione (3) have been isolated *from Strobilanthes cusia* (L. Li *et al.,* Chem. Abs., 1993, **119,** 24635). The isolation of quinazolin-4(3*H*)-one from *Dichroa febrifuga* was reported in 1948, but this compound may have been an artefact (J. R. Price, Fortschr. Chem. org. Naturst., 1956, **13**, 330). It has recently been suggested that quinazoline-2,4-dione (isolated from *Strobilanthes cusia*) should be used as a quality control standard in the HPLC analysis of Chinese drugs prepared from the same plant (L. Li *et al.,* Chem. Abs., 1993, **119**, 103455).

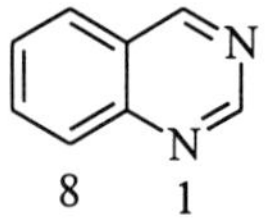

1 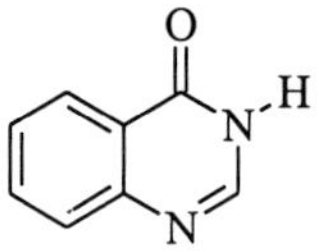 2 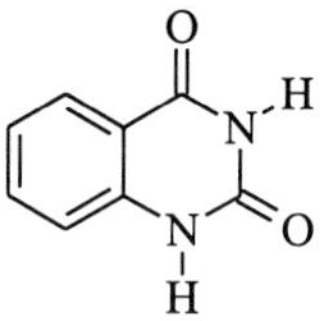 3

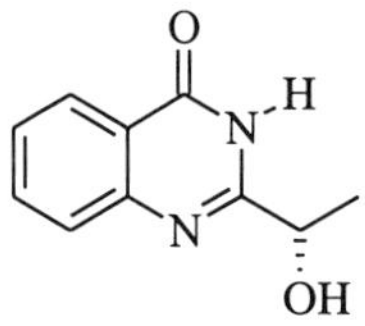

4 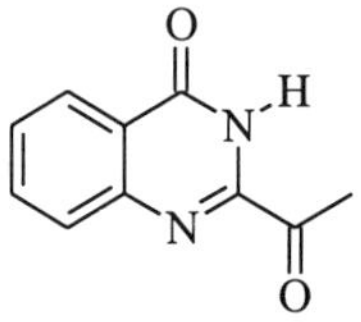 5 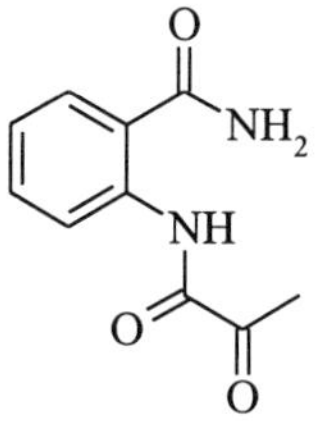 6

(-)-Chrysogine (4), 2-acetylquinazolin-4(3*H*)-one (5) and 2-pyruvylamino-benzamide (6) have been isolated as minor metabolites from cultures of *Fusarium sambucinum* (D. Niederer, C. Tamm and W. Zürcher, Tetrahedron Letters, 1992, **33**, 3997). Compound 5 maybe an artefact originating from 6. The absolute configuration of (-)-chrysogine was determined by nmr analysis of the esters formed with both (*R*) and *(S)* -α -phenylbutanoic acid. An asymmetric synthesis of (*S*)-(-)-chrysogine, starting from anthranilamide and (*S*)-(-)-2-acetoxypropanoic acid *via* the corresponding 2-acetoxyprop-anoyl chloride, has been reported (D. K. Maiti, P. P. Ghoshdastidar and P. K. Bhattacharya, J. chem. Res. (S), 1996, 306; see also J. Bergman and A. Brynolf, Tetrahedron, 1990, **46**, 1295). This synthesis confirmed the proposed *S*-configuration for the (-)-alkaloid.

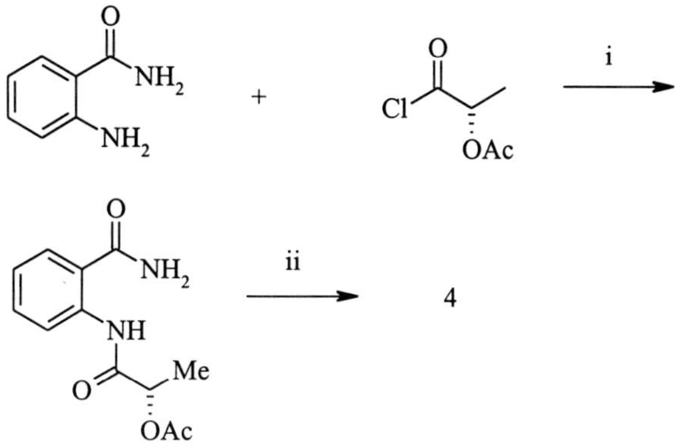

Reagents: (i) pyridine, r.t.; (ii) 1% aq. NaOH, MeOH, r.t.

2-Acetylquinazolin-4(3*H*)-one (5) has now been shown to occur in the culture broth of the mycoparasitic fungus *Cladobotryum varium* (Y. Tezuka *et al.,* Chem. pharm, Bull., 1994, **42**, 2612). *Fusarium lateritium* Nees is a natural antagonist of the plant pathogen *Eutypa armeniacae*, which is associated with sapwood necrosis in fruit trees, grapevines and ornamental plants around the world (Y. S. Tsantrizos *et al.,* Canad. J. Chem., 1993, **71**, 1362; 1994, **72**, 1415). The bioassay-guided separation of the active components of the antagonist led to the isolation of (-)-chrysogine (4) and 2-acetylquinazolin-4(3*H*)-one (5), and the enniatins A_1 , B, and B_1. Analysis of the Mosher's ester of 4 showed that the material isolated was an approximate 1.5:1 mixture of the *R* and S enantiomers. The structure of 2-acetylquinazolin-4(3*H*)-one was confirmed by X-ray crystallography. 2-Acetylquinazolinone-4(3*H*)-one (5) is an inhibitor of HIV replication; but while it inhibited syncytium formation in C8166 cells, it proved to be cytotoxic at concentrations greater than 6 mg/ml^{-1}.
Monodontamide F (7) , $C_{25}H_{27}N_5O_5$, m.p. 129 – 132 ° and five new putrescine alkaloids have been isolated from the marine gastropod mollusc

Monodonta labio (Linné) (H. Niwa *et al.*, Tetrahedron, 1994, **50**, 6805). The isolation of quinazoline alkaloids from marine sources is quite rare and this is the first isolation of a quinazolin-4(3*H*)-one from such a source. The structure of 7 follows from a detailed analysis of the ir, ^{1}H-nmr, ^{13}C-nmr, ^{1}H COSY, and ^{1}H-^{13}C COSY spectra. Monodontamide F has been synthesized from malonylated tryptamine and 4-aminobutanol. This gave an alcohol that on subsequent ozonolysis and reductive workup, yielded a formamide, which was converted into the iodide. *N*-3 Alkylation of 4-hydroxyquinazoline with the iodide afforded 7.

EtO, O, O, H, N, N, H
i
HO$(CH_2)_4$HN, O, O, H, N, N, H
ii, iii
R$(CH_2)_4$HN, O, O, H, N, O, NHCHO
iv

R = OH
R = I

O, N, N, H, N, O, O, H, N, O, NHCHO

7

Reagents: (i) 4-aminobutanol; (ii) ozonolysis in MeOH, $NaHCO_3$; (iii) 4-$MeC_6H_4SO_2Cl$, NaI, $CaCO_3$; (iv) 4-hydroxyquinazoline, KOH

For the synthesis of simple quinazoline alkaloids a new short route has been reported (B. Baudoin, Y. Ribeill and N. Vicker, Synth. Commun., 1993, **23**, 2833): the oxidation of *o*-aminobenzonitriles using sodium perborate, followed by cyclisation gives the corresponding quinazolin-4(3*H*)-ones in a 'one pot' reaction under mild, non-hazardous conditions.

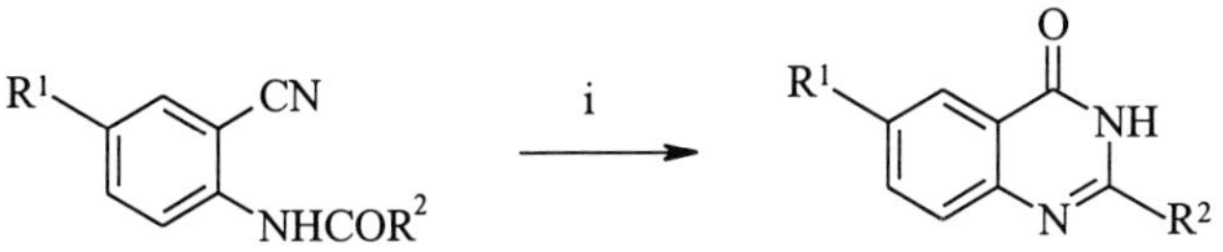

R^1 = H, I, 3,4,5-trimethoxystyryl
R^2 = Me, Ph, NMe_2, ethylcyclohexyl

Reagents: (i) $NaBH_4$, H_2O, dioxane, reflux 24 h.

2-Substituted quinazolines have been prepared (J. J. Van den Eynde *et al.*, Synthesis, 1993, 867) from aldehydes, which are converted *in situ* into *N*-(1-chloroalkyl)pyridinium chlorides using thionyl chloride and pyridine, and reacted with 2-aminobenzylamine under mild conditions. This affords tetrahydroquinazolines that can be aromatized readily by treatment with 2,3-dichloro-5,6 dicyanobenzoquinone (DDQ), or tetrachloro-1,4-benzoquinone in benzene (TCQ). This approach can be used for the synthesis of quinazoline alkaloids bearing an unbranched or a branched alkyl chain, an aryl ring (eventually substituted by an electron-withdrawing group, or an electron-releasing group), or a heteroaryl ring at the 2-position.

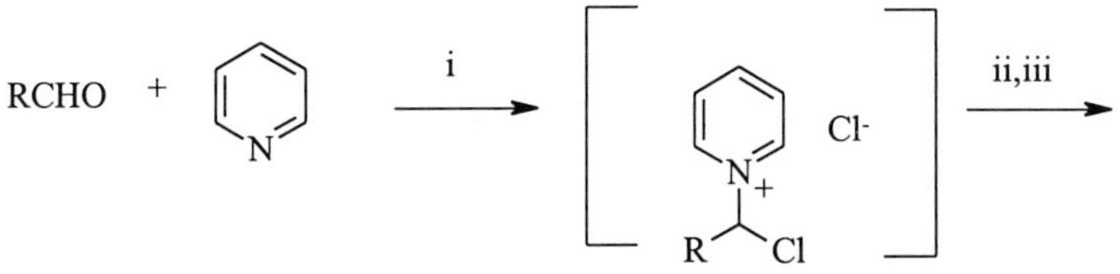

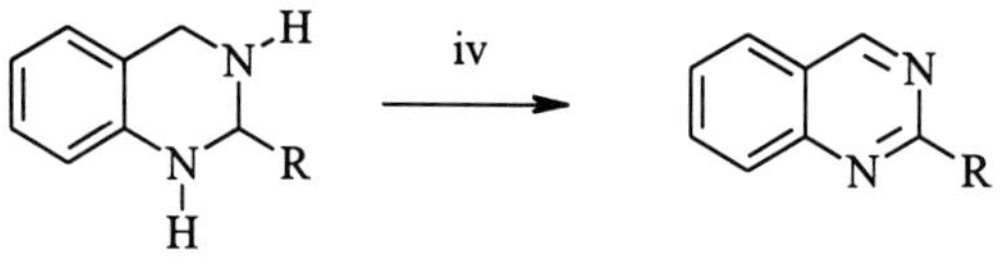

R = Et, $(Ph)_2CH$, 2-$MeOC_6H_4$, 2-furyl, 2-thienyl etc.

Reagents: (i) $SOCl_2$, DCM, 0°, 15-60 min; (ii) 2-aminobenzylamine, $NaOAc/H_2O$, 0-20°, 15 min; (ii) NaOH; (iv) DDQ, C_6H_6, r.t., 1h., or TCQ, C_6H_6, reflux 1h.

A new total stereoselective synthesis of racemic febrifugine, an alkaloid that possesses antimalarial and anticoccidial properties, has been published (L.E. Burgess, E.K.M. Gross and J. Jurka, Tetrahedron Letters, 1996, **37**, 3255). In this the silyl enol ether required for coupling with *N*-ethoxycarbonyl piperidine 2,3-epoxide was prepared from 4-hydroxyquinazoline by *N*-alkylation with chloroacetone, followed by a trimethylsilyltrifluoromethane-sulfonate entrapment of the corresponding enolate. After chromatography of the 1;1 mixture of diastereomers and *N*-deprotection of the piperidine unit of the appropriate isomer, febrifugine (8) was obtained.

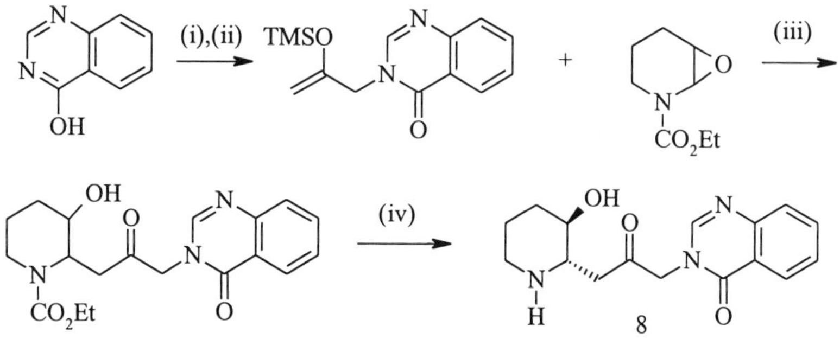

Reagents: (i) NaH, DMF 0°, then chloroacetone, 0° to r.t.; (ii) TMSOTf, $(^iPr)_2NEt$, DCM, r.t.; (iii) $TiCl_4$, DCM, 0°; (iv) flash chrom., then KOH, diethylene gylcol, H_2O, heat

Two new tryptoquivalines, CS-B (9) and CS-C (10) were isolated from the Ascomycete *Corynascus setosus*, together with fiscalin B (H. Fujimoto *et al*., Chem. Pharm. Bull., 1996, **44**, 1843). CS-B (9), $C_{28}H_{28}N_4O_7$, m.p. 236 – 238 °, $[\alpha]_D$ + 196 ° (c 0.29, $CHCl_3$), CS-C (10) : $C_{29}H_{30}N_4O_7$ m.p. 225-227 ° $[\alpha]_D$ + 138 ° (c 0.02, $CHCl_3$).
The CD spectrum of CS-B (9) resembles that of CS-C (10). It is also similar to that of nortryptoquivaline (FTD), except for the appearance of an additional positive Cotton effect at 278 nm in the spectrum of 9. This indicates that 9 is the diastereomer of FTD at position 27 [27-*epi*-nortryptoquivaline (9), 27*R*-configuration]. CS-C (10) is a diastereomer of tryptoquivaline (FTC), but not an enantiomer. This alkaloid was deduced to be 27-*epi*-tryptoquivaline (27*R*-configuration). In mice, 9 and 10, both caused weak trembling with paralysis.

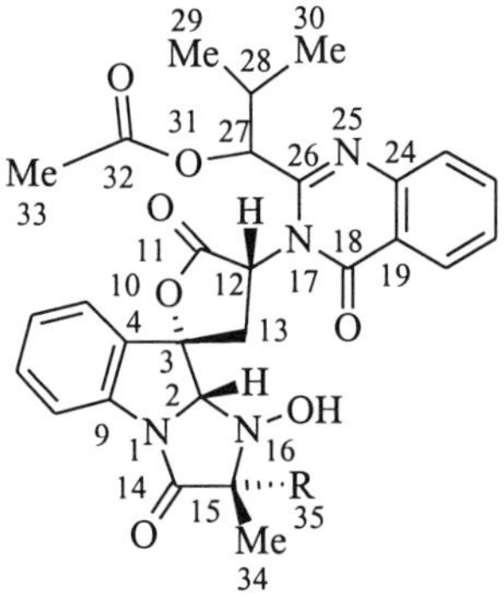

9, R = H, (*R*)-configuration at position 27
10, R = Me, (*R*)-configuration at position 27
FTD, R = H, (*S*)-configuration at position 27
FTC, R = Me, (*S*)-configuration at position 27

The addition reactions of nitrilium salts and nucleophilic isocyanates provide a simple route to quinazolinones (M. Al-Talib *et al.*, Synthesis, 1992, 697).

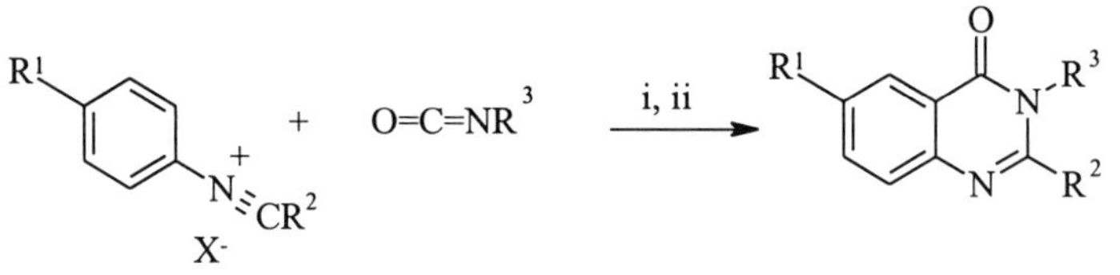

Reagents: (i) $ClCH_2CH_2Cl$, reflux: (ii) NaOH aq. DCM, r.t.

7-Hydroxyechinozolinone (11), $C_{10}H_{10}N_2O_3$ has been isolated from the flowers of *Echinops echinatus* (P.K. Chaudhuri, J. nat. Prod., 1992, **55**, 249).

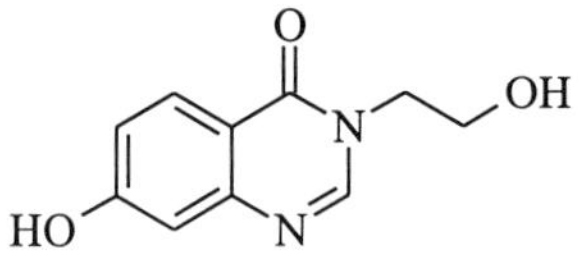

11

2. *Pyrroloquinazolines*

Recent investigations on the well-studied Indian medicinal plant, *Adhatoda vasica*, have yielded four new natural products, 3-hydroxyanisotine (1), vasnetine (2) (B.S. Joshi *et al.*, J. nat. Prod., 1994, **57**, 953), desmethoxy-aniflorine (3), and 7-methoxyvasicinone (4) (R.K. Thappa *et al.*, Phytochem., 1996, **42**, 1485). The structures have been established by ^{1}H- and ^{13}C-nmr spectral studies. The chemical shift assignments for these alkaloids were confirmed by unusually comprehensive ^{1}H homonuclear COSY, DEPT, HETCOR, selective INEPT and HMBC nmr experiments. The authors observed that a $CDCl_3$ solution of 1-vasicine kept overnight in a nmr tube gave a combined spectrum of vasicine and vasicinone. Variations of the optical rotations of vasicine reported in the literature are due to this instability. The structure of the monohydrate of 4 has been determined by X-ray crystallography, but no assignment of absolute configuration was made (D.K. Magotra *et al.*, Acta Cryst, 1996, **52C**, 1491). Other crystallographic studies include those of deoxyvasicinone, its hydrochloride salt and a tetrachlorocobaltate salt (K. K. Turgunov *et al.*, Khim. Prir. Soedin., 1995, 849).

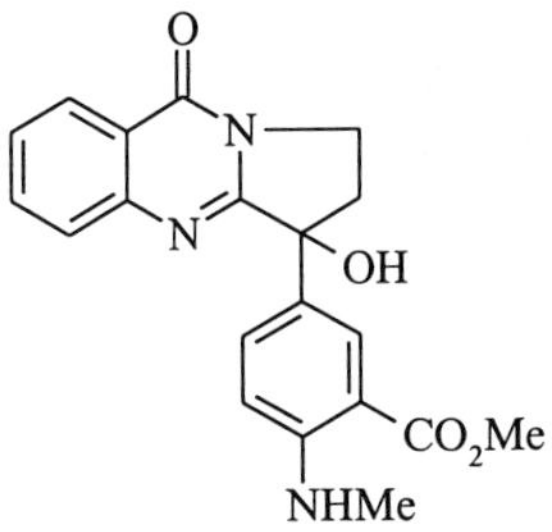

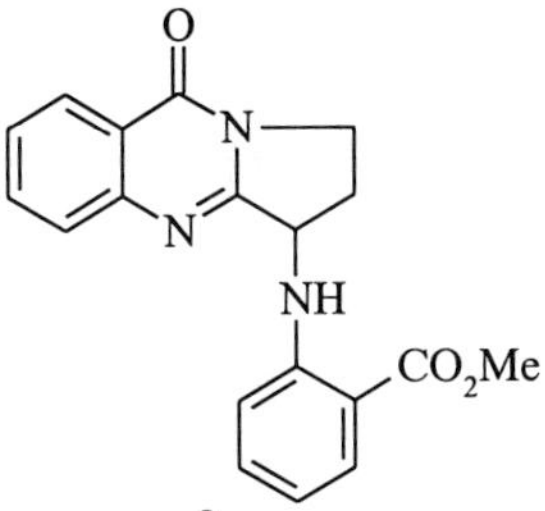

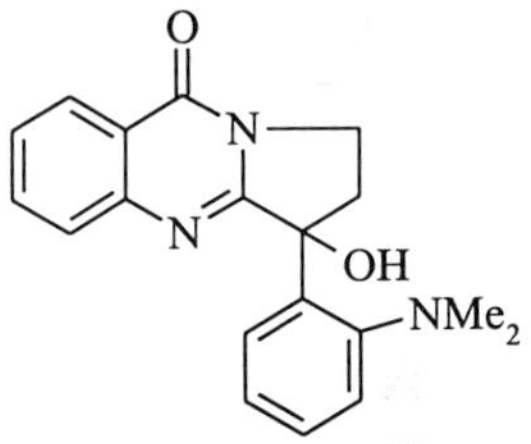

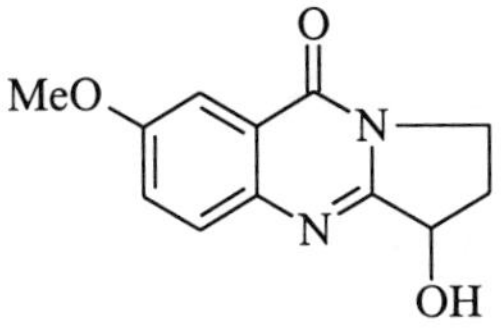

Some new pyrrolo[2,1-*b*]quinazoline alkaloids: vasicinone *N*-oxide (7), deoxyvasicinone *N*-oxide (8), deoxypeganine *N*-oxide (5) and peganine *N*-

oxide (6) have been isolated from *Nitraria komarovii* (T.S. Tulyaganov *et al.*, Chem. Nat. Comp., 1993, **29**, 73, 509; 1994, **30**, 727). Reduction of 7 with zinc and hydrochloric acid yielded a mixture of vasicinone, deoxyvasicinone, and deoxyvasicine. The similar reduction of 8 gave deoxyvasicine and deoxyvasicinone. The reduction of 5 afforded deoxypeganine, while reduction of 6 gave peganine and a small quantity of deoxypeganine. The previously assigned 3*R* configuration of (-)-vasicine, based on the results of an anomalous dispersion X-diffraction study of (+)-vasicine hydrochloride, has been reversed. On the basis of an X-ray diffraction study of the hydrobromides of (-)-vasicine (m.p. 193 –195 °, $[\alpha]_D$ + 27.5°) and of (-)-vasicinone (m.p. 178 – 180 °, $[\alpha]_D$ + 15.1°), these pyrrolo[2,1-*b*]quinzolines have been shown to have the 3 *S*-configuration (B.S. Joshi *et al.*, Tetrahedron Asymmetry, 1996, **7**, 25).

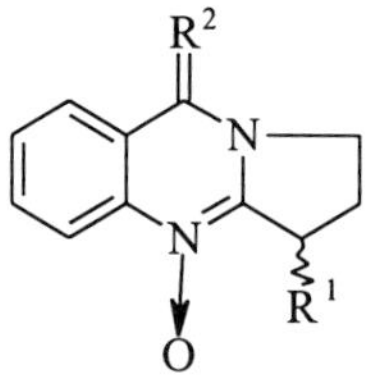

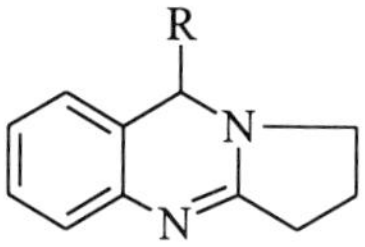

5, R^1 = H, R^2 = H_2
6, R^1 = OH, R^2 = H_2
7, R^1 = OH, R^2 = O
8, R^1 = H, R^2 = O

9, R = OH
10, R = CH_2COMe

The alkaloids (+)-vasicinol and vasicinolone, which have been interrelated, should also have the 3*S*-configuration. A study of the ^{1}H-nmr spectroscopy of (-)-vasicine by the use of Mosher's method, using MTPA [α-methoxy-α-(trifluoromethyl)phenylacetic acid] esters, indicated an exception to this protocol for establishing the absolute configuration of the alkaloids.

Peganol (9) has been converted into deoxypeganidine (10) upon heating with $CuSO_4$/acetone (M.V. Telezhanetskaya and A.L. D'yakonov, Chem. Nat. Comp. 1995, **31**, 146).
X-Ray crystallographic analyses have been reported for dehydroxypeganine, peganine and some related compounds (B. Taskhodzhaev *et al.*, Khim. Prir. Soedin., 1995, 410, 421; K.K. Turgunov *et al*, *ibid.*, 1995, 426).
In view of the pharmacological potential of vasicine, ten analogues have been synthesized employing the Schöpf-Oechler scheme (R.L. Sharma and A. Prabhakar, Orient. J. Chem., 1994, **10**, 27) and screened for their pharmacological activities. The new deoxyvasicine analogue 11 was synthesized by condensation of 2-amino-5-bromovanillin with 4-aminobutyraldehyde in phosphate buffer, followed by hydrogenation of the resulting quinazolinium complex (B. Ojo and B.K. Chowdhury, Synth. Commun., 1995, **25**, 569).

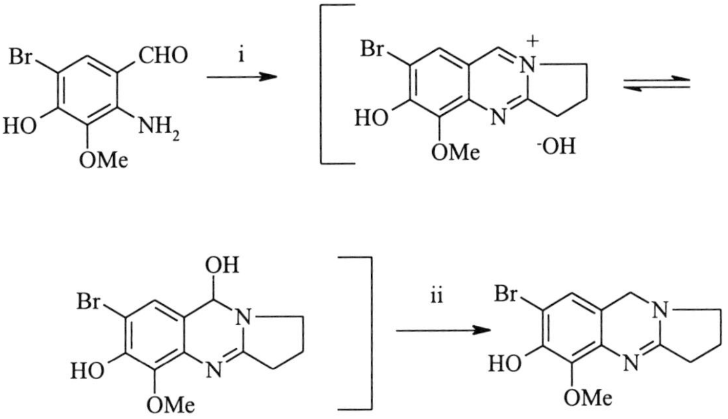

Reagents: (i) 4-aminobutanal, pH 5.8/aq. phosphate buffer, r.t., 72h;
(ii) H_2, 5% Pd/$BaSO_4$, 65^0, 1.5 h

Chloroamphenicol has been shown to stimulate the production of quinazoline alkaloids in young *Adhatoda vasica* plants (M. Rajan and G. C. Bhavsar, Chem. Abs., 1993, **119**, 45291).
Several quinazoline alkaloids are known to elicit a variety of biological responses. This has spurred the preparation and pharmacological evaluation of a great number of quinazoline derivatives and intensive research in this field is still in progress. For the treatment of Alzheimer's disease, deoxyvasicine and some synthetic analogues have been assayed for anticholinesterase activity in the rat brain and in human red blood cells. The most potent inhibitor in this series was the chloro compound 12 (J.C. Jaén *et al.*, Bioorg. Med. Chem. Lett., 1996, **6**, 737).

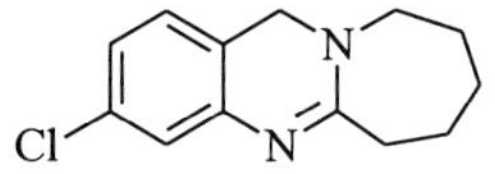

12

3. *Indolo-, pyrido- and indolopyrido-quinazolines*

Glyantrypine (1) has been isolated from the fermentation broth of *Aspergillus clavatus* (J. Penn *et al.*, J. Chem. Soc., Perkin I, 1992, 1495). Molecular modeling studies employing MOPAC with the PM3 Hamiltonian afforded an optimized geometry, where dihedral angles of 15 and 104 ° between the protons at C-11 and the adjacent NH proton of the amide were observed.
Fiscalins A (2), B (3), and C (4) occur in the fermentation broth of the fungus *Neosartorya fischeri* (S.-M. Wong *et al.*, J. Antibiotics, 1994, **46**, 545). Fiscaslin B (3) was also isolated from the ascomycete *Corynascus setosus* (H. Fujimoto *et al.*, Chem. Pharm. Bull., 1996, **44**, 1843). The compounds are similar in structure to the known fumiquinazoline A and to

glyantrypine. The absolute stereostructure has been determined by X-ray crystallographic analysis and amino acid analysis upon the products from the hydrolysis of 2.

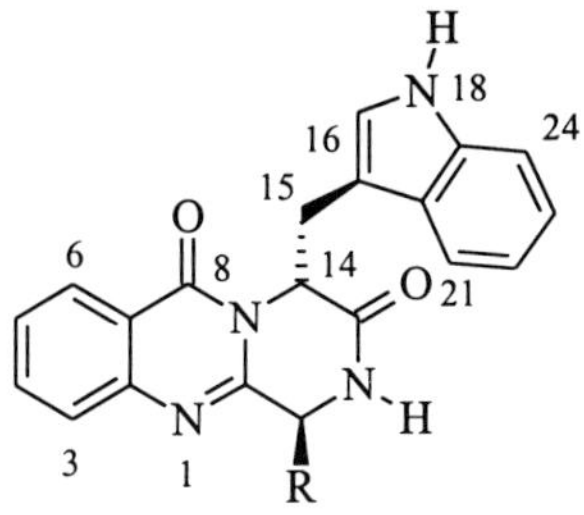

1, R = H
3, R = ^{i}Pr

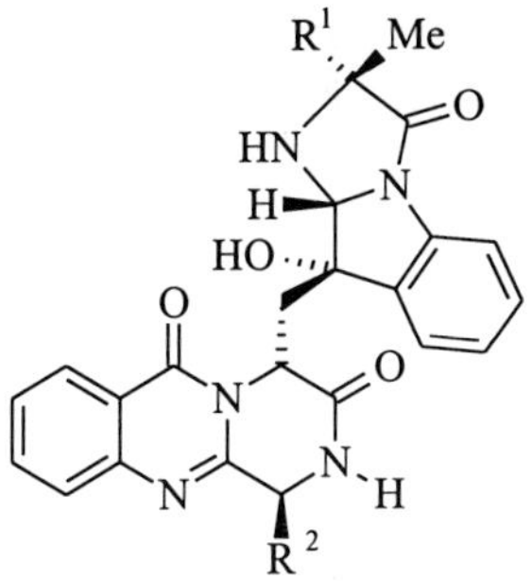

2, R^1 = H, R^2 = ^{i}Pr
4, R^1 = Me, R^2 = ^{i}Pr

The synthesis of 3 was achieved in a total of five steps and in 48% overall yield (H. Wang and A. Ganesan, J. Org. Chem., 1998, **63**, 2432) starting from D-tryptophan methyl ester. It included a Wipf dehydration of the tripeptide .

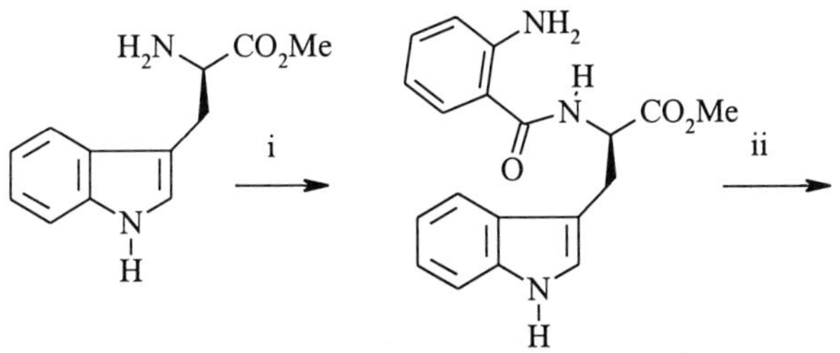

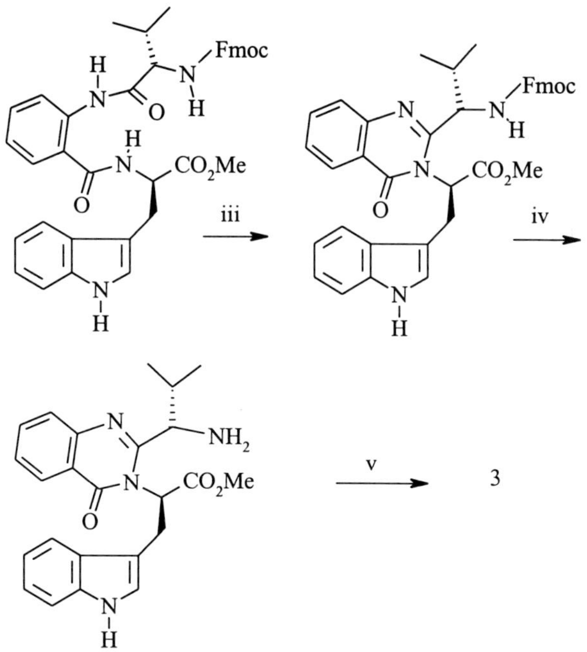

Reagents: (i) 1-ethyl-3-[3-dimethylamino)propyl]carbodiimide.HCl (2.2 eq.), anthranilic acid (2.0 eq.), MeCN, r.t., 3 h; (ii) Fmoc-D-ValCl (1.4 eq.) DCM/aq. Na_2CO_3, r.t., 2 h; Ph_3P (5.0 eq.), I_2 (4.9 eq.) $EtN(^iPr)_2$ (10.4 eq.), r.t. 8 h; (iv) 20 % piperidine in DCM, r.t., 12 min; (v) MeCN, DMAP (1.3 eq.) 19 h

Fiscalines (2) - (4) inhibited the binding of substance P, an undecapeptide neurotransmitter, to human neurokinin-1 receptors.
The new quinazolinocarboline alkaloid, (+)-7-hydroxyrutaecarpine (5), $C_{18}H_{13}N_3O_2$, $[\alpha]_D$ 187.8° (c 0.0017, MeOH) occurs in the heartwood of *Tetradium glabrifolium* (= *Evodia meliaefolia*) and in the fruit of *T. ruticarpum* (T.-S. Wu *et al.*, Heterocycles 1995, **41**, 1071). The stereo-

chemistry of H-7 was deduced by the fact that in the ^{1}H nmr spectrum its signal shows only small coupling interactions with the resonances of H-8_{ax} and H-8_{eq}. This can be accounted for if H-7 is in the equatorial site and the hydroxyl group is axial. These assignments are analogous to those of a known indolopyridoquinazoline alkaloid isolated from *Evodia rutaecarpa* (Juss) Benth. et Hook (B. Danieli, G. Lesma, and G. Palmisano, Experientia, 1979, **35**, 156). Callus tissue from *Phellodendron amurense* produced 5, 7,8-dihydrorutaecarpine, $C_{18}H_{13}N_3O_3$, m.p. 247 – 250 °, $[\alpha]_D$ – 38 ° (c 0.0016, MeOH) (A. Ikuta, H. Urabe and T. Nakamura, J. nat. Prod., 1998, **61**, 1012), and 7,8-dehydrorutaecarpine, $C_{18}H_{11}N_3O$ (A. Ikuta, T. Nakamura and H. Urabe, Phytochem., 1998, **48**, 285). The latter compound has been reported as a synthetic product from rutaecarpine (H. Möhrle, C. Kamper and R. Schmid, Arch. Pharm., 1980, **313**, 990), but this is its first isolation from a natural source.

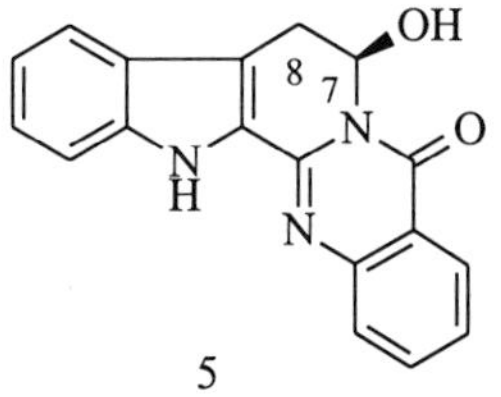

5

The bioassay-guided fractionation of a methanol extract of the fruit of *Zanthoxylum integrifoliolum* (Merr.) Merr. (Rutaceae) led to the isolation of 1-methoxyrutaecarpine (6). The structure of this alkaloid was established by its identity with the product of *O*-methylation of 1-hydroxyrutaecarpine by diazomethane. This conclusion was further confirmed by an NOE difference experiment (W.-S. Sheen *et al.*, Planta med. 1996, **62**, 175). The known alkaloid rutaecarpine has been isolated from *Tetradium glabrifolium* (T.-S. Wu *et al.*, J. Chin. Chem. Soc., 1995, **42**, 929), from *Zanthoxylum budrunga* (H. Banerjee, S. Pal and N. Adityachaudhury, Planta med., 1989, **55**, 403), and from *Z. limonella* (A.-O. Somanabandhu *et al.*, J. Sci. Soc.,

Thailand, 1992, **18,** 181). 1-Hydroxyrutaecarpine exhibited antiplatelet activity induced by arachidonic acid and showed an IC_{50} value of *ca.* 1-2 μg/ml (W.-S. Sheen *et al., loc. cit.*).

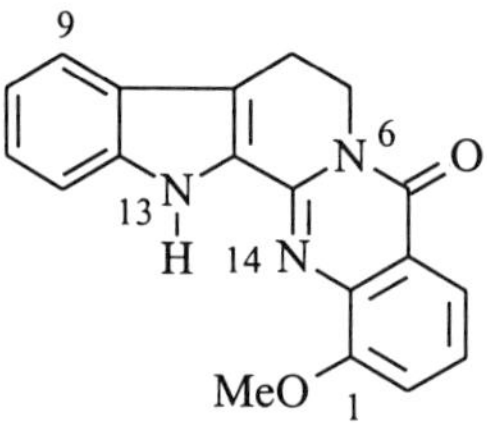

6 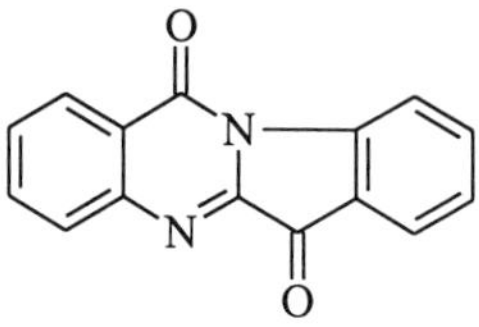 7

Tryptanthrin (7), isolated from the aerial parts of *Isatis tinctoria* L, a plant having a history of use in folk medicine, shows insecticidal and antifeedant activity against termites (*Reticulitermis santonensis*), insect preventive and control activity against larvae of the house longhorn beetle (*Hylorrupes bajulus* L.) and fungicidal activity against the brown-rot fungus *Coniophora puteana*) (K. Seifert and W. Unger, Z. Naturforsch., 1994, **49C**, 44). Dehydroevodiamine, evodiamine, and rutaecarpine, isolated from *Evodia rutaecarpa*, have been shown to produce a vasodilatory effect on the endothelium of intact rat aorta with equal potency. Rutaecarpine produced a full (100%) nitric oxide-dependent vasodilatation, whereas evodiamine and dehydroevodiamine produced a partially endothelium-dependent effect: 50% and 10%, respectively. At the same time, dehydroevodiamine and evodiamine may also act by other mechanisms, including probably an α_1-adrenoceptor blocking action and a 5-HT antagonizing action, respectively (W.-F. Chiou, J.-F. Liao and Ch.-F. Chen, J. nat. Prod., 1996, **59**, 374; W.-F. Chiou *et al., ibid.,* 1997, **60**, 708).
Bioassay-guided fractionation of preovulatory urine of the female Asian elephant, *Elephas maximus,* yielded as a sole chemical component, the well-

known antibiotic tryptanthrin (7) (L.E.L. Rasmussen *et al.*, J. chem. Ecol., 1993, **19**, 2115).
Full details of the isolation and structural elucidation of the fumiquinazolines (FQs) A – G from *Aspergillus fumitagus,* have been published (Ch. Takahashi *et al.*, J. chem. Soc., Perkin I, 1995, 2345). These compounds were originally obtained from the gastrointestinal tract of the marine fish *Pseudolabrus japonicus*, FQs A (8), B (9) and C have already been described briefly in the 1st Supplement of the Chemistry of the Carbon Compounds.
FQ D (13), m.p. 214-216 °, $[\alpha]_D$ + 68.9° (c 0.27, $CHCl_3$) has the same molecular formula as FQ C ($C_{24}H_{21}N_5O_4$). The absolute stereochemistry of 13 was deduced from a transformation of 13 into FQ E (10), the observation of NOESY cross peaks between 20-H and 16-H and 2-H, 18-H and 20-H, 27-H and 15-H_A, and 18-H and OH, and production of L-(+)-alanine from 13. The stereo structure was confirmed by a X-ray analysis.
As observed for FQ A, FQ E (10), $C_{25}H_{25}N_5O_5$, m.p. 168-172 °, $[\alpha]_D$ – 143.3 (c 0.18, $CHCl_3$), exhibits a cross peak for a W-type of long-range coupling between 14-H and 2-H in the ^1H – ^1H COSY experiment, and NOEs between 16-H and 2-H, 16-H and OMe, and 14-H and 27-H, implying that 10 exists in a twist boat conformation with the methoxy group and the C(14) – C(15) bond in a coaxial arrangement.
FQ F (11) $C_{21}H_8N_4O_2$, m.p. 88-90 °, $[\alpha]_D$ – 411.2° (c 1.36, $CHCl_3$) and FQ G (12) $C_{21}H_{18}N_4O_2$, m.p. 119-121 °, $[\alpha]_D$ – 462.8° (c 0.61, $CHCl_3$) are stereoisomers. When 7 and 8 were each treated with 0.4 % KOH in MeOH, both underwent epimerization to afford a mixture of 11 and 12 in a 3:2 ratio. This is a similar result to that described when FQ A and FQ B are reacted in this manner. The oxopiperazine rings [C(1) – C(4), N – 13 and C – 14] in 11 and 12 exist in a twist boat conformation with the 16-methyl group and the C(14) – C(15) bond, and 3-H and the C(14) – C(15) bond, respectively, in a coaxial arrangement.
All the fumiquinazolines were moderately cytotoxic in the P 388 lymphocytic leukaemia test system.

8, $R^1 = {}^{16}Me$, $R^2 = H$
9, $R^1 = H$, $R^2 = Me$
10, $R^1 = Me$, $R^2 = OMe$

11, $R^1 = {}^{16}Me$, $R^2 = H$
12, $R^1 = H$, $R^2 = Me$

13

The synthesis of *ent*-fumiquinazoline G in a total of 12 steps from Cbz-L-tryptophan has been realized (F. He and B.B. Snider, Synlett, 1997, 483). (-)-Fumiquinazoline G (12) was prepared from a tripeptide that originates from D-tryptophan methyl ester (H. Wang and A. Ganesan, J. org. Chem., 1998, **63**, 2432). Treatment of the tripeptide under Wipf's conditions (P. Wipf and C.P. Miller, *ibid*, 1993, **58**, 3604) furnished a product from which the Fmoc protecting group was then removed. Attempted chromatographic purification of the free amine induced direct intramolecular cyclization yielding (-)-fumiquinazoline G (12) in 38% overall yield for the last four steps.

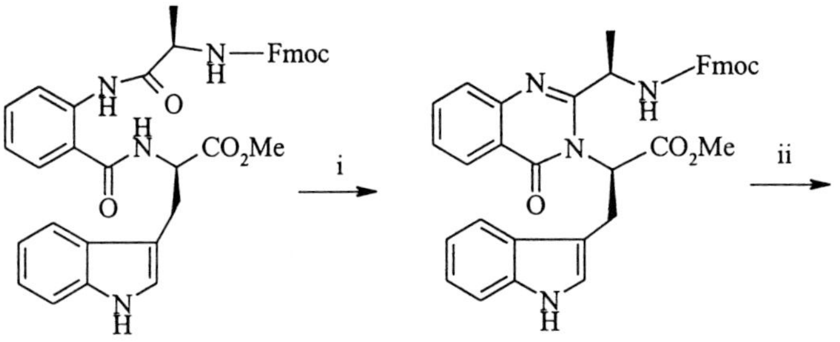

Reagents: (i) Ph_3P (5.0 eq.), I_2 (4.9 eq.), $EtN(^iPr)_2$ (10.1 eq.), r.t., 2.5 h, (65 %); (ii) 20 % piperidine in DCM, r.t., 12 min; SiO_2 (75 %)

Ardeemin (14), $C_{26}H_{26}N_4O_2$, $[\alpha]_D$ – 92 ° (c 0.24, MeOH), its 5-*N*-acetyl derivative (15), $C_{28}H_{28}N_4O_3$. $[\alpha]_D$ – 33 ° (c 0.78, MeOH), and 15b-β-hydroxy-5-*N*-acetylardeemin (16), $C_{28}H_{28}N_4O_4$, $[\alpha]_D$ 245 ° (c 0.21, MeOH) have been isolated from *Aspergillus fischeri* (J.E. Hochlowski *et al.*, J. Antibiotics, 1993, **46**, 380) designated var. *brasiliensis* by J.P. Karwowski *et al., ibid.*, 1993, **46**, 374). The structures were determined employing 1-D and 2-D homonuclear and heteronuclear nmr spectroscopy, mass spectrometry and biosynthetic considerations. An isolated acetate group with long range coupling, observed only between the acetate proton signal at δ 2.63 and carbon signals at δ 23.5 (N-5 Acetate C-2) and δ 169.9 (N-5 Acetate C-1), respectively, is positioned at N-5. An X-ray structure determination assigned the relative configuration of 15 and confirmed the structure. The most notable difference between the ^{13}C-nmr spectra of 14 and 15 is the absence in 14 of two carbon signals assigned as the N-5 acetate group in 15. The difference between the ^{1}H nmr spectra of 15 and 16 is the absence in 16 of the proton signal at δ 4.41 in 15, which was assigned to the proton on position C-15b. Additionally, signals for the two protons on C-16, which appear as two doublets of doublets in the spectrum of 15 (δ 3.01 (dd, 1 H, J = 12.7, 5.6 Hz) and δ 2.67 (dd, 1 H, J = 12.7 Hz), show in the spectrum of 16 at δ 3.18 (d, 1 H, J = 14.2 Hz) and 3.11 (d, 1H, J = 14.2 Hz) as a pair of doublets related only by a geminal coupling constant. These data are most consistent with a hydroxylation at position 15b in 16.

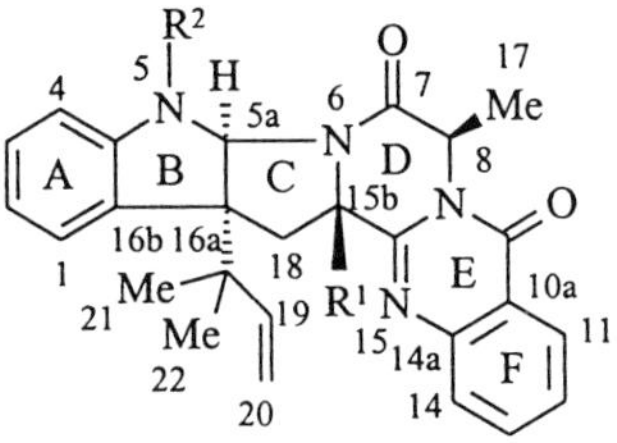

14, $R^1 = R^2 = H$
15, $R^1 = H$, $R^2 = Ac$
16, $R^1 = OH$, $R^2 = Ac$

An NOE was observed between the C-15b hydroxyl proton resonance at δ 6.44 ppm and the methyl proton signal at δ 1.51 (position 17) indicating that this C-8 methyl and the C-15 b hydroxyl group must be *cis* to one other. Alkaloids 14 - 16 are structurally related to asperlicin E with a similar skeleton. The position of oxygenation in 16 is novel, but there is precedence for oxygenation at this carbon. as supplied by the constitution of fumitremorgin B.
The total syntheses of 14 and 15 have been achieved (S.P. Marsden, K. M. Depew and S.J. Danishefsky, J. Amer. Chem. Soc., 1994, **116**, 11143) starting from (BOC)tryptophan methyl ester (17). Coupling of 18 with D-alanine methyl ester afforded 19, which after deprotection and ammonia-DMAP-induced cyclisation gave 20. An intramolecular variant of the aza-Wittig reaction was used for the efficient fusion of the quinazolin-4(3H)-one unit. Following acylation of 20 with *o*-azidobenzoyl chloride, the resultant product 21 was reacted with tributylphosphine to afford ardeemin (14) and the 5-*N*-acetyl derivative (15), respectively. The last

compound exhibits reversed multiple drug resistance in human tumour cell lines, and sensitizes cells to the anticancer agent vinblastine.

Reagents: (i) FCN, pyridine, DCM, -15⁰; (ii) D-Ala-OMe. HCl, $NaHCO_3$, H_2O, DCM; (iii) TMSI, MeCN, 0⁰, then NH_3, DMAP, MeOH; (iv) KDMDS, 2-$N_3C_6H_4COCl$, THF, -78⁰: (v) Bu_3P, benzene; (vi) LDA, THF, -78⁰ to r.t., then AcCl, reflux

4. *Other quinazoline alkaloids*

(-)-Spiroquinazoline (1), $C_{23}H_{19}N_5O_3$, m.p. 166 -168 °, $[\alpha]_D$ – 2.5° (c 0.7, MeOH) has been isolated from cultures of the fungus *Aspergillus flavipes* (C.J. Barrow and

H.H. Sun, J. nat. Prod., 1994, **57**, 471). It contains a new carbon skeleton and it is a structural relative of the fiscalins and fumiquinazolines. The relative stereochemistry of 1 was defined by the observation of an NOE from H-14 to H-23, but not to H-12. Two possible diastereoisomers exist for 1. The first diastereoisomer is that drawn as structure 1 and the second has the opposite chirality at C-14. Drawing and minimizing these structures using Macromodel clearly established that an H-14 to H-23 NOE should be observed for the isomer drawn as structure 1. The absolute stereochemistry of 1 remains undefined. The alkaloid inhibited the binding of substance P to human astrocytoma cells.
Benzomalvins A (2), $C_{24}H_{19}N_3O_2$, m.p. 105 ~ 115 °, $[\alpha]_D$ – 106 ° (c 1.0, MeOH), B (3) $C_{24}H_{17}N_3O_2$, m.p. > 266 °, $[\alpha]_D$ + 158 ° (c 0.59, MeOH), C (4) $C_{24}H_{17}N_3O_3$, m.p. 214 °, $[\alpha]_D$ + 69.7 ° (c 0.38, MeOH), and D (5), $[\alpha]_D$ + 48 °, have been isolated from the culture broth of a *Penicillium* sp. (H.H. Sun *et al.*, J. Antibiotics, 1994, **47**, 515; H.H. Sun, C.J. Barrow and R. Cooper, J. nat. Prod., 1995, **58**, 1575). The structures of these compounds were determined by physical methods including mass spectrometric measurements and nmr analyses (HRFAB-MS, ^{13}C-NMR, 2 D COSY, HMQC, HMBC, NOE, and DEPT experiments). The chirality of C-19 was assigned as *S* by comparison of optical rotation with that of a known related compound, asperlicin C (6). In solution 80 % of 5 converted into benzomalvin A (2), an atropisomer. This is the first report of natural atropisomers of 1,4-benzodiazepines. Benzomalvins (2) and (5) interconvert to form a 4:1 mixture at r.t. and are totally separable by hplc. In this case the energy barrier to interconversion is high enough to allow the isolation of the two atropisomers. Molecular modeling of 2 indicated a 3D structure 7 with an equatorial benzyl group and structure 8 with an axial group for the two isomers. The finding that 3, a molecule without asymmetric carbons, exhibits optical activity also supports the presence of two atropisomers of 2. The space-filling model of 3 showed that the bulky phenyl group, which is *cis* to the quinazoline group might prevent the inversion of the benzodiazepine ring. This compound thus adopts only one conformation and forms an atropisomer.

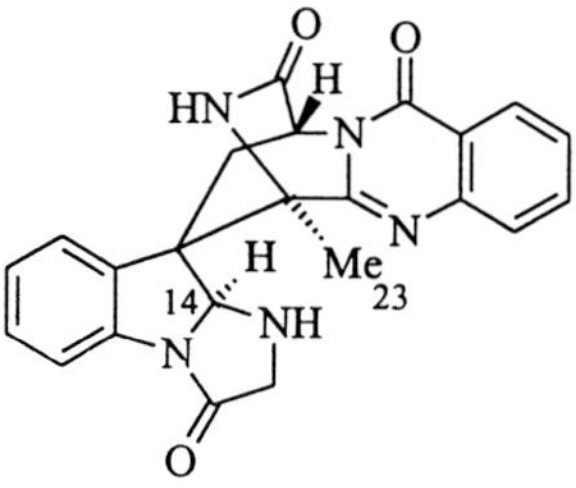

Benzomalvin C (4) differed from 2 and 3 only by the presence of an epoxide group at C-19 and C-20.

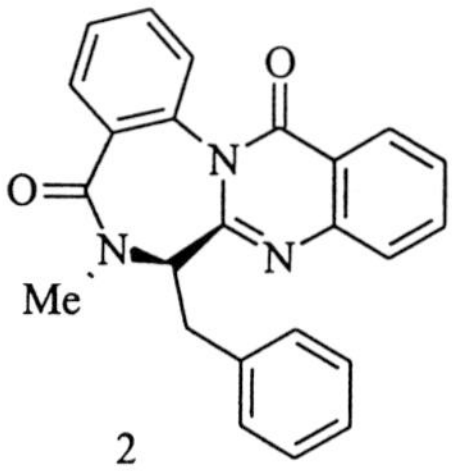

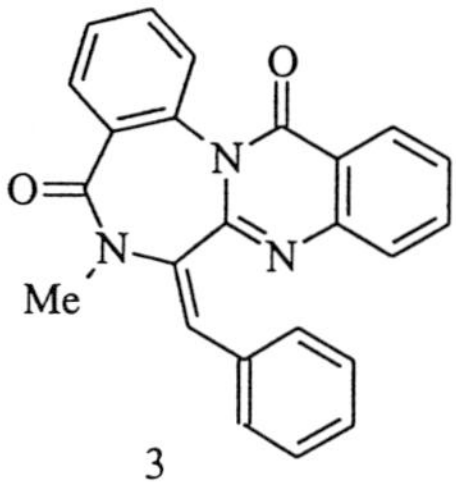

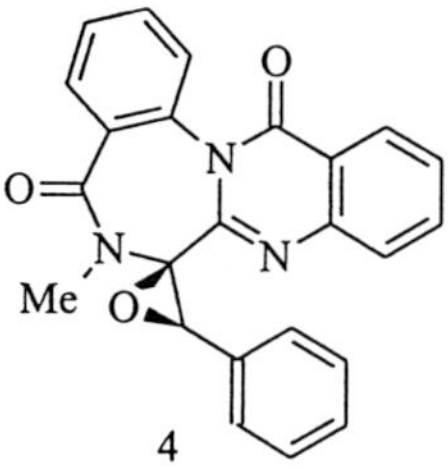

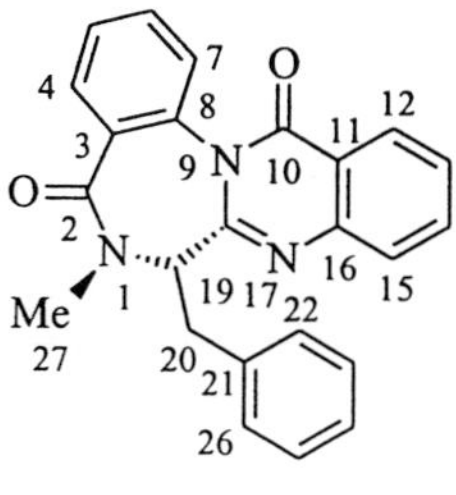

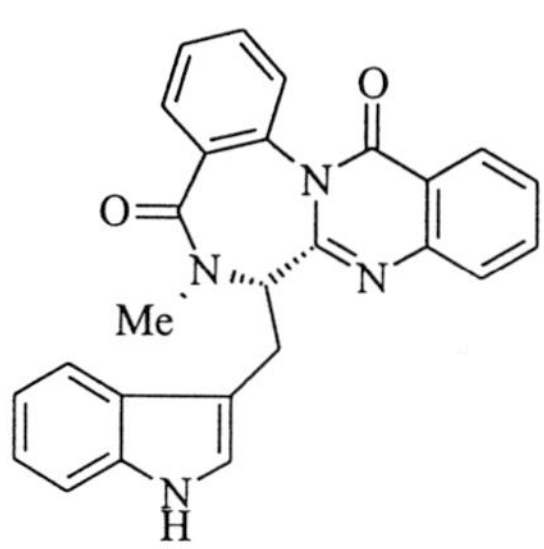

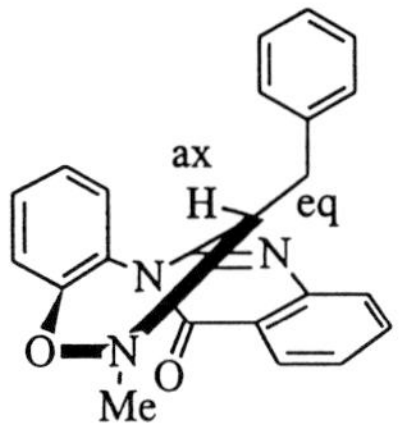

7

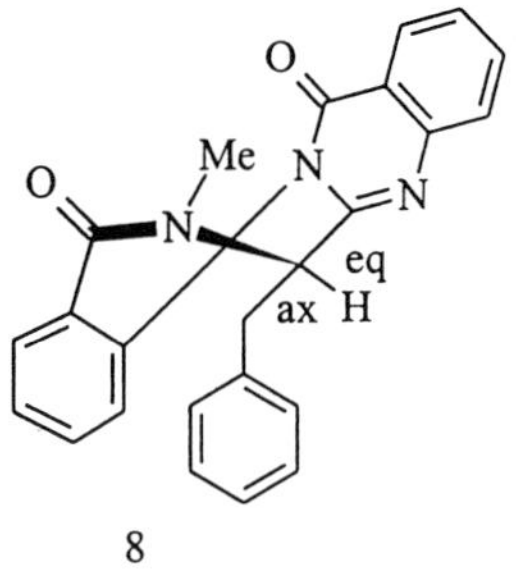

8

The total synthesis of benzomalvin A (2) was realized by the reaction of isatoic anhydride with L-phenylalanine. Without isolation the *N*-anthranoyl-L-phenylalanine was converted into a benzodiazepinedione. Further elaboration to give benzomalvin A required a regioselective annulation with anthranilic acid. This was accomplished in three steps (H. H. Sun, C.J. Barrow and R. Cooper, *loc. cit.*).

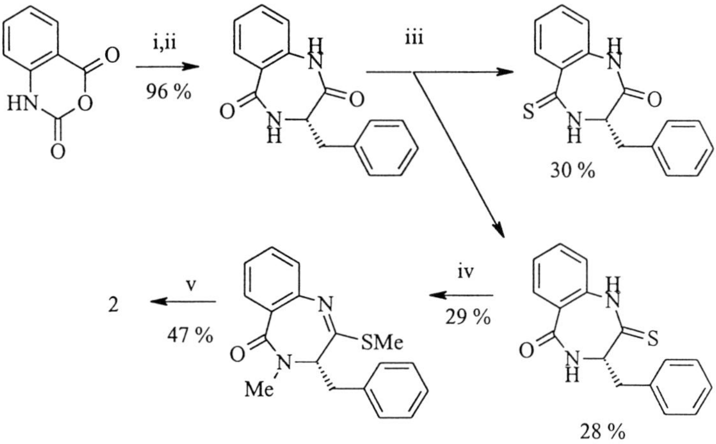

Reagents: (i) L-phenylalanine, Et_3N, H_2O, r.t.; (ii) HOAc, reflux; (iii) Lawesson's reagent THF, r.t. flash chrom. SiO_2; (iv) 40 % aq.NaOH, MeI, $(Bu)_4NHSO_4$, toluene, r.t; (v) methyl anthranilate

(-)-Benzomalvin A (2) is a good inhibitor of substance P, the endogenous ligand for the neurokinin-1 receptor, which is implicated in inflammatory response and pain. Auranthine (9), $C_{19}H_{14}N_4O_2$, m.p. > 300 °, $[\alpha]_D - 164$ ° (1 % w/v EtOH) has been isolated from sporing cultures of *Penicillium aurantiogriseum* (St. E. Yeulet *et al.*, J. chem. Soc., Perkin I, 1986, 1891). Its structure has been elucidated from spectral evidence and biosynthetic reasoning. This alkaloid is a complementary addition to the small group of fungal secondary metabolites that are biosynthesized from anthranilic acid. The closest analogy is with the microbial metabolite asperlicin (10), $C_{31}H_{29}N_5O_4$, m.p. 211 – 213 ° $[\alpha]_D - 185.3$ ° (c 1.10, MeOH) and the related asperlicins B (11), C (6), D (12) and E (13) produced by *Aspergillus alliaceus* (M.A. Goetz *et al.*, J. Antibiotics, 1985, **38**, 1633; *ibid.* 1988, **41**, 875). Asperlicin and related compounds are non-peptidal antagonists of the gastrointestinal hormone neurotransmitter cholecystokinin (CCK) and their extraction from *A. alliaceus*, represents the first isolation from microbial sources. The chemical structures were deduced spectroscopically (R.S. Chang *et al.*, Science, 1985, **230**, 177). Asperlicins (6) and (13) have been synthesized from L-tryptophan and isatoic anhydride (M.G. Bock *et al.*, J. org. Chem., 1987, **52**, 1644; J.M. Liesch *et al.*, J. Antibiotics, 1988, **41**, 878). Rose bengal sensitized photooxygenation of 6 afforded a mixture of 3-hydroperoxyindoles. The cyclisation product was reduced to give 13. This synthesis can be improved by using the Eguchi protocol for the elaboration of a quinazolin-4-one onto an amide (F. He and B.B. Snider, Synlett, 1997, 483; H. Takeuchi, S. Hagiwara and S. Eguchi, Tetrahedron, 1989, **45**, 6375; S. P. Marsden, K.M. Depew and S.J. Danishefsky, J. Amer. Chem. Soc., 1994, **116**, 11143).

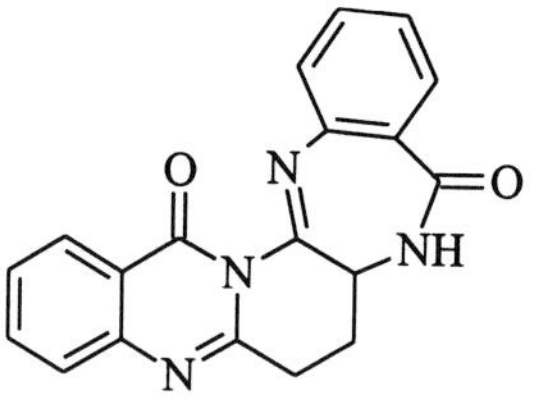

O
N
NH
O
N
N
9 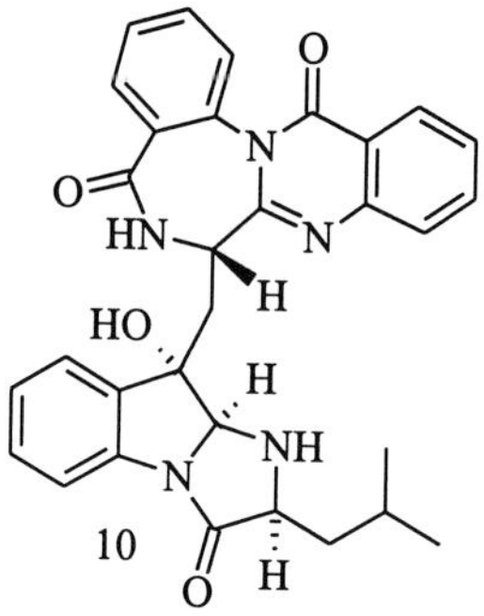
O
N
O
HN
H
N
HO
H
NH
N
O
H
10 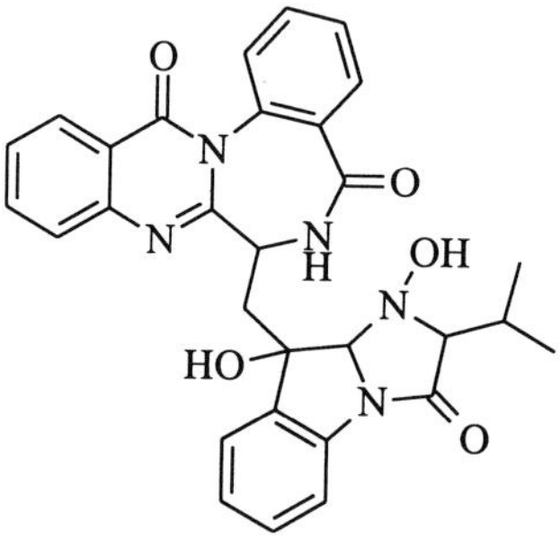
O
N
O
N
N
H
OH
N
HO
N
O
11 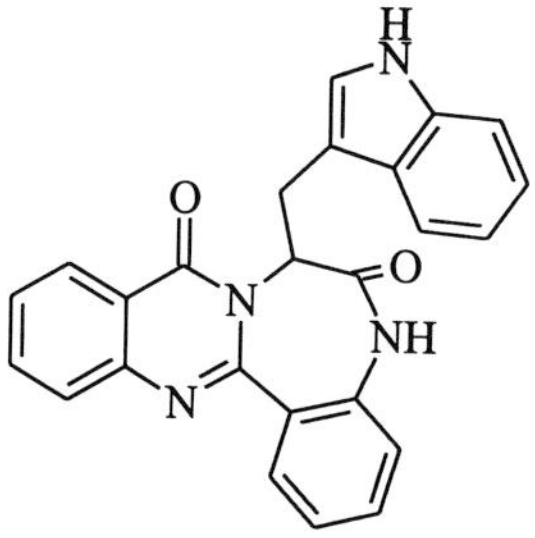
H
N
O
O
N
NH
N
12 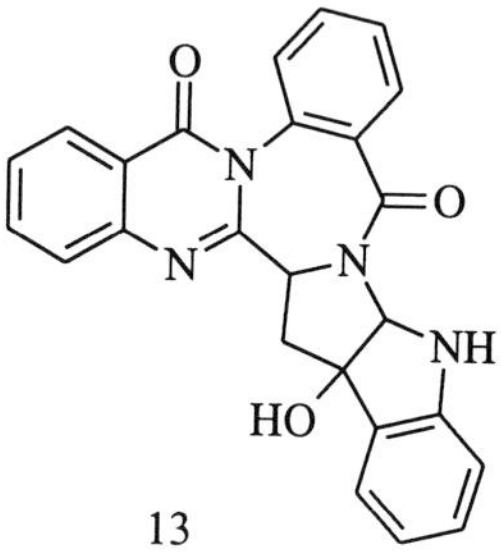
O
N
O
N
N
NH
HO
13

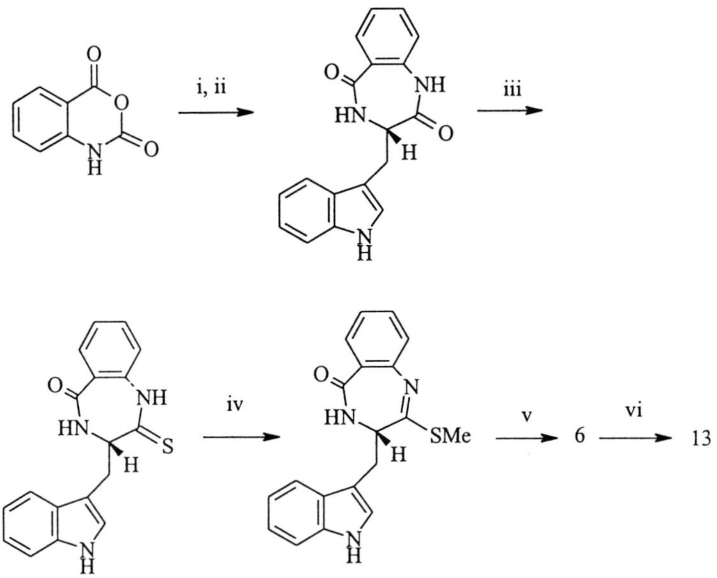

Reagents: (i) L-Trp, Et_3N, H_2O, 23^0, 5 h; (ii) HOAc, 118^0, 5 h; (iii) Lawesson's reagent, THF, 23^0, 2 h; (iv) MeI, $(Bu)_4HSO_4$, 40 % aq.NaOH, toluene, 23^0 20 min; (v) methyl anthranilate, 135^0, 1 h; (vi) $^1O_2$, rose bengal, MeOH-pyridine, 0^0, 5h

F. He, B.M. Foxman and B.B. Snider (J. Amer. Chem. Soc., 1998, **120**, 6417) have shortened and improved the synthesis of 6, developed a general route to the hydroxyimidazoindolone ring system, and applied it to the first synthesis of 10, which proceeds efficiently (15 steps, 8 % overall yield) and stereospecifically. Hinckdetine A (14) is an unusual marine alkaloid that has been isolated from the bryozoan *Hincksinoflustra denticulata*, collected off the eastern coast of Tasmania (A.J. Blackman *et al.*, Tetrahedron Letters, 1987, **28**, 5561). The structure of 14 was established by X-ray crystallography and its stereochemistry was established by detailed nmr studies.

Hinckdentine A has an unique molecular skeleton consisting of a seven membered lactam ring fused to a tribromoindolo[1,2 *c*]quinazoline. Although numerous alkaloids containing an indole unit or a quinazoline unit are known, only three alkaloids are recognized, which belong to the indoloquinazoline family: tryptanthrin, candidine, and hinckdentine A. A systematic study of the reactions of indolo[1,2-*c*]quinazoline (15) was made, as well as attempts to employ 15 as a synthon for the construction of the pentacyclic skeleton of 14 (A.D. Billimoria and M.P. Cava, J. Org. Chem., 1994, **59**, 6777). The direct bromination of 15 using one equivalent of bromine gave the 12-bromo compound (16) as a major product with smaller amounts of the 10,12-dibromo compound (17). The use of two or more equivalents of bromine resulted only in the formation of 17 as the major product, implying that the three bromine atoms in 14 are not introduced biogenetically from a fully aromatic indolo[1,2-*c*]quinazoline precursor. A review describing the progress in the chemistry of indoloquinazolines including the synthesis of tryptanthrin, candidine and 14 is available (A. D. Billimoria and M.P. Cava, Heterocycles, 1996, **42**, 453).
Tetrodotoxin (18) and approximately a dozen toxins of this type are found in marine animals of diverse classes and also in amphibia inhabiting fresh water. Other sources are nemertines, gastropod molluscs, the blue spotted octopus, marine crustacea, globe fishes, amphibian families of salamandrian animals and true salamanders, plus corals, turbellarin worms, starfish, gobies, and amphibians of the toad family. Tetrodotoxins have been detected hitherto either in algae or in higher plants. Tetrodotoxin is present in considerable concentrations in sea-bed deposits (K. Kogure, H.K. Do and E.V. Thuesen, Mar. Ecol. Prod. Ser., 1988, **45**, 303). The accumulation is the result of the activity of a whole series of tetrodotoxin-producing bacteria (Pseudomonas, Alteromonas, Bacillus, Plesiomonas, Vibrio). For further information the reader is referred to A.L. D'yakonov and M.V. Telezhenetskaya, Chem. nat. Comp., 1997, **33** 221.
Alantrypinone (19) $C_{21}H_{16}N_4O_3$, $[\alpha]_D + 37°$ (c 2.08, EtOH) and fumiquinazoline F have been isolated as the two major metabolites of *Penicillium thymicola* (Th. O. Larsen *et al.*, J. nat. Prod., 1998, **61**, 1154). The structure of 19, which contains a new ring system, was elucidated by analysis of spectroscopic data including 2 D

NMR. The absolute configuration of 19 was established by a single-crystal X-ray diffraction study.
A NOESY experiment clarified the relative configuration around the spiro-center at C-17. Interaction between the protons resonating at δ 8.63 ppm (H-2) and δ 7.18 ppm (H-24) supported the depicted geometry of the tryptophan moiety. This geometry is identical to that reported in the spiroquinazoline (1). A similar ring closure between C-3 and C-17 in fumiquinazoline C, *via* an ether linkage has been described (C. Takashashi *et al.*, J. chem. Soc., Perkin I, 1995, 2345). The absolute configuration of 19 is 3*R* and 14*R*, suggesting that the peptide incorporates L-alanine and D-tryptophan, as do fumiquinazolines F and C, and the spiroquin-azoline (1).

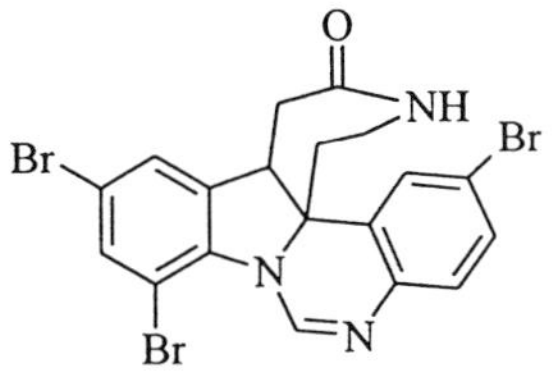

14

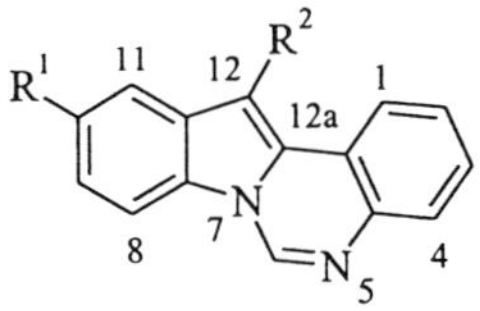

15, $R^1 + R^2 = H$
16, $R^1 = H$, $R^2 = Br$
17, $R^1 + R^2 = Br$

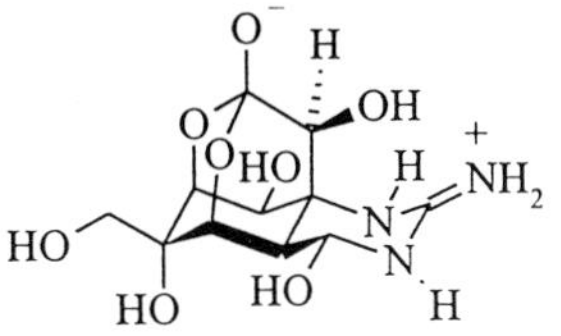

18

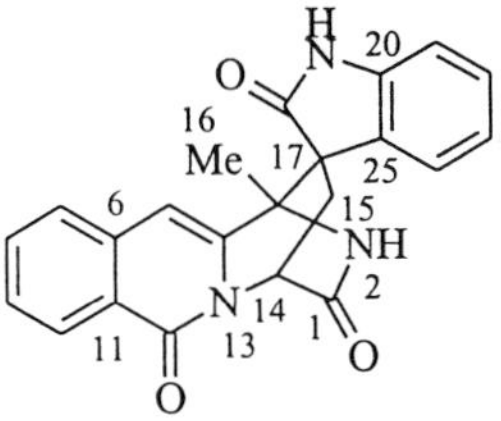

19

Second Supplements to the 2nd Edition of Rodd's Chemistry of Carbon Compounds, Vol.IV IJ, edited by M. Sainsbury

Chapter 47

SIX-MEMBERED RINGS WITH THREE OR MORE HETERO-ATOMS

DONAL F. O'SHEA

Introduction

This review covers the literature from 1993-1998. It is presented as a follow on from the two previous chapters dealing with this topic published in 1995 and 1989. The individual sections cannot be put into proper perspective unless the previous reviews are also consulted.

(I) *Nitrogen Rings*

The possible nitrogen containing rings are the triazines, 1,2,3-, 1,2,4- and 1,3,5-; the tetrazines, 1,2,3,4-, 1,2,3,5- and 1,2,4,5-; pentazine and hexazine.

(1) *Triazines*

(a) *1,2,3-Triazines*

1,2,3-Triazines have been reviewed by A. Ohsawa, T. Itoh (Compreh. Hetero. Chem. II, 1996, 6, 483).

(i) *Synthesis and Structure*

The optimised structure of the parent 1,2,3-triazine (1) (R^1, R^2, R^3 = H) determined by *ab initio* molecular orbital calculations, is consistent with the previously reported experimental X-ray diffraction study. The calculated ring bond-lengths at the MP2/6-31G* level of theory are *N*(1)-*N*(2) 1.341 Å, *N*(1)-*C*(6) 1.346 Å, *C*(5)-*C*(6) 1.388 Å with the bond angles at *N*(2) 121.1° and *C*(5) 115.3° (G. Fischer *et al.*, J. mol. Spectr., 1993, 161, 388). X-ray crystallographic data of other substituted 1,2,3-triazines has been reported, 4,5,6-tris(dimethylamino) (1) (R^1, R^2, R^3 = $N(CH_3)_2$) and 4-phenyl-6-carboxylic acid methyl ester (1) (R^1 = Ph, R^2 = H, R^3 = CO_2CH_3) (R. Bopp *et al.*, Zeit. Kristall., 1994, 209, 196; S. Eichhorn *et al.*, Zeit. Kristall., 1993, 208, 310). Substitution at the ring carbons has minor affect on the overall ring geometry. The triazine synthesis involves two previously reported procedures, the oxidation of substituted 1-aminopyrazoles and the thermal rearrangement of cyclopropenyl azides (H. Neunhoeffer, R. Bopp, W. Diehl, Ann., 1993, 367). The electronic absorption spectrum

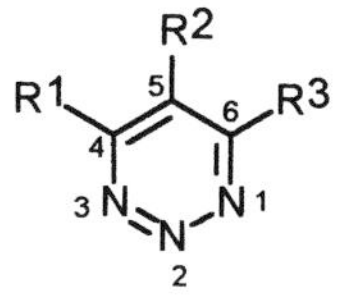

(1)

of 1,2,3-triazine has been measured in vapour and in solution. Excited state geometry and vibrational frequency calculations show the first three excited singlet states comprising $^1A_2$ (largely $a_1 \rightarrow a_2$), $^1A_2$ (largely $b_2 \rightarrow b_1$) and $^1B_1$ are energetically close, with $^1A_2$ ($a_1 \rightarrow a_2$) state being lowest. The lowest π-π* state is calculated as $^1B_2$ (G. Fischer, D.M. Smith, A.U. Nwankwoala, Chem. Phy., 1997, 221, 11). 1,6-Dihydro-1,2,3-triazines are prepared by the Bamford-Stevens reaction of *cis*-aziridinyl ketone tosylhydrazones (2) which rearrange in the presence of a slight excess of sodium hydride to afford (3) in good yields (Table 1) (M. Morioka *et al.*, Heterocycles, 1996, 43, 1759). When a large excess of sodium ethoxide is used the products isolated are the aminopyrazoles (4). The dihydrotriazines (3) are converted into aminopyrazoles (4) under identical conditions. A route to the new

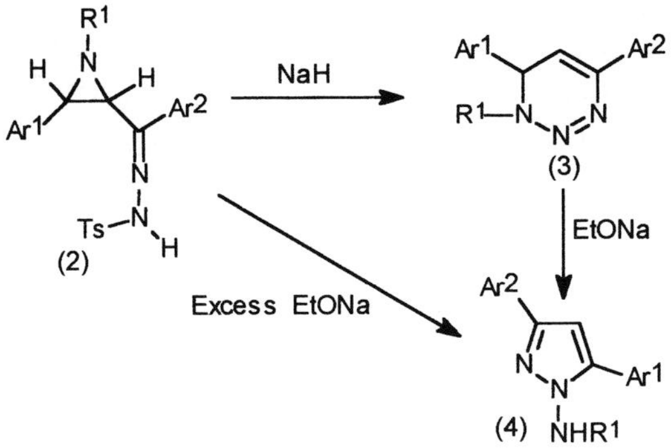

Table 1: 1,6-Dihydro-1,2,3-triazines (3).

R^1	Ar^1	Ar^2	Yield	m.p.
i-Pr	C_6H_5	C_6H_5	89	134 - 136
$C_6H_5CH_2$	C_6H_5	C_6H_5	91	167 - 170
c-C_6H_{11}	C_6H_5	*p*-$CH_3OC_6H_4$	95	120 - 123
c-C_6H_{11}	C_6H_5	*p*-$NO_2C_6H_4$	83	120 - 123
c-C_6H_{11}	*p*-$CH_3C_6H_4$	C_6H_5	90	134 - 136

phenanthro[9,10-*e*][1,2,3]triazine system from a cycloaddition-rearrangement sequence of 9,10-bisarylazophenanthrenes with dipolarophiles is described (R.N. Butler *et al.*, J. chem. Soc. Perkin 1, 1996, 1623).

(ii) *Reactions of 1,2,3-Triazines*

Direct nucleophilic substitution of 1,2,3-triazines is not possible as the highly π-deficient ring readily opens followed by dinitrogen elimination. Thus, such reactions are achieved *via* the *N*(2)-substituted triazines. 1,2,3-Triazinium-2-dicyanomethylides (5) are prepared by the reaction of 1,2,3-triazines with tetracyanoethylene oxide (TCNEO). Nucleophilic substitution with chloromethyl phenyl sulfone gives the corresponding 5-substituted derivative (6) in yields of 42-84%. The dicyanomethylene group is readily eliminated with ammonium persulfate resulting in the formation of the 5-phenylsulfonylmethyl-1,2,3-triazines (7) (T. Itoh *et al.*, Heterocycles, 1993, 35, 581; K. Nagata *et al.*, Chem. pharm. Bull., 1993, 41, 1644). Substitution reactions of 2-methyl-1,2,3-triazinium iodide salts also occur at the

TCNEO → (5); $PhSO_2CH_2Cl$, base → (6); $(NH_4)S_2O_8$ / *i*-PrOH → (7)

R^1, R^2 = Me, Me; Me, Et; Me, Ph; Et, Et; Ph, Ph

C(5)-position with allyltributyltin and silyl enol ethers (T. Itoh *et al.*, Chem. pharm. Bull., 1994, 42, 1768; T. Itoh *et al.*, J. chem. Soc. Perkin 1, 1996, 2511). The reaction of 1,2,3-triazines with silyl enol ethers in the presence of 1-chloroethyl chloroformate gives the 2,5-dihydrotriazines (9). The reaction is shown to proceed by the trapping of the unstable *N*-(1-chloroethoxycarbonyl) quaternary salts (8) with the silyl reagent. Aromatisation with ceric ammonium nitrate (CAN) in acetonitrile-water *via* an oxidation / hydrolysis mechanism gives the corresponding 5-substituted 1,2,3-triazines in yields that are dependent on ring substituents (T. Itoh *et al.*, Chem. pharm. Bull., 1995, 43, 881).

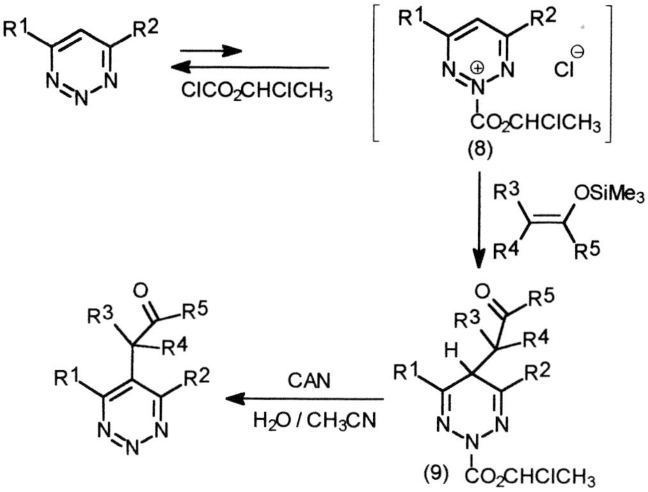

The π-deficient nature of the 1,2,3-triazine ring renders it suitable for inverse electron demand 4-π cycloaddition reactions with electron rich dienophiles. The zinc bromide catalysed reaction of 4-methyl-1,2,3-triazine in a sealed tube at 90° with a range of enamines gives the pyridines (10) (J. Koyama, T. Ogura, K. Tagahara, Heterocycles, 1994, 38, 1595). The reaction proceeds through a Diels-Alder 4+2 addition

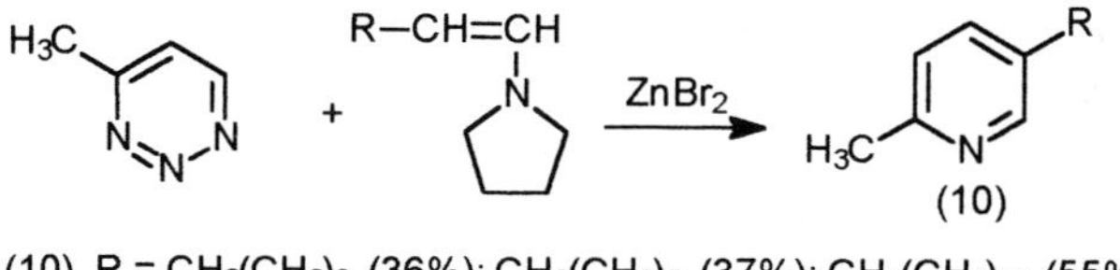

(10) R = $CH_3(CH_2)_2$, (36%); $CH_3(CH_2)_5$, (37%); $CH_3(CH_2)_{10}$, (55%)

C_6H_5, (77%); $C_6H_5CH_2$, (43%); $C_6H_5CH_2CH_2$, (32%).

across *N*(3)-*C*(6) of the triazine with subsequent elimination of dinitrogen. The vapour photolysis (290 nm) of the parent 1,2,3-triazine (1) yields acetylene, hydrogen cyanide and dinitrogen in a concerted photofragmentation mechanism. Photolysis in solution is postulated to occur through a number of different paths due to the observation of unidentified photo-products (G. Fischer, A.U. Nwankwoala, J. photochem. Photobiol., 1995, 87, 135).

(b) *1,2,4-Triazines (as-Triazines)*

1,2,4-Triazines have been reviewed by H. Neunhoeffer (Compreh. Hetero. Chem. II, 1996, 6, 507).

(i) *Synthesis and Structure*

^{13}C and ^{15}N NMR data for a series of methoxy and thiomethoxy substituted 1,2,4-triazines has been assigned. The ^{15}N NMR chemical shifts for *N*-4 of 3- and 5-methoxy or thiomethyl substituted triazines are shown to be related to the π-charge at *N*-4, calculated by the semi-empirical molecular orbital (AM1) procedure (E. Kolehmainen *et al.*, Mag. Res. Chem., 1995, 33, 690). Gas phase ultraviolet, ultraviolet photoelectron, vacuum ultraviolet and near-threshold electron energy loss spectra have been presented for the parent 1,2,4-triazine (1) (R^1, R^2, R^3 = H). The results have been interrupted with *ab initio* multi-reference configuration interaction calculations (M.H. Palmer *et al.*, Chem. Phy., 1995, 201, 381). The X-ray structures of three analogues

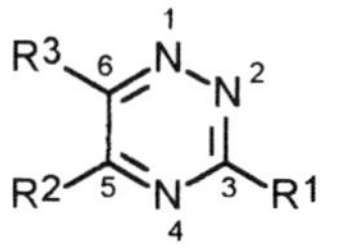

R3 6 N 1 N 2 R2 5 N 4 3 R1

(1)

of the anticonvulsant lamotrigine, 3,5-diamino-6-(aryl)-1,2,4-triazines, were determined for a structure-activity relationship study (R.W. Janes, R.A. Palmer, Acta Cryst., 1996, C52, 2627; *ibid.*, 1995, C51, 685; *ibid.*, 1995, C51, 440). Trifluoromethyl substituted triazines are of particular interest due to their pharmacological properties. A new route for the synthesis of trifluoromethyl-1,2,4-triazines derivatives has been accomplished. The reaction of 3-hydrazono-1,1,1-trifluoroalkan-2-ones (2) with aldehydes in aqueous ammonium hydroxide gives the 5-trifluoromethyl-2,3-dihydro-1,2,4-triazines (3), after an acidic work-up,

R^3CHO / NH_4OH (aq), r.t. 24 h.; 1M HCl; $R2 = CH_3$, H_2O_2 / $FeCl_2$; R2 = H, DDQ

R1 = p-$CH_3C_6H_4$; p-$NO_2C_6H_4$; n-C_6H_{13}
R2 = CH_3; H
R3 = o-$CH_3C_6H_4$; p-$CH_3C_6H_4$; CH_2CH_3; $CH(CH_3)_2$

in yields of 31-99%. The dihydotriazines (3) (R^2 = H) are readily oxidised by dichlorodicyanoquinone (DDQ) to give the 5-trifluoromethyl-1,2,4-triazines (4) in good yields (59-89%). The *N*(2)-substituted dihydrotriazines (3) ($R^2 = CH_3$) are oxidised by hydrogen peroxide to give the 5-trifluoromethyl-2,5-dihydro-5-hydroxy-2-methyl-1,2,4-triazines (5) (Y. Kamitori *et al.*, Heterocycles, 1994, 39, 155). The ability of 1,2,4,5-tetrazines to undergo inverse electron demand Diels-Alder reactions with heterodienophiles to yield triazines is exploited in the synthesis of the first 1,2,4-triazine-*C*-nucleoside. The 4+2 addition across the C=NH of 2,5-anhydro-3,4,6-tris-*O*-benzyl-*D*-allonoimidic acid methylester (6) with bis(trifluoromethyl)-1,2,4,5-tetrazine gives the *O*-benzyl protected 5-(β-D-ribofuranosyl)-1,2,4-triazine (7) (M. Richter, G. Seitz, Arch. Pharm., 1995, 328, 175). The corresponding 2,3-dideoxyribofuranosyl-1,2,4-triazine derivatives are synthesised in a similar fashion (G. Seitz, J. Seigl, Z. Naturforschung, 1997, 52b, 851). The pyrido[1,2-*b*][1,2,4]triazinium salt (8) ring opens

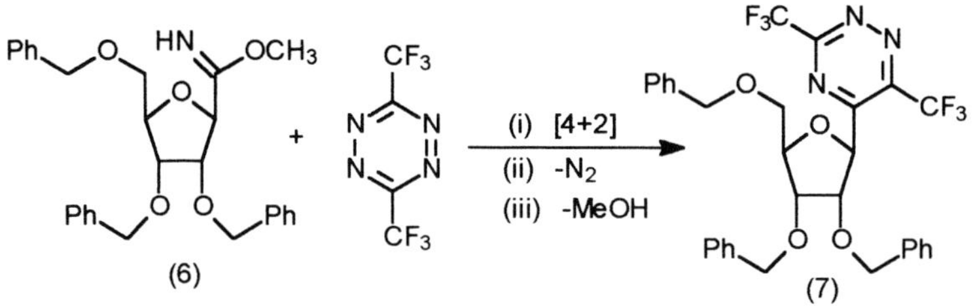

with piperidine in acetonitrile yielding 5,6-diphenyl-3-[4-(pyrrolidin-1-yl)buta-(1*Z*,3*E*)-1,3-dienyl][1,2,4]triazine (9) in an 85% yield. When the reaction is carried out in a protic solvent such as ethanol the product isolated is the thermodynamically more stable 1*E*,3*E* isomer (A. Kotschy *et al.*, J. org. Chem., 1996, 61, 4423). Two synthetic approaches to polymeric 1,2,4- triazines have been developed. The first

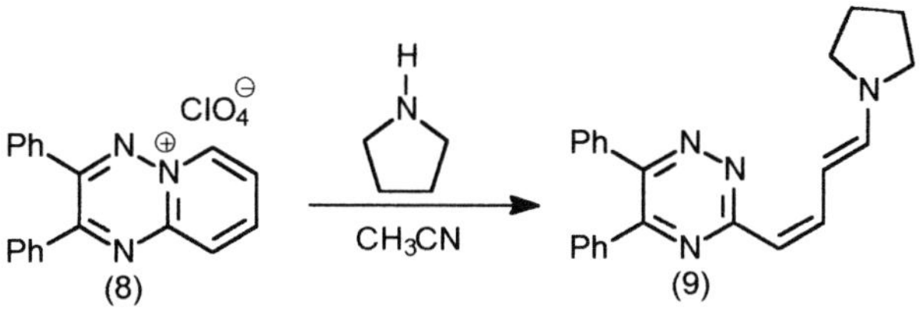

involves the synthesis of monomeric triazines and their reaction with biphenols and activated aromatic difluorides to form high molecular weight polymers. A second, more successful synthetic strategy involves first preparing the poly(aryl ether benzils) (10) and reacting them with 2-(pyrid-2-yl)semicarbazide to form the poly(aryl ether pyridyl-1,2,4-triazines) (11), (R = $C(CH_3)_2$; $C(CF_3)_2$, $C(CH_3)(C_6H_5)$). All of the polymers are amorphous and exhibit glass transition temperatures in the range 202-277°, which are significantly higher than the parent benzil polymers (10) (M. Strukelj, J.C. Hedrick, Macromolecules, 1994, 27, 7511). The synthesis and structure-activity

(10)

(i)

(ii) chlorobenene, reflux

(11)

relationships of a series of 3-substituted 5,6-bis(4-methoxyphenyl)-1,2,4-triazines as anti-platelet agents are described (A. Tanaka *et al.*, Chem. pharm. Bull., 1994, 42, 1835). 5,5',6,6'-Tetraaryl-3,3'-bi(1,2,4-triazines) are produced from the cyclo-condensation of α,β-diketones with oxamidodihydrazone (J. Breu, K.-J. Range, E. Herdtweck, Monatsh. Chem., 1994, 125, 119). A single crystal X-ray structure of the tetraphenyl derivative shows that the triazine rings are coplanar with the phenyl rings twisted 26.5° and 54.2° relative to them (J. Breu, K.-J. Range, Acta Cryst., 1993, C49, 1541). Flash vacuum pyrolysis (FVP) of α-benzotriazolyl-β-oxophosphorus ylide (12) causes the extrusion of $n\mathrm{Bu_3P}$ to generate a acetyl benzotriazolylcarbene which undergoes a 1,2-*N* migration to give the 3-acetyl-1,2,4-benzotriazine (13) in a 26% yield. A second minor product, *o*-cyanoacetophenone, was also isolated in 16% yield (R.A. Aitken *et al.*, Chem. Comm., 1993, 1517). Substituted benzo-1,2,4-triazines have been reported as

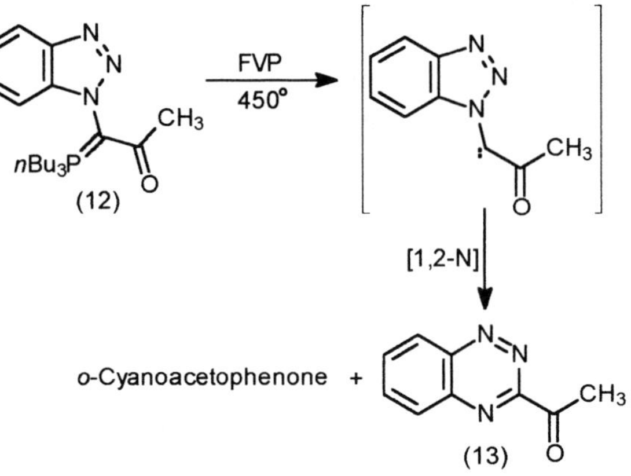

o-Cyanoacetophenone + (13)

one of the products from the reaction of nitrile oxides with *N*-aryl benzamidoximes (F. Risitano *et al.*, Tetrahedron, 1997, 53, 1089). The synthesis of 4-alkyl-3,6-diphenyl-2-tosyl-2,3,4,5-tetrahydro-1,2,4-triazines has been achieved by the ethyl ether-boron trifluoride catalysed rearrangement of *cis-N*-alkylaziridinyl ketonetosylhydrazones (M. Morioka, T. Ogata, Heterocycles, 1998, 48, 769). 1,4,5,6-Tetrahydro-1,2,4-triazin-3(2*H*)-ones are discussed as azacyclic peptide mimetics (J. Gante, H. Neunhoeffer, A. Schmidt, J. org. Chem., 1994, 59, 6487). 4,5-Dihydro-1,2,4-triazin-6-one derivatives are also described (R. Zupet, M. Tišler, J. org. Chem., 1994, 59, 507). The synthesis and annelation effect of 1,2,4-triazino[3,2-*a*]isoquinolinium and 1,2,4-triazino[2,3-*b*]isoquinolinium salts are illustrated (G. Hajós, Z. Riedl, A. Messmer, Acta Chem. Scand., 1993, 47, 296). A new route to phenanthro[9,10-*e*][1,2,4]triazines has been developed (R.N. Butler *et al.*, J. chem. Soc. Perkin 1, 1997, 1047).

(ii) *Reactions of 1,2,4-Triazines*

The chemistry of 1,2,4-triazines is determined by the π-deficient nature of the ring system. Nucleophilic substitution reactions generally occur at *C*-5 and inverse electron demand Diels-Alder reactions are central to their reactivity. Recently research interest in the metal complexes of

1,2,4-triazines has developed. The reaction of 3-substituted 1,2,4-triazines with nitromethane in KOH / DMSO gives the 5-formyl-1,2,4-triazine oximes (14). The reaction occurs exclusively at the *C*-5 position. Dehydration of the oxime with phenyl isocyanate in benzene, under reflux, converts it to the nitrile (15). The cyano substituent at the *C*-5 position is an efficient leaving group and provides an entry into a range of 5-substituted 1,2,4-triazines (16) (Table 1) (A. Rykowski *et al.*, J. heterocyclic Chem., 1996, 33, 1567; *ibid.*, 1993, 30, 413).

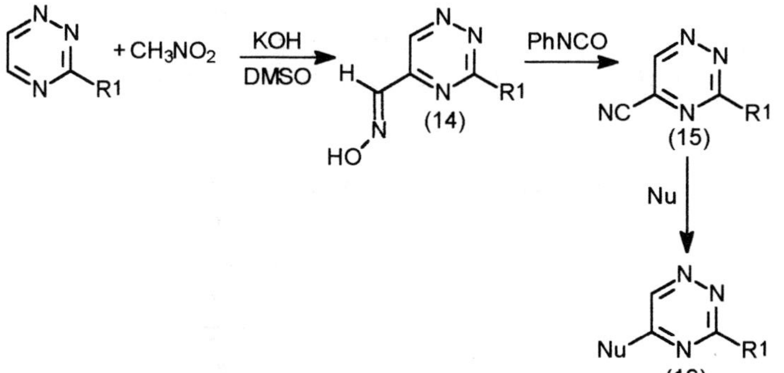

Table 1: Substituted 1,2,4-triazines (16).

R^1	Nu	Yield	m.p.
Ph	NH_2	90	252 - 253
Ph	CH_2NO_2	83	104 - 105
Ph	OH	90	244 - 246
$SCH(CH_3)_2$	NH_2	91	156 - 157
$SCH(CH_3)_2$	$CH(CN)CO_2Et$	75	204 - 205

3-Methoxy-1,2,4-triazine (1) (R^1 = OMe, R^2, R^3 = H) undergoes substitution with morpholine and piperidine at *C*-5 in 95% ethanol (S.G. Alekseev, S.V. Shorshnev, B.V. Rudakov, Chem. heterocyclic Comp., 1997, 33, 621). The reaction of 3-chloro-5,6-diphenyl-1,2,4-triazine (1) (R^1 = Cl, R^2, R^3 = Ph) with chloroacetonitrile in pyridine yields 3-cyanochloromethyl-5,6-diphenyl-1,2,4-triazine (1) (R^1 = CH(Cl)CN, R^2, R^3 = Ph). This is converted in a three steps to 9-cyano 6-methyl-2,3-diphenyl-8*H*-[1,2,4]triazino[4,5-*b*][1,2,4]triazine (R.M. Abdel-Rahman, I.E. Islam, Indian J. Chem., 1993, 32B, 526). Nucleophilic substitution of 3-chloro-6-phenyl-1,2,4-triazine with α-chlorocarbanions gives rise to the pyrazoles (18) and not to the expected 5-substituted 3-chloro-6-phenyl-1,2,4-triazines. The reaction mechanism is discussed in terms of an initial addition of carbanion to *C*-5, followed by breaking of the *N*(4)-*C*(5) bond to generate the ring opened intermediate (17) which undergoes intramolecular ring closure followed by hydrolysis and decarboxylation of the cyano group (A. Rykowski, D. Branowska, Heterocycles, 1996, 43, 2095). In a similar

R = *p*-$CH_3C_6H_4$ (93%); C_6H_5 (67%); $N(CH_3)_2$ (74%); OC_6H_5 (47%).

reaction with malononitrile ($CH_2(CN)_2$) under basic conditions the products isolated are 3-aminopyridazines (A. Rykowski, E. Wolińska, Tetrahedron Letters, 1996, 37, 5795). The first reported metalation of a 1,2,4-triazine derivative has been described. 3-Aryl-5-methoxy-1,2,4-triazines are lithiated with lithium 2,2,6,6-tetramethylpiperidide to form the 3-aryl-6-lithio-5-methoxy-1,2,4-triazines which are trapped with aldehydes or iodine. The iodo-1,2,4-triazines undergo palladium catalysed cross-coupling reactions with olefins (N. Plé *et al.*, Ann., 1993, 583). 1,2-Dihydro-3-thioxo-1,2,4-benzotriazine condenses with prop-2-ynyl bromide to afford the corresponding 3-(prop-2-ynylsulfanyl)-1,2,4-benzotriazine. Palladium catalysed cyclisation yields 3-methylidene-2,3-dihydrothiazolo[2,3-*c*][1,2,4]benzotriazine (M.M. Heravi *et al.*, Monatsh. Chem., 1997, 128, 1143). The Diels-

Alder reaction of a range of substituted 1,2,4-triazines with cyclic vinyl ethers gives a new route to substituted pyridines with hydroxyalkyl and oxoalkyl side chains. The unstable cycloadduct (19) is derived from addition of the vinyl ether dienophile across the *C*-3 and *C*-6 positions of the 1,2,4-triazine followed by the elimination of dinitrogen. The opening of the furan or pyran ring generates the substituted pyridine. The reactivity is enhanced by electron withdrawing groups on the triazine (A.M. d'A. Rocha Gonsalves, T. Pinho e Melo, T.L. Gilchrist, Tetrahedron, 1993, 49, 5277). 1,2,4-Triazines are utilized in the

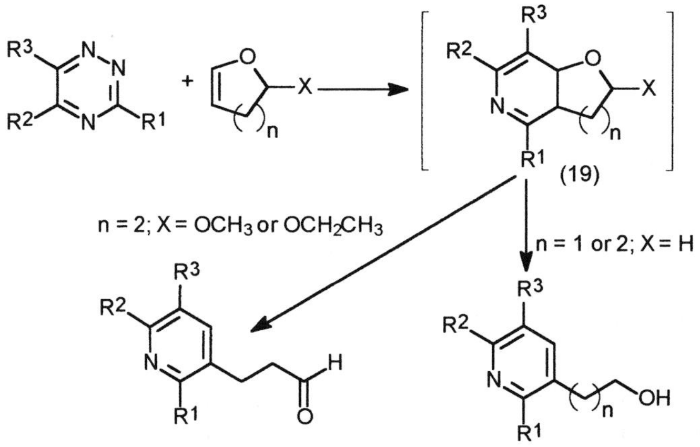

(19)

n = 2; X = OCH_3 or OCH_2CH_3

n = 1 or 2; X = H

synthesis of pyridines by the [4+2] cycloaddition with enamines (G. Seitz, J. Siegl, Z. Naturforschung, 1997, 52b, 851; H. Neunhoeffer *et al.*, Heterocycles, 1993, 35, 1089; A. Rykowski *et al.*, J. heterocyclic Chem., 1996, 33, 1567). The reaction of 5,6-bis(4-methoxyphenyl)-3-(4-pyridyl)-1,2,4-triazine with bicyclo[2.2.1]hepta-2,5-diene in xylene under reflux gives 2,3-bis(4-methoxyphenyl)-6-(4-pyridyl)pyridine in 94% yield (A. Tanaka, Y. Motoyama, H. Takasugi, Chem. pharm. Bull., 1994, 42, 1828). The key step in the synthesis of mappicine ketone alkaloid analogues (21) is the intramolecular Diels-Alder cycloaddition of *N*-2-propenyl-5,6-dimethyl-1,2,4-triazine-3-carboxamide (20) (I. Pendrak *et al.*, J. org. Chem., 1994, 59, 2623).

Derivatives of the β-carboline alkaloid class are prepared by the intramolecular cycloaddition of indole with 1,2,4-triazines (W.-H. Fan, M. Parikh, J.K. Snyder, Tetrahedron Letters, 1995, 36, 6591; J.-H. Li, J.K. Snyder, *ibid.*, 1994, 35, 1485). The first non-synchronous reaction of 3-ethoxycarbonyl-1,2,4-triazine (1) (R^1 = CO_2Et, R^2, R^3 = H) and an enamine dipolarophile has been demonstrated (J.E. Macor, W. Kuipers, R.J. Lachicotte, Chem. Comm., 1998, 983). The products from the 1,3-dipolar cycloaddition reaction of 1,2,4-triazine 1-oxides (22) with benzyne are determined by the *C*(6) substituent on the triazine ring.

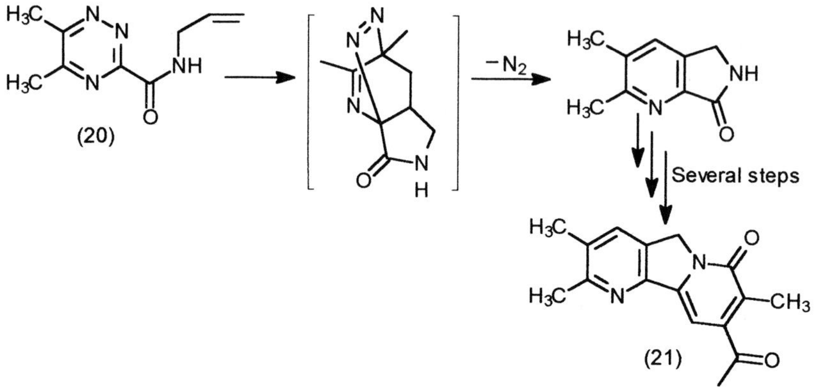

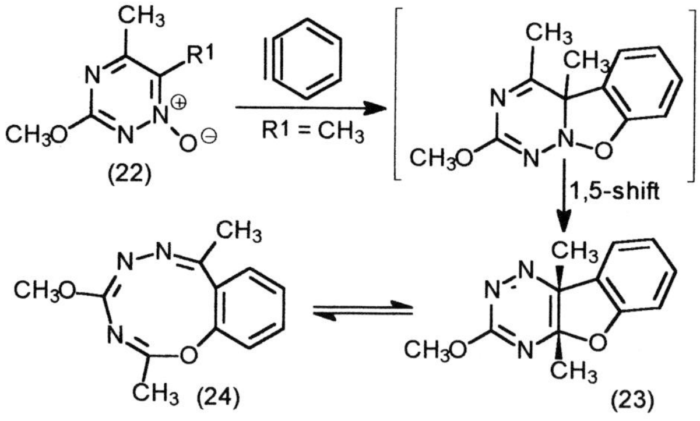

If the $C(6)$ position is unsubstituted (e.g. 22, R^1 = H) the reaction yields 1,3-benzoxazepines and 6-(*o*-hydroxyphenyl)-1,2,4-triazines *via* an unstable 1,3-dipolar cycloadduct. If $C(6)$ is substituted (e.g. 22, R^1 = CH_3) the initial unstable adduct undergoes a 1,5-shift to give a benzofuranotriazine (23). Heating of (23) in toluene promotes a thermal equilibrium with the ring opened 2,7-dimethyl-4-methoxy-1,3,5,6-benzoxatriazonine (24) (N. Kakusawa *et al.*, Heterocycles, 1996, 43, 2091). The [3+2] cycloaddition of the singlet vinylcarbene / three carbon 1,3-dipole (generated from the thermal ring opening of cyclopropenone ketal) with 1,2,4-triazines gives rise to the pyrrolo[1,2-*d*][1,2,4]triazines (25). When R^2 = H, the cycloadduct is unstable and undergoes an *in situ* 1,4- elimination of methanol to yield the aromatic

R1 R2 N N N R3 + OCH3 OCH3 50° H3CO OCH3 H3CO OCH3

R1 R2 N N N R3 OCH3 OCH3 R2 = H -CH3OH R1 N N N R3 OCH3

(25)

R1 = CO_2Et, R2 = CF_3, R3 = SCH_3 (27%)

R1 = CO_2Et, R2 = CF_3, R3 = C_6H_5 (35%)

R1 = H, R2 = CF_3, R3 = SCH_3 (34%)

R1 = CF_3, R2 = CF_3, R3 = SCH_3 (28%)

(26)

R1 = CF_3, R3 = SCH_3 (20%)

R1 = C_6H_5, R3 = C_6H_5 (51%)

R1 = H, R3 = SCH_3 (21%)

R1 = C_6H_5, R3 = SCH_3 (17%)

pyrrdotriazines (26) (G. Frenzen, M. Rischke, G. Seitz, Ber., 1993, 126, 2317). 5,6-Diamino-3-thiomethyl-1,2,4-triazines (1) (R^1 = SCH_3, R^2, R^3 = NH_2) react with phenylglyoxal or 2-bromoacetophenone in methanol to form dihydropyrazino[2,3-*e*][1,2,4]triazines (C.-C. Tzeng *et al.*, Heterocycles, 1996, 43, 1007; K.-H. Lee *et al.*, *ibid.*, 1993, 36,

2577). Novel camphor-1,2,4-triazines fused with a variety of heterocycles have been synthesised and evaluated as central nervous system stimulants (S. Nagai *et al.*, J. heterocyclic Chem., 1998, 35, 293; S. Nagai *et al.*, Heterocycles, 1997, 44, 117). The copper(II) and ruthenium(II) complexes of 5,6-diphenyl-3-(2-pyridyl)-1,2,4-triazine and their spectroscopic data have been reported (R. Uma, M. Palaniandavar, R.J. Butcher, J. chem. Soc. Dalton, 1996, 2061; T.E. Keyes *et al.*, *ibid.*, 1995, 2705).

(c) *1,3,5-Triazines (s-Triazines)*

1,3,5-Triazines have been reviewed by D. Bartholomew, (Compreh. Hetero. Chem. II, 1996, 6, 575). V.V. Dovlatyan (Chem. heterocyl. Comp. 1996, 32, 375) has reviewed the synthesis and rearrangement of chloroalkyl-substituted 1,3,5-triazines.

(i) *Synthesis and Structure*

As a result of their varying industrial applications, 1,3,5-triazines remains the most prevalent of the triazine series. The vibrational / rotation spectra of the parent 1,3,5-triazine and its halogenated

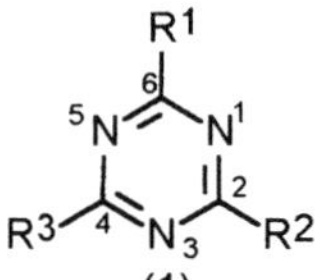

R1
6
5N N1
2
R3 4 N3 R2
(1)

derivatives have been further studied (A. Navarro *et al.*, J. mol. Struct., 1996, 376, 353; W. Bodenmuller, A. Ruoff, J. mol. Spectr., 1995, 173, 205; R. Tripolt, E. Nachbaur, Monatsh. Chem., 1998, 129, 139). Symmetrised quantum-mechanical force-fields and inelastic neutron scattering spectra have been reported for *s*-triazine and trichloro-*s*-triazine (G.J. Kearley *et al.*, Chem. Phy., 1997, 216, 323). The theoretical prediction for 2,4,6-trinitro-1,3,5-triazine (1) (R^1, R^2, R^3 =

NO_2) shows it to have sufficient energy and density to be of interest as a high-energy molecule, but its synthesis has yet to be achieved (A.A. Korkin, R.J. Bartlett, J. Am. chem. Soc., 1996, 118, 12244). The applications of *s*-triazines vary from agricultural / medicinal to light-stabilisers and chiral synthesis. The established routes to the ring system which are summarised in the previous editions of this chapter continue as the main synthetic routes. Cyanuric chloride (1) (R^1, R^2, R^3 = Cl) made by the trimerisation of cyanogen chloride (Cl-CN) is the principal starting point for substituted *s*-triazine synthesis. The immense quantity of industrially produced *s*-triazines is demonstrated by the fact that *Atrazine* (1) (R^1 = Cl, R^2 = $NHCH(CH_3)_2$, R^3 = NHC_2H_5) is one of the most widely utilised herbicide in the United States (R.N. Lerch, W.W. Donald, J. agric. Food Chem., 1994, 42, 922). Environmental concerns about triazine herbicides have lead to a study of their degradation processes and their ability to contaminate groundwater (Ph. Schmitt *et al.*, Anal. Chem., 1997, 69, 2559; Ph. Schmitt *et al.*, J. Chromatogr., 1995, 709, 215; C.D. Adams, S.J. Randtke, J. environ. Sci. Technol., 1992, 26, 2218). Successive replacement of chlorine in cyanuric chloride by a nucleophile is possible because the electron deficient nature of the ring is being progressively reduced, thereby making the next nucleophilic displacement more difficult. A robust industrial synthesis of 2-chloro-4,6-dimethoxy-1,3,5-triazine (1) (R^1 = Cl, R^2, R^3 = OCH_3) from cyanuric chloride is described (J.S. Cronin *et al.*, Syn. Comm., 1996, 26, 3491; H.-W. Lee *et al.*, *ibid.*, 1998, 28, 1339). This triazine is utilised as a chiral carboxylate activating agent in peptide synthesis (P.A. Hipskind *et al.*, J. org. Chem., 1995, 60, 7033; R. Menicagli, C. Malanga, P. Peluso, Syn. Comm., 1994, 24, 2153). A procedure for the sequential, selective derivatisation of cyanuric chloride that allows for the incorporation of carbohydrates and peptides has been elucidated. The authors claim that a combinatorial array of over 40,000 individual derivatives in 50 μmol quantities has been synthesised using automated parallel solution phase synthesis (G.R. Gustafson *et al.*, Tetrahedron, 1998, 54, 4051). A library of 1,920 1,3-dialkylamino-5-aryloxytriazines is obtained by the parallel solution phase, nucleophilic aromatic substitution of cyanuric chloride. The purification of the end products is achieved by solid support liquid extraction (C.R. Johnson *et al.*, Tetrahedron, 1998, 54, 4097). Tri- and di-amino substituted 1,3,5-triazines which are synthesised from cyanuric chloride have significant aromatase-inhibitory activity (T. Matsuno *et al.*, Chem. pharm. Bull., 1997, 45, 291) and cholesteryl ester transfer protein inhibitory activity (Y. Xia *et al.*, Bioorg. Med. Chem. Letters, 1996, 6, 919). A general two-step synthesis of trisubstituted 1,3,5-triazines that does not involve substitution of cyanuric chloride has been developed. *N'*-Acyl-*N,N*-dimethylamidines (3) are prepared in quantitative yields by heating amides (2) with dimethylformamide dimethyl acetal. Subsequent condensation with amidines gives the 1,3,5-triazines (4) (Table 1) (C. Chen, R. Dagnino Jr., J.R. McCarthy, J. org. Chem., 1995, 60, 8428).

Table 1: 1,3,5-Triazines (4).

R^1	R^2	R^3	m.p.	Yield
CH_3	*p*-$CH_3OC_6H_4$	$CH_2CH_2C_6H_5$	oil	81
CH_3	*p*-$CH_3OC_6H_4$	C_6H_5	142 - 144	76
H	$N(CH_2)_4O$	*p*-BrC_6H_4	160 - 161	80
H	C_6H_5	*m*-$(Cl)_2C_6H_4$	105 - 106	87
H	$(CH_3)_2N$	*m*-$(Cl)_2C_6H_4$	71	67

The reaction of benzonitrile with metallic lithium in THF results in the formation of the 2,4,6-triphenyl-1,3,5-triazine radical anion which on treatment with water gives the triazine, in a 37% yield (A.S. Kurbatova, Y.V. Kurbatov, Russ. J. org. Chem., 1997, 33, 1050). Triaryl-*s*-triazines are the products from the ring expansion of triarylimidazoles (C.S. Panda, B. Patro, A.K. Padhy, J. Indian Chem. Soc., 1996, 73, 283). *S,S'*-Dimethyl *N*-cyanocarbonimidodithioate reacts with amides or thioamides (R^1 = alkyl or aryl) in the presence of sodium hydride to yield the corresponding *N*-acyl-*N'*-carbamoyl-*S*-methylisothioureas (5).

CH_3S CH_3S N CN R1 X NH_2 NaH X H N SCH_3 R1 N NH_2 (5) O Δ SCH_3 N N R1 N X H (6)

$POCl_3$ $C_6H_5N(C_2H_5)_2$ X = O X = S MeI / NaOH

SCH_3 N N R1 N Cl $R2R3NH$ SCH_3 N N R1 N $NR2R3$ $R2R3NH$ SCH_3 N N R1 N SCH_3

Thermal cyclisation of these products affords the 4-methylthio-1,3,5-triazin-2(1*H*)-ones (6) which are useful intermediates for the synthesis of a range of trisubstituted 1,3,5-triazines (S. Kohra, K. Ueda, Y. Tominaga, Heterocycles, 1996, 43, 839). A new cyclic transformation of the sodium salts of 1,3,4-oxadiazol-2(3*H*)-ones (7) with two

H N–N R1 O O (7) (i) NaH, DMF (ii) R2—N=C=O O R2 N N R2 O N O H N O (8) R1 10% HCl O R2 N N R2 O N O NH_2 (9)

equivalents of either aryl or alkyl isocyanates affords the 1,3,5-triazine-2,4,6-trione derivatives (8) in yields of 15-74%. Acidic hydrolysis of (8) provides the 1-amino-1,3,5-triazine-2,4,6-trione derivatives (9) (F.

Chau, J.-C. Malanda, R. Milcent, J. heterocyclic Chem., 1997, 34, 1603). A refinement in the synthesis of (9) is achieved by the treatment of the ethoxycarbonylhydrazones of aromatic aldehydes or ketones with either aryl or alkyl isocyanates (F. Chau, J.-C. Malanda, R. Milcent, J. heterocyclic Chem., 1998, 35, 261). 1,3,5-Triazine-2,6-dione and 1,3,5-triazine-2-one-6-thione are available from the reaction of isocyanates with *N*-trimethylsilylimines (J. Barluenga, C. del Pozo, B. Olano, Synthesis, 1995, 1529). 1-Chloro-1-aryl-2,2,2-trifluoroethylheterocumulenes (10) (isocyanates X = O; carbodiimides X = NPh) undergo regioselective cyclisation with diphenylguanidine to give 2-aryl-2-trifluoromethyl-4-oxo-6-imino- and 2-aryl-2-trifluoromethyl-4,6-diiminoperhydro-1,3,5-triazines (12). A proposed mechanism for the reaction is that it proceeds through formation of the alkylation product of the more nucleophilic imine nitrogen of diphenylguanidine giving rise to the intermediate (11). This undergoes

CF_3-C(R1)(Cl)-N=C=X (10) —[Ph–NH–C(=NH)–NHPh]→ [PhNH–C(NHPh)=N–C(R1)(CF_3)–N=C=X] (11) → (12)

R1 = Ph, X = O; (65%)
R1 = *p*-$CH_3C_6H_4$, X = O; (69%)
R1 = *p*-$CH_3OC_6H_4$, X = O; (69%)
R1 = *p*-ClC_6H_4, X = O; (68%)
R1 = Ph, X = NPh; (63%)
R1 = *p*-$CH_3C_6H_4$, X = NPh; (58%)

a rapid *in situ* cyclisation by attack of the imino group at the electrophilic heterocumulene moiety (M.V. Vovk, I.F. Tsymbal, Chem. heterocyl. Comp., 1997, 33, 614). The reaction of two equivalence of arylamidoxime with ethyl orthoacetate in the presence of an acid catalyst gives substituted 6-aryl-1,3,5-triazine 1-oxides *via* a Beckmann type rearrangement (G. Kaugars, W. Watt, J. heterocyclic Chem., 1993, 30, 497).

(ii) *Reactions of 1,3,5-Triazines*

The 1,3,5-triazine ring is an electron deficient system. An *ab initio* study of 1,3,5-triazines calculated with the 6-31G**, 6-311++G** and cc-pVTZ basis sets shows that the lowest energy decomposition mechanism is the concerted triple dissociation (*s*-triazine → 3 HCN) (S.V. Pai, C.F. Chabalowski, B.M. Rice, J. phys. Chem., 1996, 100, 5681; *ibid.*, 1996, 100, 15368). It has been proposed that the photo-dissociation of 1,3,5-triazine is not a symmetric triple dissociation but rather a two-step concerted process, where concerted refers to the correlation between the asymptotic velocity vectors of the three ejected HCN fragments (T. Gejo, J.A. Harrison, J.R. Huber, J. phys. Chem., 1996, 100, 13941). A significant feature of *s*-triazine chemistry is facile acidic hydrolysis. Kinetic measurements of the acidic hydrolysis of unsubstituted 1,3,5-triazine show that the rate determining step is ring opening (M. Privman, P. Zumam, Heterocycles, 1994, 37, 1637). A comprehensive study of the scope of the amidine Diels-Alder reaction with 1,3,5-triazines is described (D.L. Boger, M.J. Kochanny, J. org. Chem., 1994, 59, 4950). The reaction of amidines with the

EtO_2C N CO_2Et N N CO_2Et (13) + R NH.HCl NH_2 ⇅ NH.HCl R NH_2 → [CO_2Et N N H_2N CO_2Et R N H_2N CO_2Et] (14) ↓ H_2N N CO_2Et N R (15) CO_2Et

symmetrical 1,3,5-triazine (13) proceeds with an *in situ* amidine to 1,1-diaminoethene tautomerisation, [4+2] cycloaddition with the 1,3,5-triazine yielding (14), loss of ammonia from the initial Diels-Alder adduct with imine generation, imine to enamine tautomerisation and retero Diels-Alder loss of ethyl cyanoformate to provide the substituted 4-aminopyrimidines (15). This reaction has been utilised as a key step in the total synthesis of bleomycin A_2, and (+)-P-3A (D.L. Boger, T.J. Honda, J. Amer. chem. Soc., 1994, 116, 5647; *ibid.*, 1994, 116, 5619; *ibid.*, 1994, 116, 82). Utilising 5-aminopyrazoles as dienophiles in a similar reaction yields the pyrazolopyrimidines (16) in yields of 30-70% (Q. Dang, B.S. Brown, M.D. Erion, J. org. Chem., 1996, 61, 5204). The reaction of unsubstituted 1,3,5-triazine (1) (R^1, R^2, R^3 = H)

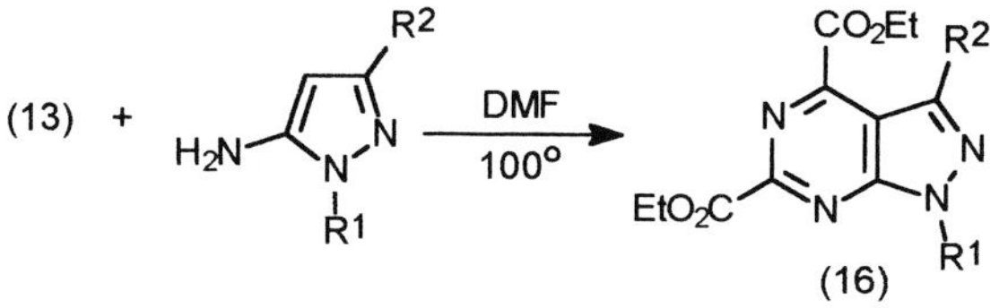

(13) + [5-amino-1-R1-3-R2-pyrazole] → DMF, 100° → (16)

with organolithium compounds yield stable mono-substituted 1,4-dihydrotriazines (W.M. Boesveld, P.B. Hitchcock, M.F. Lappert, Chem. Comm., 1997, 2091). An efficient triple addition of allyl groups to unsubstituted 1,3,5-triazines is achieved with allyltributyltin to give 2,4,6-triallyl-1,3,5-triazacyclohexanes (R. Yamaguchi, S. Kobayashi, S. Kozima, Chem. Letters, 1994, 2349).

(2) *Tetrazines*

(a) *1,2,3,4- and 1,2,3,5-Tetrazines*

(i) *Synthesis and Structure*

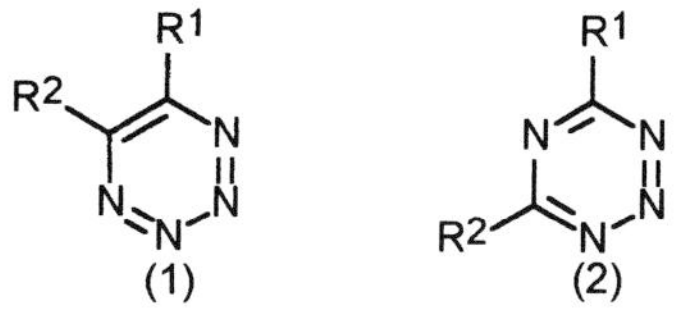

(1) (2)

The 1,2,3,4- and 1,2,3,5-tetrazine systems are uncommon and difficult to synthesise although interesting derivatives of them have been reported. The unsubstituted systems have been the subject of extensive theoretical calculations. *Ab initio* theoretical methods have been utilised to determine the equilibrium geometry of the unsubstituted 1,2,3,4- and 1,2,3,5-tetrazines. Relative energy differences show that the 1,2,3,5-tetrazine is the lowest energy isomer, with 1,2,4,5-tetrazine 8.0 kcal mol^{-1} higher and 1,2,3,4-tetrazine 22.2 kcal mol^{-1} above the 1,2,3,5-tetrazine (J.R. Thomas, G.E. Quelch, H.F. Schaefer III, J. org. Chem., 1991, 56, 539). The synthesis, structure and reactivity of cyclopenteno-1,2,3,4-tetrazines show that the 1,2,3,4-tetrazine unit embedded in a 10-π system has unexpectedly high stability (P.J. Mackert *et al.*, Ber., 1994, 127, 1479). These systems are generated by the reaction of diazocyclopentadienes (3) with arenediazonium salts. They exist in solution equilibrium with aryl-azodiazocyclopentadienes (4), the nature of which depends on the substituents (Table 1). Single

Table 1: Solution equibrium ratios of (4) ⇆ (5)

R^1	R^2	Ratio 4:5	Yield (4), (5)
H	C_6H_5	3 : 97	21
H	$4\text{-}NO_2C_6H_4$	13 : 87	74
H	$2\text{-}CH_3\text{-}4\text{-}NO_2C_6H_3$	94 : 6	38
$C(CH_3)_3$	$4\text{-}NO_2C_6H_4$	5 : 95	35
$C(CH_3)_3$	$2\text{-}CH_3\text{-}4\text{-}NO_2C_6H_3$	96 : 4	32
$C(CH_3)_3$	CH_3	0 : 100	6

crystal X-ray analysis of (5) (R^1 = $C(CH_3)_3$, R^2 = CH_3), demonstrates the overall planar and aromatic nature of the bicyclic system. These compounds undergo electrophilic substitution reactions such as bromination, formylation and trifluoroacetylation of the cyclopentadiene ring. Photolysis of the solution equilibrium mixtures is dominated by the reactivity of the diazo group of the ring opened isomer (4), which undergoes a loss of dinitrogen and rearrangement to generate ketenimines. 5,7-Dibromobenzo-1,2,3,4-tetrazine 1-oxide is obtained by the intramolecular reaction of *o*-(*t*-butylazoxy)phenyl-diazonium tetrafluoroborate (A.M. Churakov *et al.*, Mendeleev. Comm., 1994, 122).

(b) *1,2,4,5-Tetrazines (s-Tetrazines)*

1,2,4,5-Tetrazines have been reviewed by J. Sauer (Compreh. Hetero. Chem. II, 1996, 6, 901).

(i) *Synthesis and Structure*

The 1,2,4,5-tetrazine isomer continues to be the most widely studied of the tetrazine class. The unsubstituted 1,2,4,5-tetrazine (1) (R^1, R^2 = H) (m.p. 99°) is a planar ring with bond lengths of 1.334 Å (*C-N*) and 1.321 Å (*N-N*). The angle at *C* is 127° 22' and at *N* 115° 57'. The ^{1}H-nmr chemical shift is 11.05 ppm and the carbon shift is 161.2 ppm in deuteriated acetone. The hydrogen bonding and solvent polarity effects

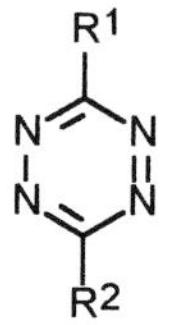

R1

N N

N N

R2

(1)

on the ^{14}N-nmr shielding of 1,2,4,5-tetrazines is discussed (M. Witanowski *et al.*, J. mag. Reson., 1997, 124, 127). The molecular and electronic states of the parent *s*-tetrazine have been reinvestigated by vacuum ultraviolet and near-threshold energy loss spectra, which has led to the identification of several new excited states (M.H. Palmer *et al.*, Chem. Phy., 1997, 214, 191). The electronic states of a series of symmetrically disubstituted *s*-tetrazines have been studied by linear dichroism spectroscopy in the UV-visible region, magnetic circular dichroism spectroscopy with theoretical calculations of structures and spectra (J. Waluk, J. Spanget-Larsen, E.W. Thulstrup, Chem. Phy., 1995, 200, 201). The general methods utilised for the synthesis of symmetrically substituted *s*-tetrazines are described in the previous two editions of this chapter. The synthesis of unsymmetrically substituted 1,2,4,5-tetrazines in reasonable yields continues to be problematic. A new route to such systems involves the reaction of the *tert*-alkyl substituted iminium chlorides (2) with *S*-methylisothiocarbonohydrazide iodide (3) in ethanol / triethyl amine, followed by an *in situ* oxidation with acetic acid / sodium nitrite. The yields obtained (20-30%) are an improvement on other procedures (Table 1) (S.C. Fields *et al.*, J. org. Chem., 1994, 59, 8284). These compounds can be readily

Table 1: 3-Alkyl-6-methylthio-1,2,4,5-tetrazines (4).

R^1	R^2	R^3	Yield	m.p.
CH_3	CH_3	CH_3	27	60-61
CH_3	$R^2 = R^3$	$(CH_2)_2$	30	oil
CH_3	n-C_3H_7	n-C_3H_7	22	oil
CH_3	C_2H_5	n-C_4H_9	19	oil

converted to 3-alkyl-6-aminotetrazines (5) in 90% yields, by their reaction with ammonium hydroxide in ethanol. 3-Aryl-6-cyano-*s*-tetrazines are generated by the introduction of the cyano group after the formation of the tetrazine ring (C.M. Hudson *et al.*, Liquid Crystals, 1995, 19, 871). The synthesis of 1-(tetrazol-5-yl)-4-aryl-1,4-dihydro-

1,2,4,5-tetrazines (6) is achieved in yields of 57-61%. An X-ray structure shows the 1,4-dihydrotetrazine ring to be boat shaped as expected for an 8π-six-atom ring (R.N. Butler *et al.*, J. chem. Res. (S), 1995, 224). An X-ray structure of 3,6-di(2-pyridyl)-1,4-dihydro-1,2,4,5-tetrazine also shows a boat structure for the tetrazine ring (L.-S. Liou *et al.*, Acta Cryst., 1996, C52, 1841). Hydrazones of aliphatic aldehydes and ketones react with nitrile imine 1,3-dipoles to yield acyclic addition products which on oxidation with palladium-carbon cyclise to the substituted 1,6-dihydro-1,2,4,5-tetrazines (7) (A.Q. Hussein, J. chem. Res., (S), 1996, 174).

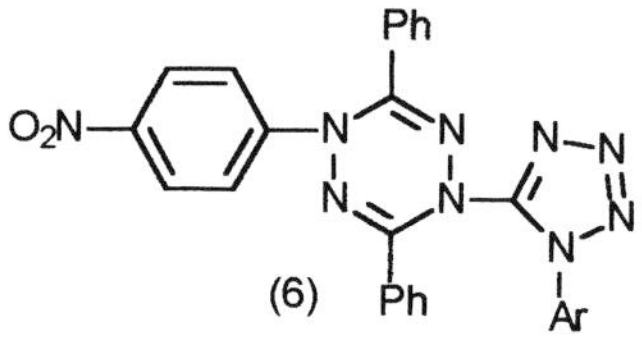

(6)

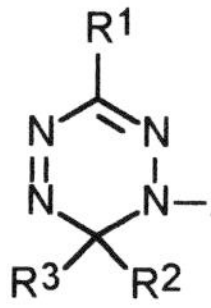

(7)

(ii) *Reactions of 1,2,4,5-Tetrazines*

The 1,2,4,5-tetrazines stands as the most efficient and widely utilised 4π-electron deficient azadiene system for inverse electron demand Diels-Alder reactions with electron-rich dienophiles. The tetrazines have facilitated the preparation of heterocyclic systems not readily attainable from other methods. The regiochemistry of the cycloaddition reaction is investigated with the unsymmetrical 3-methoxy-6-methylthio-1,2,4,5-tetrazine (8). The [4+2] cycloaddition of (8) with a dienophile gives an initial unstable adduct (9) which eliminates dinitrogen, followed by loss of HX with aromatisation to generate the diazines (10) or (11). The cycloaddition of (8) with electron-rich 1,1-disubstituted enamines and enol ethers provides regiospecific cycloadducts of type (10) only. Neutral and conjugated mono substituted alkynes undergo [4+2] cycloaddition to provide both regioisomeric isomers (10) and (11), as a 1:1 mixture (S.M. Sakya, K.K. Groskopf, D.L. Boger, Tetrahedron Letters, 1997, 38, 3805). The kinetics of the Diels-Alder reaction of di(2-pyridyl)-1,2,4,5-tetrazine

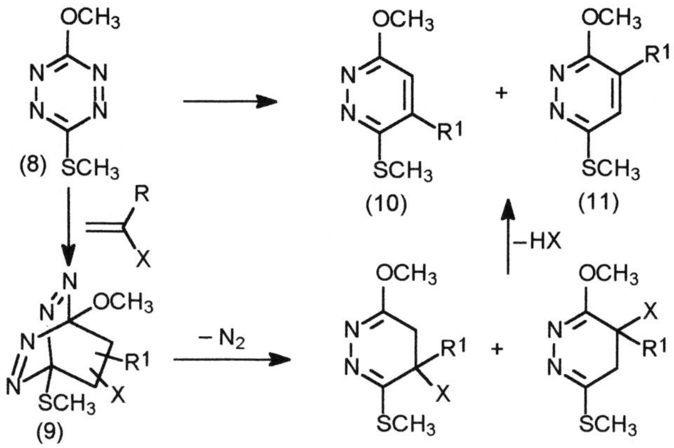

with substituted styrenes in aqueous media was examined. The second-order rate constants for the reaction increase dramatically in water-rich media. The solvent effects seen are very similar to those for HOMO controlled Diels-Alder reactions (J.W. Wijnen *et al.*, J. org. Chem., 1996, 61, 2001). The utilisation of this reaction procedure in solid phase synthesis has been accomplished. The polystyrene resin immobilised 3,6-disubstituted tetrazines (12) requires a four-step procedure to generate, with a loading efficiency of 90%. Reaction with

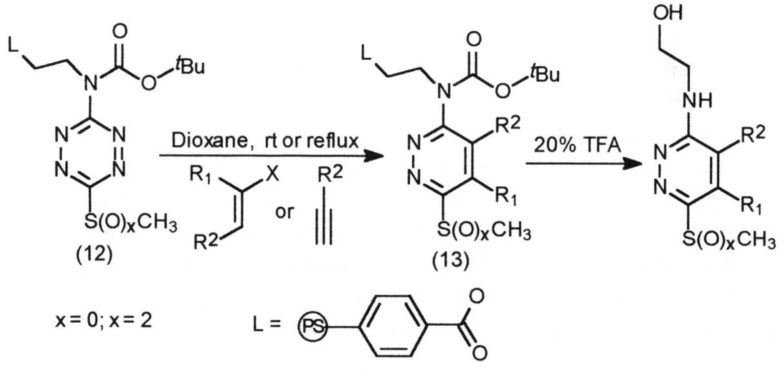

a variety of olefins and alkynes gives the diazines (13), which can be cleaved from the resin with 20% trifluoroacetic acid (TFA) (J.S. Panek, B. Zhu, Tetrahedron Letters, 1996, 37, 8151). The reactions are regioselective with the reactivity of (12) (x = 0) being less than that of (12) (x = 2) due to the strong electron withdrawing effect of the sulfone. Cycloaddition with substituted 1,5-benzothiazepine and 1,4-benzothiazines yield the fused pyridazino[1,5]benzothiazepine (14) (n = 1) and pyridazino[1,4]benzothiazine (14) (n = 0) respectively (S. Pippich, H. Bartsch, Heterocycles, 1996, 43, 1967). Intramolecular cycloadditions between tethered *s*-tetrazines and indoles are utilised to produce the complex fused compounds (15) (S.C. Benson, L. Lee, J.K. Snyder, Tetrahedron Letters, 1996, 37, 5061). Further cycloaddition reactions of *s*-tetrazines include reaction with azaazulenes (G. Frenzen *et al.*, Ber., 1993, 126, 441) and pseudoazulenes (U. Reimers, G. Seitz, Ber., 1993, 126, 2143). The products from the reaction with the

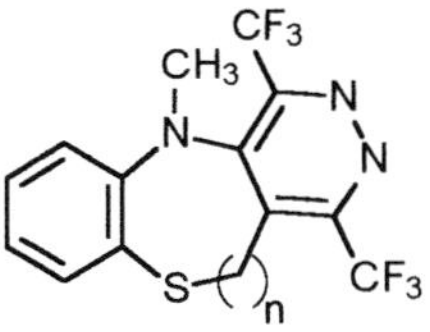

(14) 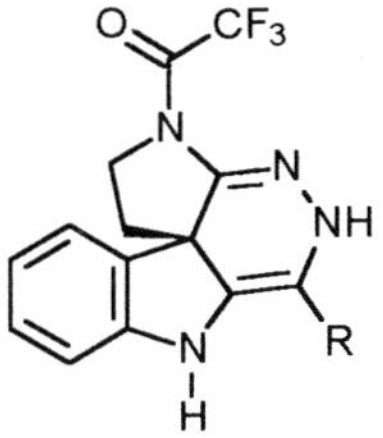 (15) 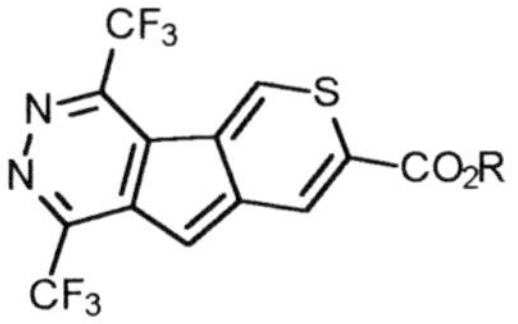 (16)

pseudoazulene, cyclopenta[*c*]thiopyrane are the thiopyranocyclopenta-[1,2-*d*]pyridazines (16). The versatility of this reaction is demonstrated by the wide range of unusual dienophiles utilised, 2,3-dioxabicyclo[2.2.2]oct-5-ene (M. Balci *et al.*, Tetrahedron Letters, 1996, 37, 921), 1-methyl-2-methylthio-2-pyrroline (N. Haider *et al.*, Heterocycles, 1994, 38, 1845), [60]fullerene olefins (R.N. Warrener *et al.*, Chem. Comm., 1995, 1417), cyclopropanobenzene (J. Laue *et al.*, Z. Naturforschung, 1996, 51b, 348), hetaryl-substituted dienamines (A. Kotschy *et al.*, J. org., Chem., 1995, 60, 4919), and ethynylboranes (G. Seitz *et al.*, Arch. Pharm., 1994, 327, 673). Cycloadditions with heterodienophiles (C=N) yield 1,2,4-triazines as products (M. Richter, G. Seitz, Arch. Pharm., 1995, 328, 175; G. Seitz, J. Seigal, Z. Naturforschung, 1997, 52b, 851). Polymerisation chemistry *via* this reaction is also possible (C. Glidewell *et al.*, J. chem. Soc. Perkin 2, 1997, 1167; G.A. McLean *et al.*, J. chem. Res., (S), 1996, 448). *s*-Tetrazines undergo (4+1) cycloadditions with nucleophilic singlet carbenes. 3,6-Trifluoromethyl-1,2,4,5-tetrazine reacts with the singlet carbenes (17) to give the unstable adducts (18) which undergo an *in situ* dinitrogen loss to form the tetraazaspiro compounds (19). This is

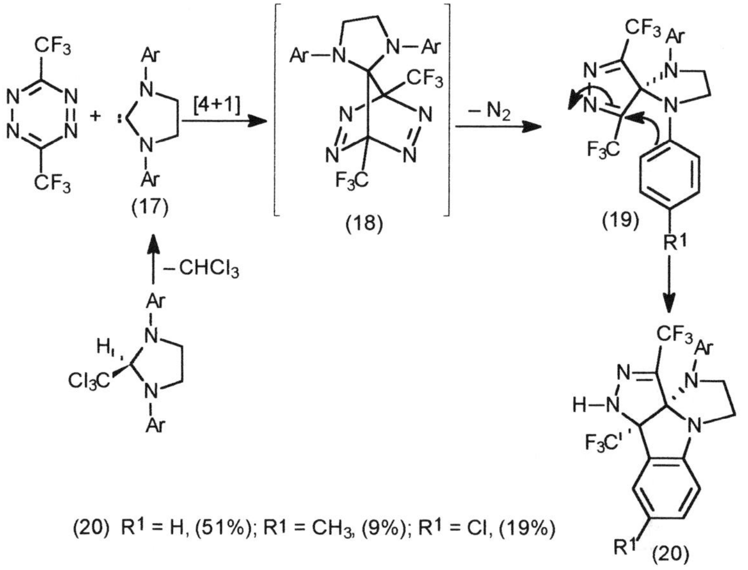

(20) R^1 = H, (51%); R^1 = CH_3, (9%); R^1 = Cl, (19%)

followed by an intramolecular electrophilic aromatic substitution to yield the chiral tetracyclic imidazo[1,2-*a*]pyrazolo[4,3-*b*]indoles (20), with high diastereoselectivity (G. Frenzen *et al.*, Ber., 1994, 127, 1803). The first regioselective oxidation of the *N*-1 atom of 3-aryl-1,2,4,5-tetrazines (21) can be achieved with methyl(trifluoromethyl)dioxirane in high yields. The structure (22) was comfirmed by ^{1}H, ^{13}C and ^{15}N nmr. The proposed mechanism is a S_N2 attack by the nitrogen lone pair

(21) $\xrightarrow[CH_2Cl_2 / 0°]{}$ (22)

Ar = C_6H_5 (68%)
Ar = p-$CH_3C_6H_4$ (90%)
Ar = p-$CH_3OC_6H_4$ (73%)
Ar = p-ClC_6H_4 (74%)

of the tetrazine on the dioxirane peroxide bond (W. Adam, C. van Barneveld, D. Golsch, Tetrahedron, 1996, 52, 2377).

(c) *Verdazyls*

Verdazyls are stable substituted 2,3,4-trihydro-1,2,4,5-tetrazin-1-yl free radicals which are commonly used for the investigation of single electron transfer reactions. The main route to these systems is condensation of formaldehyde with 1,3,5-triarylformazans (23) with air oxidation generating the stable radical. A modification of the procedure is a two-phase reaction to generate the triarylverdazylium salt (24) with reduction to the verdazyl (25) using ascorbic acid (A.R. Katritzky, S.A. Belyakov, Synthesis, 1997, 17; *ibid.*, 1995, 577). This

(23) $\xrightarrow[CHCl_3 / H_2O]{CH_2O / HClO_4}$ (24) $\xrightarrow[\text{ascorbic acid}]{NH_4OH}$ (25)

approach has been adapted for the synthesis of new bisverdazyls (26), from their corresponding bisformazans (A.H.M. Elwahy, A.A. Abbas, Y.A. Ibrahim, J. chem. Res. (S), 1998, 184). The study of stable biradicals with closely spaced conjugated radical centres provides a useful model for studing intramolecular electronic interactions. The esr and magnetic properties of the diradical 1,1',5,5'-tetramethyl-6,6'-dioxobisverdazyl (27) reveal that the molecule has a singlet ground state with a thermally populated triplet lying 0.094 eV above it (D.J.R.

Brook *et al.*, J. phys. Chem., 1996, 100, 2066). The X-ray structure shows that the neutral diradicals can distort geometrically and form strong intermolecular interactions through π-stacking. Copper(I)

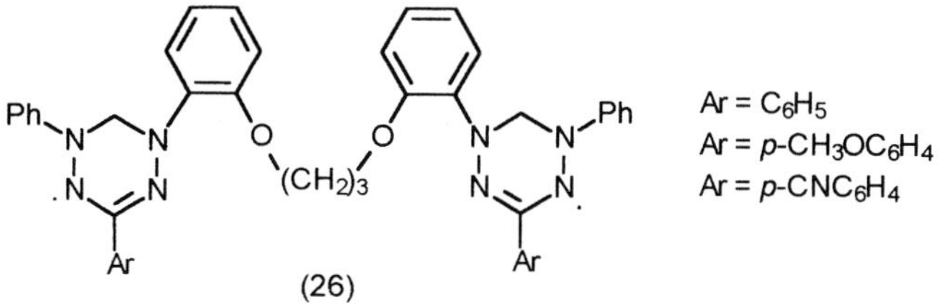

Ar = C_6H_5
Ar = *p*-$CH_3OC_6H_4$
Ar = *p*-CNC_6H_4

(26)

coordination polymers of composition $[Cu_2X_2(27)]_x$ are generated from the reaction of the diradical (27) with copper(II) halides (X = Cl, Br) in methanol (D.J.R. Brook *et al.*, J. Amer. chem. Soc., 1997, 119, 5155).

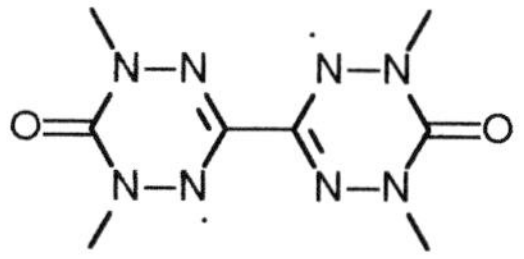

(27)

(3) *Pentazine and Hexazine*

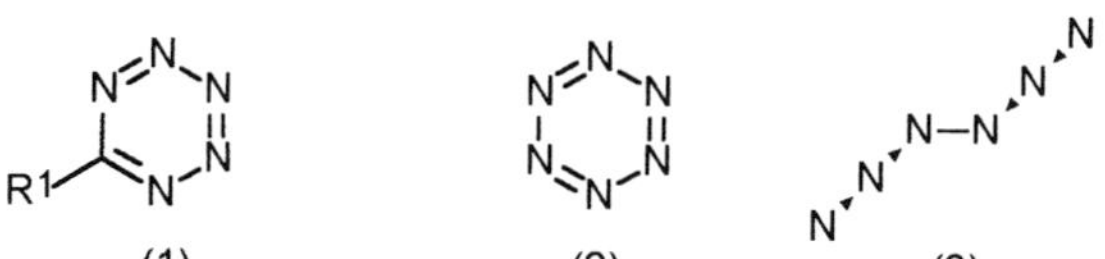

(1) (2) (3)

The pentazine structure (1) is unknown but remains the object of extensive calculations. The sodium salt of 2-(tetrazol-5-yl)-6-imino-1,2,3,4,5-pentazine mesoion (5) is formed by the acid tautomerisation of the disodium salt of azotetrazole (4). The reverse reaction is also possible by treatment of (5) with sodium hydrogencarbonate. The structure was elucidated by IR spectra, ^{15}N isotope labelling and chemical properties (A.G. Mayants *et al.*, Chem. heterocyl. Comp., 1993, 395). The full nitrogen analogue of benzene, hexazine (2), can

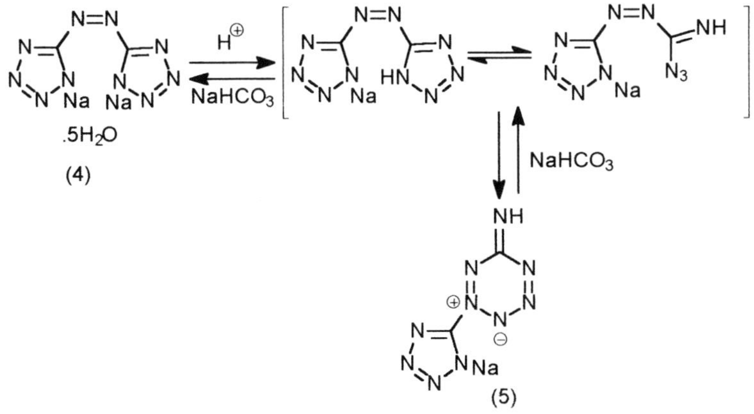

N=N N=N N=N NH H⊕ NaHCO3 .5H2O (4) NaHCO3 N3 NH (5)

be classified as either a heterocycle or an allotrope of nitrogen. It has not been isolated and extensive calculations predict that structure (2) cannot exist because the D6h structure would give two imaginary frequencies for out-of-plane vibrations. Though it is generally agreed that the structure (2) is unfavourable, several open-chained structures have been proposed. A computational study of the all possible structures for N_6 reveals that the most stable is the C_2 open-chain structure (3). However, this structure is calculated to be unstable thermodynamically by 188.3 kcal mol^{-1} relative to the dissociation into three nitrogen molecules (M.N. Glukhovtsev, P. von Ragué Schleyer, Chem. Phy. Letters, 1992, 198, 547; *ibid.*, 1993, 204, 394). The combination of azidyl radical with azide generates the transient hexazine radical anion species ($N_3^{\bullet} + N_3^{-} \leftrightarrows N_6^{\bullet -}$). UV-visible and time-resolved infrared spectroscopy show that the species is characterised by a broad, featureless visible absorption at 700 nm and an IR band at 1842 cm^{-1} (M.S. Workentin *et al.*, J. Chem. Phy., 1995, 99, 94; J. Amer. chem. Soc., 1994, 116, 1141). The most stable structure predicted by *ab initio* calculations is a planar, symmetric dimer consisting of two symmetric N_3 fragments connected by two long bonds to form a rectangle.

(II) *Nitrogen with Sulphur and/or Oxygen*

(1) *Sulphur-Nitrogen Rings*

The synthesis of 2-methylthio-5,6-dihydro-4*H*-1,3,4-thiadiazines is achieved in three steps starting from an aldehyde or a ketone. *S*-Methyl alkylidenehydrazinecarbodithioates (1) are readily obtainable from the reaction of an aldehyde or ketone with methyl dithiocarbazate ($CH_3SCSNHNH_2$). Alkylation at sulphur results in the formation of the alkylthiols (2) which when heated in the presence of a catalytic amount of base yield the 5,6-dihydro-1,3,4-thiadiazines (3) as a mixture of isomers (K. Shimada *et al.*, Bull. Chem. Soc. Jpn., 1996, 69, 1043; Chem. Letters, 1995, 925). The oxidation of the amino(thiocarbonyl) amidines (4) with iodine generates the 1,3,5-thiadiazine ring system (5).

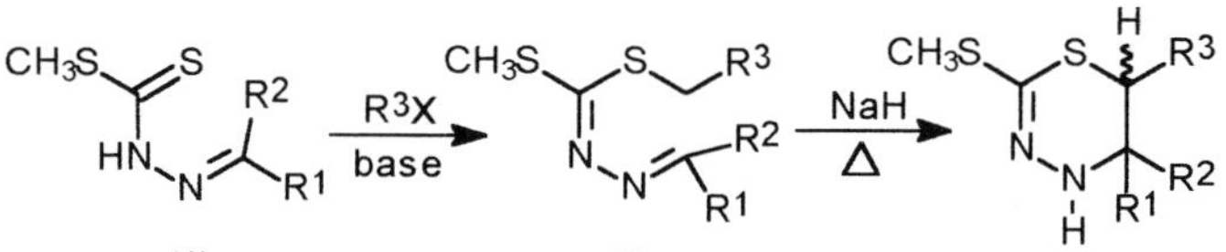

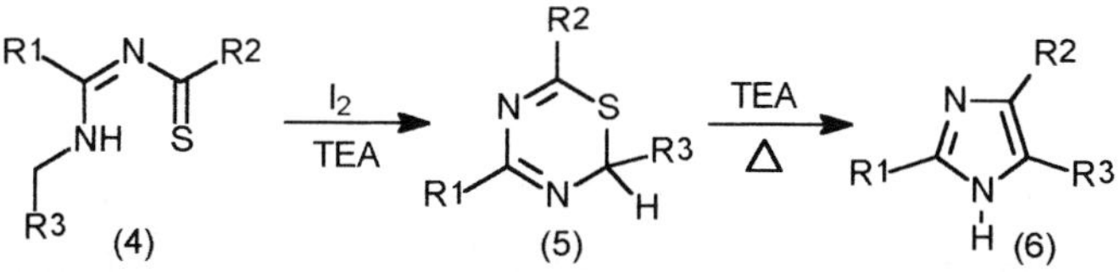

The thiadiazines are readily desulphurised to provide a new route to the imidazoles (6) (A. Rolfs, J. Liebscher., J. org. Chem., 1997, 62, 3480).

A new facile high-yield route to the 1,3,4,5-thiatriazine ring system (8) is the reaction of hydrogen sulfide with 2,4,5-triaryl-1,2,3-triazolium N^1-arylimides (7). The thiatriazine ring can be desulphurised in

H_2S, CH_2Cl_2 ; *m*CPBA or Δ

(7) (8) (9)

(8): R1 = Ph, R2 = Ph, (82%); R1 = Ph, R2 = *p*-BrC_6H_4, (83%)
R1 = Ph, R2 = *p*-$NO_2C_6H_4$, (80%); R1 = Ph, R2 = *p*-ClC_6H_4, (79%)
R1 = Ph, R2 = *p*-$CH_3C_6H_4$, (80%); R1 = *p*-ClC_6H_4, R2 = Ph (77%)

quantitive yield by thermolysis or by oxidation with *m*-chloroperbenzoic acid (*m*CPBA) to produce the 1,2,3-triazoles (9) (R.N. Butler, D.F. O'Shea, J. chem. Res. (S), 1994, 350). The use of 1,2,3-triazolium-N^1-ylides for the synthesis of a range of azimines has been reviewed (R.N. Butler, D.F. O'Shea, Heterocycles, 1994, 37, 571). The 1,4,2-dithiazine ring system is accessible from the ring expansion of 1,3-dithiolium salts (10) with ammonia and iodine. A X-ray crystal structure of the 8π-1,4,2-dithiazine (11) (R^1 = *p*-$CH_3OC_6H_4$) shows that the ring adopts a boat conformation (M.R. Bryce *et al.*, J. chem. Soc. Perkin 1, 1994, 2571; *ibid.*, 1992, 2295; Chem. Comm., 1992, 478).

I_2 / NH_3 aq

(10) (11)

The reaction of 5,6-dimethyl-3-(4-bromophenyl)-1,4,2-dithiazine (11) (R^1 = *p*-BrC_6H_4) with norbornene in *o*-dichlorobenzene, under reflux, affords the dithiine derivative (14), while the reaction with dimethyl acetylenedicarboxylate (DMAD) gives the tetra-substituted thiophene. The suggested mechanism for both reactions is *via* a zwitterionic

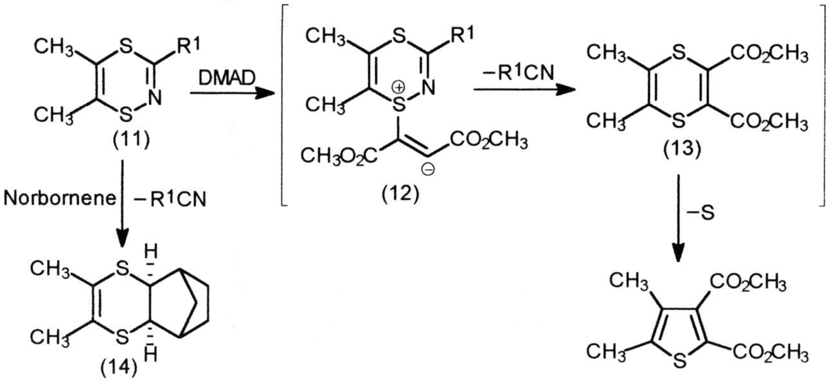

intermediate such as (12) which eliminates a nitrile to generate the dithiine derivatives (13) or (14). In the case of (13) this then undergoes an *in situ* desulphurisation to generate the corresponding thiophene (M.R. Bryce *et al.*, J. chem. Soc. Perkin 1, 1997, 1157). The oxidation of the dithiazine (11) ($R^1 = p$-BrC_6H_4) with peracetic acid results in the formation of 1,4,2-dithiazine 1,1-dioxide (15) in a 23% yield. The structure is confirmed by X-ray analysis which shows the ring to be planar with increased π-delocalisation in comparison with (11). The 1,4,2-diselenazine (16) is generated in a similar fashion to its sulphur

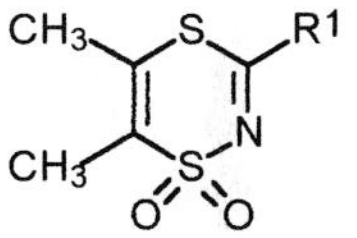

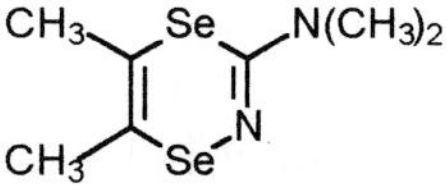

analogues from the ring expansion of the appropriate 1,3-diselenolium salt (S. Yoshida, M.R. Bryce, A. Chesney, Chem. Comm., 1996, 2375). This type of approach is further extended by the synthesis of the

1,4,2,5-dithiadiazines (18) by the ring expansion of 3-aryl-1,4,2-dithiazolium salts (17). The X-ray crystal structure of (18)

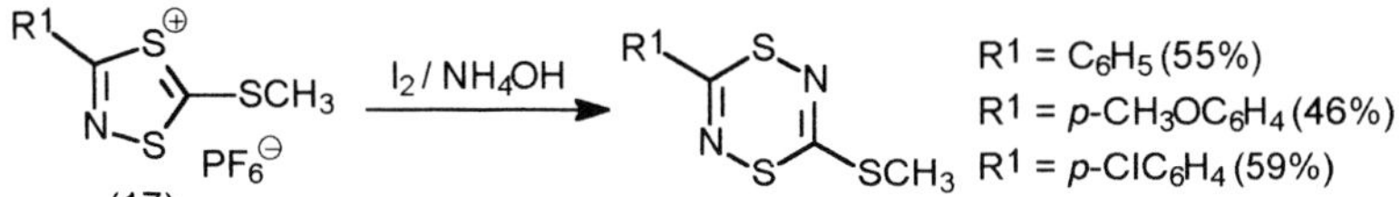

I_2 / NH_4OH

(17) (18)

R1 = C_6H_5 (55%)
R1 = *p*-$CH_3OC_6H_4$ (46%)
R1 = *p*-ClC_6H_4 (59%)

(R^1 = 4-$CH_3OC_6H_4$) shows the dithiadiazine ring to be folded along the *S*(1)•••*S*(4) vector by 48.3° (M.R. Bryce *et al.*, J. chem. Soc. Perkin 1, 1997, 1157). The structure of the alkene adducts of 1,3,2,4,6-dithiatriazine (19) have been investigated by ^{1}H-nmr and X-ray crystal

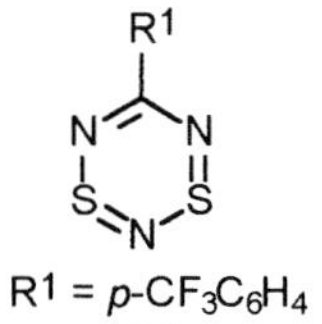

R1 = *p*-$CF_3C_6H_4$
(19)

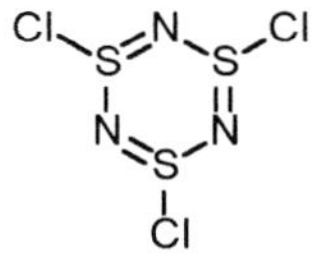

(20)

analysis (R.T. Boeré *et al.*, Can. J. Chem., 1994, 72, 1171). Trithiazyl trichloride (20) is proving to be a useful reagent in the synthesis of sulphur-nitrogen heterocycles. 1,2,5-Thiadiazoles are obtainable in a one step synthesis from the reaction of (20) with alkenes or alkynes (X.-G. Duan *et al.*, J. chem. Soc, Perkin 1, 1997, 2597). Isothiazoles are the products from its reaction with furans (X.-L. Duan, R. Perrins, C.W. Rees, J. chem. Soc. Perkin 1, 1997, 1617). The general utility of trithiazyl trichloride (20) for the synthesis of heterocyclic rings containing *S-N*, *N-S-N* and *S-N-S* units has been reviewed (X.-G. Duan *et al.*, J. heterocyclic Chem., 1996, 33, 1419).

(2) *Oxygen-Nitrogen Rings*

The Diels-Alder [4+2] cycloaddition of aza-1,3-butadienes (1) with aryl sulphonyl nitrosite (2) is utilised in the synthesis of the 1,2,3-

oxadiazines (3). Using this method the 2-arylsulphonyl-3,6-diaryl[1,2,3]-oxadiazines (3) are formed in yields of 30-50% (K.K. Singal, B. Singh, B. Raj, Syn. Comm., 1993, 23, 107). The 2-methyl-3,4-diaryl-1,2,5-oxadiazol-2-ium salts (4) are easily obtained from the treatment of 3,4-diarylfurazan with dimethyl sulfate. The unstable furazan-*N*-methanides (5), which are generated by deprotonation of (4) with base, rapidly ring-open to (6) followed by an *in situ* electrocyclisation to the 3,4-diaryl-6*H*-1,2,5-oxadiazines (7) in high yields (R.N. Butler *et al.*, J. chem. Soc. Perkin 1, 1995, 1083; 1997, 2919; 1997, 1047). Theoretical calculations using the 6-31G basis set

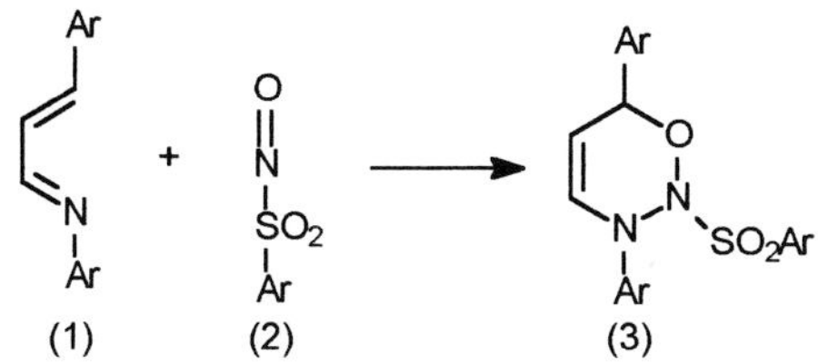

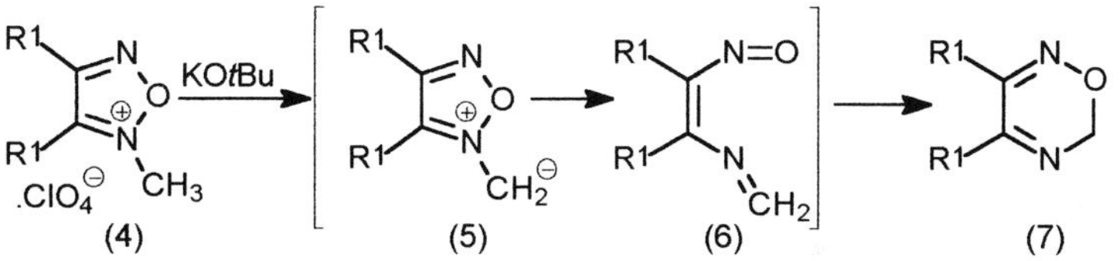

(7): R1 = Ph (91%); R1 = *p*-ClC_6H_4 (97%); R1 = *p*-BrC_6H_4 (86%).

confirms that the species (5) would rapidly ring open and that there is a large gain in stability (-53.6 kcal mol^{-1}) for the transformation of (5)→(6)→(7). Attempts to use the chiral 2-alkenyl-1,3,4-oxadiazines (8) as cyclic 1-aza-1,3-dienes in Diels-Alder reactions failed. The only products isolated were linear addition products to the exocyclic double bond (D. Batty, Y. Langlois, Tetrahedron Letters, 1993, 34, 457). An X-ray crystal structure of the new diprotonated 1,3,5-oxadiazine (9) shows the ring to be planar with interatomic distances that suggest *C-N* double bond character (C.H.L. Kennard *et al.*, Chem. Comm., 1996, 1731). This ring system is synthesised from the cycloaddition of 1,1-dimethylcyanamide ($Me_2NC{\equiv}N$) with acetone in the presence of trifluoromethanesulphonic acid. The oxazolo[4,5-*d*][1,2,3]triazoles

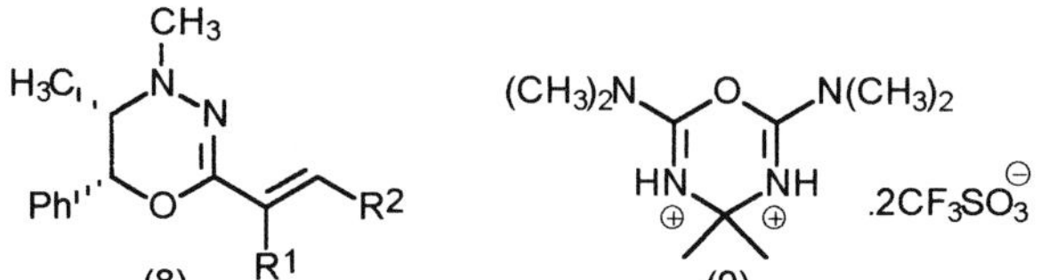

(8) (9) $.2CF_3SO_3^{\ominus}$

(10) undergo an acid catalysed ring expansion to the 2,4,6-triaryl-1,3,4,5-oxatriazines (11) and an imine. Kinetic studies of the reaction indicate the presence of a delocalised carbocation intermediate from

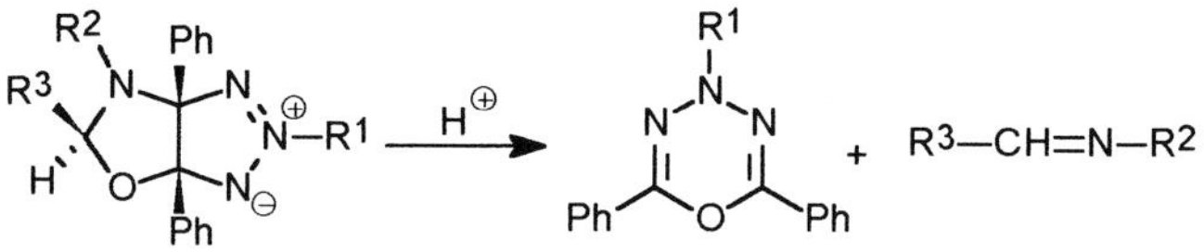

(10) $\xrightarrow{H^{\oplus}}$ (11) + R3—CH=N—R2

(11): R1 = Ph (93%); R1 = *p*-BrC_6H_4 (78%); R_1 = *p*-$NO_2C_6H_4$ (85%).

which the mechanism of ring expansion is explained (R.N. Butler, D.F. O'Shea, J. chem. Soc. Perkin 1, 1994, 2797). The *O*-allyl hydroxamic acids (12) are readily converted to phenylseleno-substituted 1,4,2-dioxazines (13) by an organoselenium-induced cyclisation (M. Tiecco

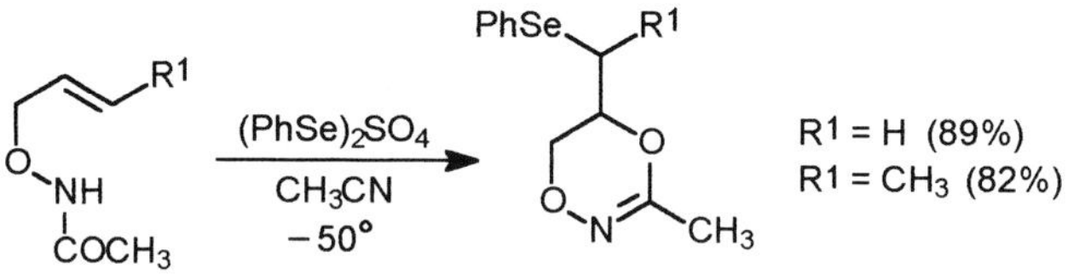

(12) $\xrightarrow[CH_3CN,\ -50°]{(PhSe)_2SO_4}$ (13)

R1 = H (89%)
R1 = CH_3 (82%)

et al., Chem. Comm., 1995, 237). The nature of the substituents on the hydroxamic acid and the reaction temperature are critical to the formation of the dioxazine. The oxygen atom of the carbonyl group acts as the nucleophile in the formation of the dioxazine but in some cases the nitrogen atom behaves as nucleophile generating a five membered *N*-acylisoxazolidine. The reaction of nitrones (14) with benzaldehyde *O*-oxide (15) is utilised to generate 1,2,4,5-trioxazines (16) (K.J. McCullough *et al.*, J. chem. Soc. Perkin 1, 1995, 41). *N*-Methyl-1,3,5,2-trioxazinane is proposed to be a possible spontaneous

Ph
=N(O)R2 + PhCH=O
R1 (14) (15)
Ph O–O Ph
H O–N R1
(16) R2

ignition sensitiser (W.-T. Chan, D. Shen, H.O. Pritchard, Chem. Comm., 1998, 583).

(3) *Oxygen-Nitrogen-Sulphur Rings*

α,*N*-Alkyl- or arylsulphonamides (1) are dialkylated *via* their corresponding dianions (2) with gaseous formaldehyde for the synthesis of the tetrahydro-1,4,3-oxathiazine 4,4-dioxides (3) (E. Grunder-Klotz,

R1 NH–R3 R2 (1)
n-BuLi, THF −78°
R1 N–R3 R2 (2)
CH_2O (g)
R1 R2 R3 (3)

P. Humbert, J.-D. Ehrhardt, Heterocycles, 1993, 36, 733). 3-(Benzo[*b*]-thien-2-yl)-5,6-dihydro-1,4,2-oxathiazine 4-oxide (4) and compounds of this class are reported to have uses as agricultural fungicides and herbicides. The X-ray structure of (4) shows that the 1,4,2-oxathiazine ring system adopts a slightly distorted half-chair conformation (J.F. Gallagher, G. Ferguson, W.G. Brouwer, Acta Cryst., 1997, C53, 823).

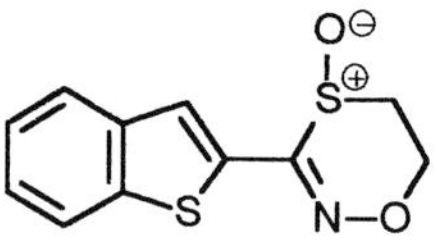

(4)

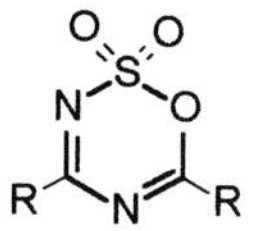

(5)

The chemistry of the oxadiazine and thiadiazine ring systems have been reviewed (R.K. Smalley, Compreh. Hetero. Chem II, 1996, 645). The synthesis, structure and properties of 1,2,3,5-oxathiadiazine 2,2-dioxides (5) have been reviewed (A.A. Michurin, A.V. Shishulina, Chem. heterocyl. Comp., 1997, 33, 1028).

(III) *Oxygen, Sulphur Rings*

(1) *Trioxanes and Oxygen-Sulphur Rings*

The 1,2,4-trioxane unit occurs naturally in the anti-malarial drug artemisinin (1). As a result of the increasing resistance to traditional quinoline-based anti-malarial drugs, a significant interest in the synthesis of trioxane derivatives has developed (W.-S. Zhou, X.-X. Xu, Acc. chem. Res., 1994, 27, 211; J.N. Cumming, P. Ploypradith, G.H. Posner, Adv. Pharmacol., 1996, 37). The synthesis of (1) and some

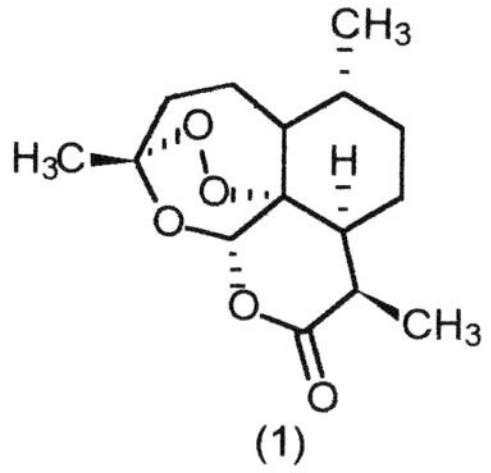

(1)

derivatives together with an analysis of mechanism is discussed (S.C. Vonwiller *et al.*, J. Amer. chem. Soc., 1995, 117, 11098; R.K. Haynes *et al.*, Tetrahedron Letters 1995, 36, 4641; R.K. Haynes *et al.*, J. org. Chem., 1994, 59, 4743). The mechanism of action of artemisinin and other endoperoxide anti-malarials has been investigated (S. Bharel *et al.*, J. Chem. Soc. Perkin 1, 1998, 2163; A. Robert, B. Meunier, J. Eur. Chem., 1998, 4, 1287). The design, synthesis and structure-activity relationships of simplified 1,2,4-trioxane analogues as potential replacements for artemisinin (1) has attracted considerable interest (J.N. Cumming *et al.*, J. med. Chem., 1998, 41, 952; C.W. Jefford *et al.*, Helv. Chem. Acta, 1993, 76, 2775). The synthesis of these analogues (4) involves the treatment of (2) with arylbromide and butyllithium to generate (3) in yields of 41-91%. The trioxane ring is formed by singlet oxygenation of (3). The fluorobenzyl ether derivative of (4) (Ar = *p*-(*p*'-$FC_6H_4CH_2OCH_2)C_6H_4$) is twice as potent as artemisinin (1) (G.H. Posner *et al.*, J. med. Chem., 1998, 41, 940).

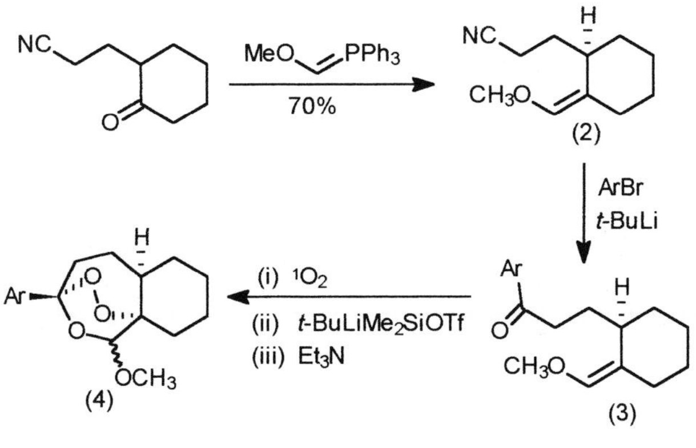

(2) *Tri- and Tetrathianes*

Treatment of 1,2,4-trithiolane (1) with $Cp_2Ti(CO)_2$ affords the titanocene chelate complex (2) by insertion into the *S-S* bond. A wide variety of sulphur heterocycles are attainable from this precursor, one of which is the 1,2,3-5-tetrathiane (3), in 13% yield (R. Steudel *et al.*, Ber., 1997, 130, 757). The silylation of 1,3,5-trithiane-1,1,3,3,5,5-

hexaoxide was investigated for the synthesis of sulphoxonium ylides (W. Sundermeyer, A. Walch, Ber., 1996, 129, 161).

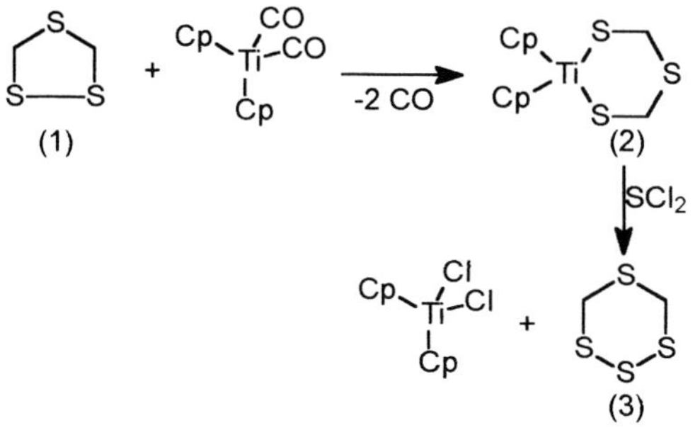

Thiobenzophenone and sulphur react, with sodium thiophenoxide as a catalyst, in acetone to furnish 3,3,6,6-tetraphenyl-1,2,4,5-tetrathiane (4) in 95% yield (R. Huisgen, J. Rapp, Heterocycles, 1997, 45, 507). The thermolysis of (4) in mesitylene at 150° leads to the formation of

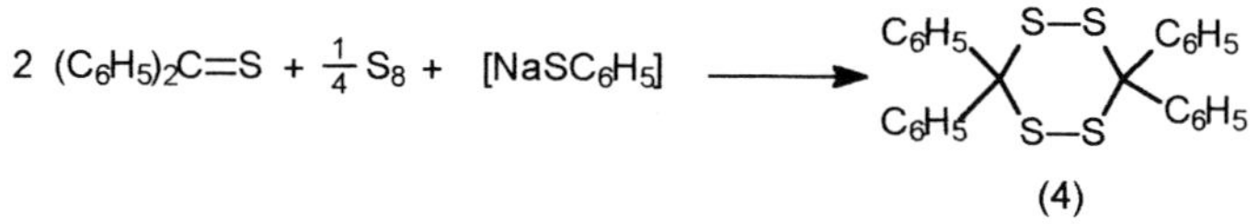

$$2\ (C_6H_5)_2C{=}S + \frac{1}{4}S_8 + [NaSC_6H_5] \longrightarrow (4)$$

thiobenzophenone in quantitative yield. When the thermolysis is carried out using dimethyl acetylenedicarboxylate (DMAD) as a trapping agent, the Diels-Alder product (5) of thiobenzophenone and DMAD is isolated. Reduction of (4) with lithium aluminium hydride and subsequent acetylation yields diphenylmethyl thioacetate (6).

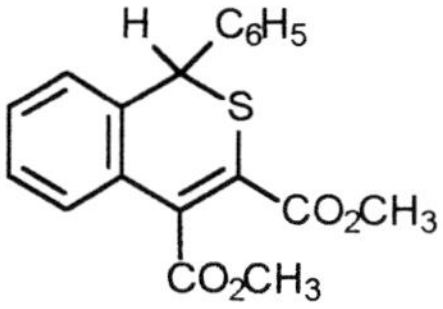

$$(C_6H_5)_2CH{-}S{-}C(=O){-}CH_3 \quad (6)$$

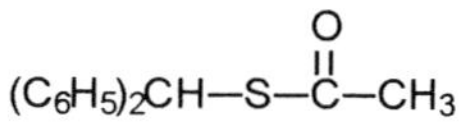

The dianions of α,α-disubstituted alkanedithioic acids (7) are oxidised with [bis(trifluoroacetoxy)iodo]benzene to give the 1,2,4,5-tetrathianes (8) in moderate yields (K. Hartke, U. Wagner, Ber., 1996, 129, 663).

R1, R2 >CH–C(=S)SH —BuLi→ (7) —[O]→ (8)

(8) R1, R2 = CH_3 (35%); R1, R2 = Ph (51%).

Guide to the Index

This index is constructed in a similar manner to the volume indexes of the first edition of the Chemistry of Carbon Compounds. However, to make the index easier to use, more descriptive entries have been made for the commonly occurring individual, and groups of chemicals.

The indexes cover primarily the chemical compounds mentioned in the text, and also include reactions and techniques, where named, and some sources of chemical compounds such as plant and animal species, oils, etc.

Chemical compounds have been indexed alphabetically under the names used by authors, editing being restricted to ensuring uniformity of entries under the same heading. In view of the alternative nomenclature that can often be used, a limited amount of cross-referencing has been done where it is considered to be helpful, but attention is particularly drawn to Convention 2 below.

For this and the succeeding volumes, the indexing conventions listed below have been adopted.

1. Alphabetisation

(a) A letter by letter alphabetical sequence is followed for entries, firstly for the main entry, followed by the descriptive entry.

(b) The following prefixes have not been counted for alphabetising:

n-	*o-*	*as-*	*meso-*	*C-*	*E-*
	m-	*sym-*	*cis-*	*O-*	*Z-*
	p-	*gem-*	*trans-*	*N-*	
	vic-			*S-*	
		lin-		*Bz-*	
				Py-	

Some prefixes and numbering have been omitted in the index, where they do not usefully contribute to the reference.

(c) The following prefixes have been alphabetised:

Allo	Epi	Neo
Anti	Hetero	Nor
Bis	Homo	Pseudo
Cyclo	Iso	

2. Cross references

In view of the many alternative trivial and systematic names for chemical compounds, the indexes should be searched under any alternative names which may be indicated in the main body of the text. Only a limited amount of cross-referencing has been carried out, where it is considered that it would be helpful to the user.

3. Derivatives

Simple derivatives are not normally indexed if they follow in the same short section of the text.

4. Collective and plural entries

In place of "– derivatives" the plural entry has normally been used. Plural entries have occasionally been used where compounds of the same name but differing numbering appear in the same section of the text.

5. Main entries

The main entry of the more common individual compounds is indicated by heavy type. Multiple entries, such as headings and sub-headings over several pages are shown by "–", e.g., 67–74, 137–139, etc.

Index

Acesulfame, 189
2-Acetoxypyrazines, 115
5-Acetoxyquinoxaline-2,3-diones, 151
4-Acetylamino-5-amino-2-phenylpyridazin-3(2*H*)-ones, 20, 21
3-Acetyl-4-aryl-2-pyrazolines, rearrangement, 13
3-Acetyl-1,2,4-benzotriazines, 241
3-Acetylcinnolinones, 48
5-Acetyl-5,10-dihydrophenazine, Friedel–Crafts alkylation, 167
5*N*-Acetyl-15b-β-hydroxyardeemin, 221–223
5-Acetyl-2-methyl-4-nitro-6-phenylpyridazin-3(2*H*)-one, reaction with 1,3-dienes, 23, 24
2-Acetylquinazolin-4(3*H*)-one, 204, 205
Acyclovir, pyrazinone analogues, 122
3-Acylamino-6-chloropyridazine, metalation, 21
5-Acylamino-6-hydroxymethyl-3-phenylpyrimidin-4(3*H*)-ones, 87
3-Acylaminopyridazines, metalation, 21
3-Acylaminoquinazolin-4(3*H*)-ones, 96
N-Acyl-*N*′-carbamoyl-*S*-methylisothioureas, 251
1-Acyl-3-(2-chloroethyl)-2-pyrazolines, 13
N′-Acyl-*N*,*N*-dimethylamidines, synthesis from amides/dimethylformamide dimethyl acetal, 250
N-Acylisoxazolidines, 271
2-Acylpyrazines, 104
1-Adamantylphthalazine, 60
Aequorins, 109
6-(Alk-2-enyl)amino-5-formyl-1,2,3,4-tetrahydro-1,3-dimethylpyrimidine-2,4-dione, 88
2-Alkenyl-1,3,4-oxadiazines, 269
2-Alkenylpyrazines, intramolecular cycloadditions, 105
N-(α-Alkoxyalkyl)benzotriazoles, 93
2-Alkoxycarbonyl-1-ethynylphthalazine, 59
Alkoxypyrazines, 112
Alkoxypyrimidines, alkyl-O to -N-rearrangements, 87
3-Alkyl-6-amino-1,2,4,5-tetrazines, 257
Alkylation of 4-hydroxycinnoline, reaction with ethyl bromoacetate, 49
2-Alkyl-5-hydroxymethyl-4-phenylcarbainoyl-1*H*-imidazole, 87
3-Alkyl-6-methylthio-1,2,4,5-tetrazines, 257
N-Alkylphenazinium salts, amination, 167, 168
Alkylpyrazines, synthesis from ethylenediamines, 102, 103
1-Alkyl-1,2,5,6-tetrahydro-4,6-diphenyl-2-tosyl-1,2,3-triazines, synthesis from *cis*-*N*-alkylazidinyl ketone tosylhydrazones, 235
Alkylthiopyrazines, 112
3-Alkynylpyridazines, 19
Alloxan, condensations with 2-amino-*N*,*N*-dimethylanilines, 160, 161
1-Allyl-2-alkoxycarbonylphthalazines, 59
Allyldihydropyridazines, 28
1-Allyl-2,5-dimethylpiperazine, 130
Amethyst violet, 186
2-Amino-6-(aminoethyl)aminopyrimidine-4,5-diones, 89
4-Amino-3-arylcinnolines, 44
7-Amino-2-arylthieno[3,4-*d*]pyridazin-1(2*H*)-ones, cycloaddition reactions, 55
2-Amino-3-bromopyrazine, Stille coupling, 108
9-Aminocarbazoles, 200
4-Amino-5-chloro-2-phenylpyridazin-3(2*H*)one, *N*-acetylation, 20, 21
3-Amino-5-chloropyrazinones, cycloaddition reactions, 124, 125

4-Aminocinnoline 3-carboxamides, 42, 43
3-Aminocinnoline-4-carboxylic acid *N*-oxides, 46
4-Amino-8-cyanoquinazolines, 93
2-Amino-3,5-dibromopyrazine, Stille coupling, 108
3-Amino-3,4-dihydro-2-methyl-quinazolin-4-one, 97
6-Amino-2,3-dimethylquinoxaline, reaction with acetaldehyde, 148
2-Amino-3-ethoxypyrazine, 111
α-Amino-3-hydroxy-5-methylisoxazole-4-propanoic acid, 151
6-Amino-3-methylpyridazin-4-ones, 43
2-Aminophenazinium salts, 167, 168
4-Aminophenol, 201
N-(2-Aminophenyl)-3,6-dichloro-4-pyridazinecarboxamide, 15
3-Amino-6-phenylpyridazines, antidepressant activity, 34
3-Amino-6-phenylthiopyridazine, oxidation, 27
Aminopyrazine, alkylation, 110
2-Aminopyrazine, 111
Aminopyrazines, 111
Aminopyridazines, 15
–, as muscarinic agonists, 33
3-Aminopyridazines, 244
–, synthesis from 3-chloro-6-phenyl-1,2,4-triazine, 13
2-Aminopyrimidines, 83
–, reaction with heteroaromatic aldehydes, 83
4-Aminopyrimidines, 254
1-Aminopyrrolidines, 5
2-Aminoquinazolines, synthesis from guanidine carbonate/2-fluorobenz-aldehydes, 93
3-Aminoquinazolin-4(3*H*)-ones, 96
3-Aminoquinoxaline-2-carbonitrile, 154, 155
3-Aminoquinoxaline-2-carbonitrile 1,4-dioxides, 154
5-Aminoquinoxaline-2,3-diones, 151
7-Aminoquinoxaline-5,8-diones, 157
Aminoquinoxalines, 147, 148
6-Aminoquinoxalines, 148
3-Aminoquinoxalin-2(1*H*)-ones, 148, 149
Amino-1,3,5-triazines, synthesis from cyanuric chloride, 250
1-Amino-1,3,5-triazine-2,4,6-triones, 251
1-Amino-2,6,6-tricyanocyclohexadienes, coupling with aryldiazonium salts, 1
Aminovinylpyrazines, 117
Amphotericin B, 178
2,5-Anhydro-3,4,6-tris(*O*-benzyl)-allonoimidic acid methyl ester, reaction with bis(trifluoromethyl)-1,2,4,5-tetrazine, 239
Anilines, electropolymerisation, 178
3-Anilino-5-hydroxy-4-(2-nitroaryl)-pyrazoles, 46
3-Anilino-4-(2-nitrophenyl)isoxazolin-5(2*H*)-ones, 46
Aposafranine, 186
Arachidonic acid, 218
Arbuzov reaction, 135
Ardeemin, 221–223
Arglecin, 108
Aroylpyridazines, 23
2-Aroylpyridazin-3(2*H*)-ones antimicrobial activity, 36, 37
Artemisinin, 272, 273
4-Arylamino-6-aryl-2,3,4,5-tetrahydropyridazin-3-ones, 5, 6
5-Aryl-4-carbamoyl-3-chloro-6-methyl-pyridazine *N*-oxides, antimicrobial activity, 36
6-Aryl-4-carboxypyridazin-3(2*H*)-ones, 33
6-Aryl-3-chloropyridazine carboxamides, 36
2-Aryl-4-cyanopyridazine-3(2*H*)-ones, 1
6-Aryl-2,3-dihydroimidazo[2,1-*b*]thiazoles, 59
6-(4-Aryl)-4,5-dihydropyridazin-3(2*H*)-ones, 39
4-Aryl-1,4-dihydro-1-(tetrazol-5-yl)-1,2,4,5-tetrazines, 257, 258
3-Aryl-1,4,2-dithiazolium salts, 268
6-Aryl-3-ethoxycarbonyl-4-hydroxypyridazines, synthesis from ethyl (*Z*)-5-aryl-2-diazo-5-hydroxy-3-oxopent-4-onoates, 9, 10

6-(3-Aryl-2-hydroxypropyl)-4,5-dihydro-3(2*H*)-pyridazinones, reactions with trimethylsilyltriflate, 19
Arylidene-4,5-dihydropyridazines, structural correction, 30
5-Arylidene-5,6-dihydrouracils, 74
5-Arylidenetetrahydropyridazin-3(2*H*)-ones, 37
2-Arylindoles, oxidation, 191, 192
3-Aryl-6-lithio-5-methoxy-1,2,4-triazines, 244
3-Aryl-5-methoxy-1,2,4-triazines, lithiation, 244
5-Aryl-3-methylpyridazin-4(1*H*)-ones, 13
6-Aryl-2-methylpyridazin-3(2*H*)-ones, as platelet aggregation inhibitors, 39
4-(Aryloxymethyl)phthalazin-1(2*H*)-ones, 58
Aryloxypyridazinyl-*N*-(4-arylsulfonyl)-acetainides, as endothelin receptor antagonists, 34, 35
2-Arylphenazine-5,10-dioxides, synthesis from quinoxaline-1,4-dioxides, 165, 166
4-Arylphthalazinones, 53
1-Arylpiperazines, synthesis from bis(2-haloethyl)amines/anilines, 129
–, – from ethanolamines/anilines, 129
–, – from piperazines/haloarenes, 129
N-Arylpiperazinones, 133
3-Arylpiperazinyl-5-benzylpyridazines, analgesic activity, 37
N-Arylpyridazines, 1
6-Aryl-3-pyridazin-2-ones, as platelet aggregation inhibitors, 39
5-Arylquinoxalines, 138
2-Arylsulfonyl-3,6-diaryl[1,2,3]-oxadiazines, 269
3-Aryl-1,2,4,5-tetrazines, synthesis from methyl(trifluoromethyl)dioxirane, 261, 262
6-Aryl-1,3,5-triazine 1-oxides, 252
2-Aryl-2-trifluoromethyl-4,6-diimino-perhydro-1,3,5-triazines, 252
2-Aryl-2-trifluoromethyl-4,6-diimino-perhydro-1,3,5-triazin-4-ones, 252
Aspericins, 224, 225, 227, 228
Astechrome, 113, 114
Aurantine, 227, 228
Aza-1,3-butadienes, reactions with arylsulphonyl nitrosite, 268, 269
1-Azafagomine, 39
Azelastine, 69
3′-Azido-2′,3′-dideoxypyridazine nucleosides, 39
2-(2-Azidophenyl)ethylamine, photolysis, 45
1-(2-Azidophenyl)pyrazoles, flash vacuum pyrolysis, 142
Azidopyrazines, 110
2-Azidopyrazines, 114
Azidopyridazines, 15
Azidyl radical, reaction with azide ion, 264
Azotetrazoles, tautomerisation, 263, 264
Azur B, 184
Azur C, 184

Bamford–Stevens reaction, 234
Barbiturates, 80
Barbituric acid, reaction with 5-formyl-4,6-dihydroxypyridinine 5-Formyl-4,6-dihydroxypyrimidinine, 83
Beckmann rearrangement, 252
Beirut reaction, 152, 153, 169
Benzil polymers, 241
Benzimidazoles, 140, 141
2,1-Benzisoxazolines, pyridazinyl, 22
Benzocinnolines, methods of synthesis, 49, 50
Benzo[c]cinnolines, heat of formation, 50
–, nitration, 51
–, physical properties, 50, 51
–, reactions, 51, 52
–, spectra, 50, 51
–, synthesis from 2,2′-dinitrobiphenyl, 49
–, – from 2-ethylamino-2′-azidobiphenyl, 50
–, transition-metal complexation, 51, 52
Benzo[*f*]cinnolines, 52
Benzo[*h*]cinnolines, 52, 53
Benzo[c]cinnoline 5-oxide, deoxygenation, 51
Benzocinnoline *N*-oxide, synthesis from 2,2′-dinitrobiphenyl, 50

1,4-Benzodiazepines, 224
–, pyridazinyl, 22
1,5-Benzodiazepines, flash vacuum pyrolysis, 142
Benzomalvins, 224–229
Benzo[*b*]phenazine, oxygenation, 170
–, photodimerisaiton, 170
Benzo[*g*]phthalazin-1,4(2*H*,3*H*)-dione, 66, 67
Benzo[*g*]phthalazino[1,2-*b*]pyridazin-6,13-diones, 66, 67
Benzopyranopyrazines, synthesis from benzopyrans, 8
Benzopyrazines, 137–161
3-(Benzo[*b*]thien-2-yl)-5,6-dihydro-1,4,2-oxathiazine 4-oxide, 271, 272
Benzo-1,2,4-triazines, synthesis from nitrile oxides/*N*-arylbenzamidoximes, 242
2-(Benzotriazol-1-yl)enamines, 93
1,3-Benzoxazepines, 247
6-[4-(Benzylamino)-4,5-dihydro-5-methyl-7-quinazolinyl]-pyridazin-3(2*H*)-ones, cardiotonic activity, 38
5-Benzyl-2-[4-(3-chlorophenyl)-6-methylpiperazin-1-yl]methyl-pyridazin-3(2*H*)-one, 34
4-(α-Benzyl-α-hydroxybenzyl)-quinazolines, 97
5-Benzyloxyacetyl-4-methyloxazole, 78
4-Benzyloxymethyl-5-hydroxy-6-methyl-pyrimidine, 78
1-Benzyloxypyrazinones, photolysis, 122
1-Benzyloxypyrimidin-2(1*H*)-ones, reactions with hydroxylamine, 87
1-Benzylpiperazin-2-ones, 133
5-Benzyl-2-[4-(3-trifluoromethylphenyl)-6-methylpiperazin-1-yl]-methyl-pyridazin-3(2*H*)-one, 34
Benzyl viologen, 186
Biosensors, 179
Bi(phthalazinylidenes), 63
Bipyrazine, synthesis from 2-chloropyrazine, 108
Bis(benzal)ethylenediamine reaction with tetrachloro-1,2-benzoquinone, 142
2,4-Bis(benzyloxy)-6-methylpyrimidine, rearrangement, 87
2,3-Bis(bromomethyl)quinoxaline, debromination, 144
3,6-Bis(diphenylphosphino)pyridazine, 29
4,6-Bis(ethoxycarbonylmethylthio)-2-methylthipyrimidine-5-carbonitrile, 81
Bisformazans, 262
2,5-Bis(4-hydroxyphenyl)pyrazine, synthesis from α-amino-4-hydroxyacetophenone, 105
2,3-Bis(4-methoxyphenyl)-6-(4-pyridyl)-pyridine, 245
5,6-Bis(4-methoxyphenyl)-3-(4-pyridyl)-1,2,4-triazine, reaction with bicyclo[2.2.1]hepta-2,5-diene, 245
5,6-Bis(4-methoxyphenyl)-1,2,4-triazines synthesis and structure-activity, 241
2,4-Bis(methylthio)pyrimidines, 75
2,5-Bis(trichloromethyl)pyrazine, 104
2,4-Bis(trimethylsilyl)oxypyrimidine, reaction with aldehydes, 84
1,4-Bis(tri-^{i}propylsilyl)-1,4-dihydro-pyrazine, reaction with carbon dioxide, 122
–, – with carbon oxysulfide, 122
–, – with [60]fullerene, 122
Bisverdazyls, synthesis from bisformazans, 262
Bleomycin A_2, 254
Brilliant cresyl blue, 180, 188
5-Bromo-2-chloropyrimidine, 85
5-Bromo-2-iodopyrimidine, 79
2-Bromo-5-methylpyrimidine, 86
3-(4-Bromophenyl)-5,6-dimethyl-1,4,2-dithiazine, reaction with dimethyl acetylenedicarboxylate, 266, 267
–, – with norbornene, 266, 267
6-(4-Bromophenyl)pyridazin-3(2*H*)-ones, 31
3-Bromopiperazine-2,5-dione, methylenation, 135
Bromothymol blue, 187
4-Bromo-2-trimethylstannylpyrimidine, 85
6-[2-(*N*-^{t}Butylamino-2-hydroxypropoxy)-phenyl]-3-pyridazinyl hydrazine hemihydrate, 32

2-(^{t}Butylazoxy)phenyldiazonium tetrafluoroborate, intramolecular cyclisation, 256
2-^{t}[Butyl-4-chloro-5-(silylmethylthio)]-pyridazin-3(2*H*)-ones, protodesilylation, 27
3-^{t}Butyl-8-methyl-1,7-dinitrobenzo[c]-cinnoline, 51
2-Butylphthalazin-1(2H)-ones, 65
2-^{t}-Butylsulfinylpyrazine, 112
2-^{t}-Butylsulfonylpyrazine, 112
6-(2-Butynylaminomethyl)pyrazinones, 125

Camphor-1,2,4-triazines, 247
Candidine, 230
Carbadox, 155
β-Carboline alkaloids, 246
1-Carboxy-7-(*N*,*N*-dimethylamino)-3,4-dihydroxyphenoxazin-5-ium chloride, 184
Celestine blue, 188
Cephalostatins, 99, 100
Chloroamphenicol, 214
1-Chloro-1-aryl-2,2,2-trifluoroethyl-heterocumulenes, 252
3-Chlorocinnoline, reaction with lithium amide/aldehydes, 47, 48
4-Chlorocinnolines, 43
2-Chloro-3,6-dialkylpyrazines, 104
6-Chloro-(*N*,*N*-dimethylamino-methylidenehydrazine)-2-phenyl-pyridazin-3(2*H*)-one, 32
6-Chloro-4-(*N*,*N*-dimethylamino-methylidenehydrazino)-2-phenyl-3(2*H*)-pyridazinones, 17
6-Chloro-5-(*N*,*N*-dimethylamino-methylidenehydrazino)-2-phenyl-3(2*H*)-pyridazinones, 17
3-Chloro-5,6-diphenyl-1,2,4-triazine, reaction with chloroacetonitrile, 243
3-Chloro-6-hydrazinopyridazine, 16
3-Chloro-6-methoxypyridazine metalation, 22
3-Chloromethyl-5,6-dihydro-1,5,5-trimethylpyrazin-2(1*H*)-one self condensation, 123
6-Chloro-2-(1-methylhydrazino)-quinoxaline 1-oxide, 152
6-Chloro-2-(1-methylhydrazino)-quinoxaline 4-oxide, 152
3-Chloro-6-methylpyridazine, homolytic substitution, 20
–, oxidation, 22, 23
2-Chloro-5-methylpyrimidine, 86
4-Chloro-5-nitropyrimidines, reactions with alkylisothioureas, 88
3-Chloro-6-phenylpyridazine, 20, 22
3-Chloro-6-phenyl-1,2,4-triazine, substitution by α-chlorocarbanions, 244
6-Chloro-3-phenyluracil, 82
1-Chlorophthalazines, 60
3-Chloropyrazine 1-oxides, 114
2-Chloropyrazines, synthesis from pyrazine *N*-oxides, 106
3-(5-Chloro-3-pyrazolyl)-2(1*H*)quinazolinone, 15
3-Chloropyridazine-6-carboxylic acid hydrazide, synthesis from methyl levulinate, 7
α-(6-Chloropyridazin-3-yl)-phenylacetonitrile, oxidative decyanation, 17
6-Chloropyrimidines, nucleophilic substitution, 80
Chloroquinazolines, 95, 96
2-Chloroquinoxaline, reaction with potassium 2-nitrophenoxide, 145
–, solvolysis, 145
Chloroquinoxaline sulfonamide, 137
3-Chloro-5-tosylpyridazine, 32
Cholecystokinin, 227
Chrysogine, 204, 205
C.I. Basic blue 9, 175
C.I. Basic blue 24, 182
C.I. Basic blue 25, 191
C.I. Basic blue 122, 189
C.I. Basic blue 141, 189
C.I. Basic blue 148, 189
C.I. Basic green, 185
C.I. Mordant blue, 184, 185, 188
Cinnoline 4-carboxylic acids, 45
Cinnolines, methods of synthesis, 42–47
–, ortho-directed lithiation, 47, 48
–, physical properties, 46–48
–, reactions, 47–49
–, – with ynamines, 48, 49

Cinnolines (*cont'd*)
–, synthesis from *o*-aminophenylalkynes, 43
–, – from ketoesters, or cyclohexan-1,3-diones, 46
–, – from phenylhydrazones, 43
–, tetracyclic, 43
Cinnolinium 3-oxides, 44, 45
Cinnolinones, bond parameters, 47
4-Cinnolinones, 43
4-Cinnolones, X-ray crystallography, 47
Cresyl violet, 191
Crystal violet, 176
3-Cyanochloromethyl-5,6-diphenyl-1,2,4-triazine, 244
9-Cyano-6-methyl-2,3-diphenyl-8*H*-[1,2,4]triazino[4,5-*b*][1,2,4]triazine, 244
2-Cyanomethylphenazine, 166
1-Cyanophthalazine, 60
2-Cyanopyrazine, 112, 116
Cyanopyridazines, 1
5-Cyano-1,2,4-triazines, nucleophilic substitutions, 243
Cyclamates, 189
Cycloalkanone 1-phthalylhydrazones, thermolysis, 62, 63
Cyclopentano-1,2,3,5-tetrazines, ring-chain tautomerism, 255, 256

DABCO, 193
Deferoxamine, 177, 178
Dehydroevodiamine, 218
7,8-Dehydrorutacarpine, 217
Deoxyapergillic acid, 111
Deoxypeganidine, 213, 214
Deoxypeganine, 211–213
Deoxypeganine *N*-oxide, 211
2-Deoxypyrazine *C*-nucleosides, 119, 120
Deoxyvasicine, 211–214
Deoxyvasicinone, 211, 212
Deoxyvasicinone *N*-oxide, 211
Desmethoxyaniflorine, 210
2,5-Diacetyl-3,6-dimethylpyrazine, 104
1,4-Diacetyl-1,2,3,4-tetrahydro-5,6-diphenylpyrazine, photo-oxidation, 128
6,7-Dialkoxyquinoxaline-2,3-diones, attempted nitration, 151
2,2-Dialkyl-1,3-dibenzyl-2,3-dihydrobenzimidazoles, thermolysis, 160
3,5-Diamino-6-(aryl)-1,2,4-triazines, 238
2,5-Diamino-3,6-dicyanopyrazine, 116
2,3-Diamino-3-(phenylthio)acrylonitrile, 117
5,6-Diamino-3-thiomethyl-1,2,4-triazines, reactions with 2-bromoacetophenone, 247
–, – with phenylglyoxal, 247
2,4-Diaminotoluene, 201
3,6-Dianilinopyridazine, 41
4,6-Diaryl-3-arylpiperazinylpyridazines, analgesic activity, 38
2,5-Diaryl-4-(chloromethylthio)-imidazolines, 77, 78
4,6-Diaryl-3-ethoxycarbonyl-1,4-dihydropyridazines, synthesis from 3,5-diaryl-2,5-diketopentanoates/hydrazine, 9
3,4-Diarylfurazan, reaction with dimethyl sulfate, 269
3,4-Diaryl-2-methyl-1,2,5-oxadiazol-2-ium salts, 269
3,4-Diaryl-6*H*-1,2,5-oxadiazines, 269
Diaryloxadiazolo[3,4-*d*]pyridazines, 14
4,6-Diarylpyridazin-3(2*H*)-ones, antidepressant activity, 34
5,6-Diarylpyridazin-3(2*H*)-ones, 22
2,3-Diarylquinoxalines, 140, 141
Diazaacridones, pyridazinyl, 22
1,4-Diaza[2,2,2]bicyclooctane, 193
Diazahelicenes, 51
1,2-Diazines, synthesis from 1,2,4,5-tetrazines/alkynes, 259, 260
1,4-Diazines, 99–137
Diazocyclopentadienes, reactions with arenediazonium salts, 255
3,6-Dibenzoylpyridazine, synthesis from 3,6-dichloropyridazine, 17
5,7-Dibromobenzo-1,2,3,4-tetrazine 1-oxide, 256
3,8-Di-^{t}butylbenzo[*c*]cinnoline, 51
3,7-Di-^{t}butylphenazin-5(10*H*)-yl radicals, 164
4,5-Dichloro-1-(ω-bromoalkyl)pyridazin-6-ones, 18

2,3-Dichloro-5,6-dicyanopyrazine, 117, 118
–, reactions with *N*-acyl-*N*-trimethylsilylmethylanilines, 118
4,5-Dichloro-3-(*N*,*N*-dimethylamino)-pyridazin-6-one, 20
4,5-Dichloro-1-hydroxymethylpyridazin-6-one, 18
4,5-Dichloro-2-methyl-3(2*H*)-pyridazinone, 30
4,6-Dichloro-2-methylthiopyrimidine-4-carbonitrile, 81
2,4-Dichloro-4-nitrophenol, reaction with sodium sulfide, 200
2,3-Dichloro-6-nitroquinoxaline, 148
6,7-Dichloro-5-nitroquinoxaline-2,3-dione, 138, 151
1,4-Dichlorophenazines, synthesis from 3,3,6,6-tetrachlorocyclohexane-1,2-dione/1,2-phenylenediamines, 164, 165
2,4-Dichlorophenol, reaction with sodium sulfide, 201
3-(3,4-Dichlorophenyl)-1,4,5,6-tetrahydro-1-(4-nitrobenzyl)-pyridazine, 36
1,4-Dichlorophthalazines, 61
3,5-Dichloropyrazin-2(1*H*)-ones, 122, 123
4,5-Dichloropyridazine, 20
3,6-Dichloropyridazine, Raman spectrum, 32
4,5-Dichloropyridazin-6-one, alkylations, 18
3,6-Dichloro-4-pyridazinyl phenyl ketone, amination, 21
2,3-Dichloroquinoxaline, reaction with disodium glycolates, 145
6,7-Dichloroquinoxaline-5,8-dione, reactions with arylamines, 157
2,3-Dichloro-6-(trifluoromethyl)-quinoxaline, reactions with ammonium carbonate, 147
2,3-Dicyano-5-methylpyrazine, alkylation, 117
2,3-Dicyanopyrazines, ketenesilylacetals, 118, 119
–, photochemical reactions, 118, 119
–, reactions with allylsilanes, 118, 119
–, – with benzylsilane, 118, 119
2,3-Dicyanoquinoxaline, reaction with silicon tetrachloride, 156
2′,3′-Dideoxy-2′,3′-dideoxypyridazine nucleosides, 39
2,3-Dideoxyribofuranosyl-1,2,4-triazines, 239, 240
1,2-Di(ethoxycarbonyl)-1,2,3,4-tetrahydropyridazine, hydroethoxycarbonylation, 27
1,2-Di(ethoxycarbonyl)-1,2,3,6-tetrahydropyridazine, hydroformylation, 27
2,7-Diethoxyphenazine, 166
5,6-Difluorobenzofuroxan, reactions with enamines, 154
–, – with 1-aminocyclohexene, 169, 170
6,7-Difluoroquinoxaline, ethylation, 146
1,2-Diformylhexahydropyridazine, 11
3,6-Diformylpyridazine, 29
4-{2-(4-*N*,*N*-Dihexadecylaminophenyl)-ethenyl}-*N*-methylpyridazinium halides, 32
5,6-Dihydrobenzo[*c*]cinnoline, 51
5,6-Dihydrobenzo[*h*]cinnolin-3-ones, 52, 53
5,7-Dihydrobenzo[*a*]phenazin-5-ones, 170
2,3-Dihydrocinnolines, 45
5,6-Dihydrocyclobuta[*d*]pyrimidines, 77
(3*s*)-2,5-Dihydro-3,6-dimethoxy-2-^{i}propylpyrazine, synthesis and reactions, 126, 127
5,6-Dihydro-3,7-diphenyl-4*H*-diazepines, 13
3,6-Dihydrofuro[3,4-*c*]pyridazinones, 2, 3
1,6-Dihydro-5-hydroxy-6-oxo-3-(trifluoromethyl)-4-pyridazinecarboxylates, synthesis from 2-(alkylamino)-3-(trifluoroacetyl)butene dioates, 4
4,5-Dihydro-4-hydroxypyridazin-3(2*H*)-ones, synthesis from 4,5-dihydroisoxazoles, 9
1,4-Dihydro-1-hydroxyquinoxaline-2,3-diones, 159

2,3-Dihydroimidazo[2,1-*a*]phthalazin-4-ium-6-olates, ring transformations, 64, 65
1,3-Dihydro-1-4′-[2″-(methacryloxy)-ethoxycarbonyl]benzyl-3,3 -dimethylspiro{2*H*-indole-2,3′-naphth[2,1-*b*][1,4]oxazine}, 197
1,2-Dihydro-2-methylenequinazolin-4(3*H*)-ones, 92
1,2-Dihydro-2-methyl-1-phenyl-3,6-pyridazinedione, reactions with phosphate esters, 26
5,6-Dihydro-2-methylthio-4*H*-1,3,4-thiadiazines, 265
Dihydro-1,2′,3,3,5-pentamethylspiro-{2*H*-indole-2,3′[3*H*]-naphth[2,1-*b*][1,4]oxazine}, 196
1,3-Dihydro-1,3,3,4,5-pentamethyl-spiro{2*H*-indole-2,3′[3*H*]naphth[2,1-*b*]oxazine}, 193
5,10-Dihydrophenazine, 166
Dihydrophenazine radical, 164
5,10-Dihydrophenazines, acylated, 169
1,2-Dihydro-1-phenyl-3,6-pyridazindione, reactions with phosphate esters, 26
1,2-Dihydro-4-phenylquinazoline-2-carboxylic acids, 91
Dihydropyrazines, 121, 127, 162–171
2,3-Dihydropyrazines, dimerisation, 121
2,5-Dihydropyrazines, use in stereo-controlled reactions, 127
Dihydropyrazino[2,3-*e*][1,2,4]triazines, 247
1,4-Dihydropyridazines, synthesis from tricarbonyl compounds, 9
4,5-Dihydropyridazin-3(2*H*)-ones, 5, 7
–, synthesis from Meldrum's acids/ hydrazine, 6
Dihydropyridazino[4,5-*c*]pyridazin-5(1*H*)-ones, 17, 18
Dihydropyrido[3,4-*b*]quinoxalines, 152
1,4-Dihydropyrimidine-2-thiones, 76
7,8-Dihydroquinazoline alkaloids, 217, 218
3,4-Dihydroquinazolines, 93, 94
3,4-Dihydroquinoxalin-2(1*H*)-ones, 160
1,3′-Dihydrospiro[1-benzopyran-2,2′-indole], 196
Dihydrospiro[indole-3,2′-(1′,2′,3′,4′-tetrahydroquinoline)]-2,4′-diones, 194
2,9-Dihydro-6-(2,3,4,5-tetrahydro-5-methyl-3-oxo-pyridazin-6-yl)pyrazolo[4,3-*b*][1,4]benzoxazine, cardiotonic activity, 38
2,3-Dihydro-2,2,5,6-tetramethyl-pyrazine, 121
1,4-Dihydrotetrazines, X-ray crystallography, 258
1,6-Dihydro-1,2,4,5-tetrazines, synthesis from hydrazones/ nitrilimines, 258
5,6-Dihydro-4*H*-1,3,4-thiadiazines, 265
1,2-Dihydro-3-thioxo-1,2,4-benzotriazine, reaction with prop-2-ynyl bromide, 244
Dihydro-1,2,3-triazines, conversion into aminopyrazoles, 234
1,6-Dihydro-1,2,3-triazines, 235
–, synthesis, 234
2,5-Dihydro-1,2,3-triazines, 236
4,5-Dihydro-1,2,4-triazin-6-ones, 242
1,3-Dihydro-1,3,3-trimethylspiro{2*H*-indole-2,3′[3′H]naphth[2,1-*b*][1,4]-oxazine}, 193
1,3-Dihydro-1,3,3-trimethylspiro{2*H*-indole-2,3′[3*H*]pyrido[3,2-*f*][1,4]-benzoxazine}, 193
3,4-Dihydroxybenzaldehyde, polymerisation, 178
2,6-Dihydroxy-3,5-dimethylpyrazine, 112
4,5-Dihydroxy-3-(hydroxymethyl)-hexahydropyridazine, 38
4,6-Dihydroxypyrimidine, 83
–, 5-substitution, 80
3,6-Dimethoxypyridazine amination, 15
5,6-Dimethoxyquinoxalines, 158
1,3-Dimethylbarbituric acid, 80
1,1-Dimethylcyanamide, reaction with acetone, 270
S,S′-Dimethyl *N*-cyanoarbonimido-dithioate, reactions with amides, 250, 251
3,8-Dimethyl-1,7-dinitrobenzo[*c*]-cinnoline, 51

2,2-Dimethylhexahydro-1,3-dioxo[4,5-d]pyridazine, 15
2,3-Dimethylhexahydroquinoxalines, 161
2,7-Dimethyl-4-methoxy-1,3,5,6-benzoxatriazonine, 247
1,4-Dimethylpiperazine-2,5-dione, 131
2,5-Dimethylpyrazine, exhaustive chlorination, 104
–, photo-irradiation, 104
2,3-Dimethylquinoxaline, standard enthalpy of formation, 138
2,3-Dimethylquinoxaline 1,4-dioxides, 155
3,6-Di-(β-naphthylamino)pyridazine, 41
1,7-Dinitrobenzo[c]cinnoline, 51
Diphenazithionin, 162
6,7-Diphenylphthalazines, 61
3,6-Diphenylpyridazine, 13, 14
2,6-Diphenylpyridazin-3(2*H*)-ones, herbicidal action, 41
5,6-Diphenyl-3-(2-pyridyl)-1,2,4-triazine, copper(II) and ruthenium(II) complexes, 248
5,6-Diphenyl-3-[4(pyrrolidin-1-yl)buta-(1*Z*,3*E*)-1,3-dienyl][1,2,4]triazine, 240
2,3-Diphenylquinoxaline 1-oxides, 151, 152
2,3-Diphenylquinoxaline radical anion, 139
2,3-Diphenylquinoxalines, spectra, 139
3,6-Di(2-pyridyl)-1,4-dihydro-1,2,4,5-tetrazine, structure, 258
Di(2-pyridyl)-1,2,4,5-tetrazine, reaction with styrenes, 258, 259
1,4,2-Diselenazines, 267
1,3-Diselenolium salts ring-expansion, 267, 268
2,6-Distilbenylazobenzene, 45
Dithiadiazines, structure, 268
1,4,2,5-Dithiadiazines, synthesis from 3-aryl-1,4,2-dithiazolium salts, 268
1,4,2-Dithiazine 1,1-dioxides, 267
1,4,2-Dithiazines, boat conformation, 266
–, synthesis from 1,3-dithiolium salts, 266
3,6-Di-(trifluoromethyl)-1,2,4,5-tetrazine, reaction with carbenes, 261
Dragmacidin B, 131

Elsaminose, 126
Emorfazone, 37
Enniatins, 205
β-(4-Ethoxybenzoyl)-α-(4,5-dihydro-5-oxo-1,3-diphenylpyrazol-4-yl)-propionic acid, 7
6-[*N*-(3-Ethoxycarbonylpropyl)amido]-tetrahydropyridazin-3(2*H*)-one, 23
3-Ethoxycarbonyl-1,2,4-triazine, cycloaddition reactions with enamines, 246
(*E*)-5-(2-Ethoxycarbonylvinyl)-2,4-dimethoxypyrimidine, 83
Ethyl 2-aminoquinoxaline-2-carboxylate, 156
Ethyl 2-chloro-4-trifluoromethyl-pyrimidine-5-carboxylate, 90
Ethyl 6-ethyl-4-hydroxycinnolin-3-carboxylate, 48
Ethyl 3-hydroxy-5,6-diphenyl-pyridazine-4-carboxylate, hydrazinolysis, 16
Ethyl 3-methylquinoxaline-2-carboxylate, 156
Ethyl 5-(phenoxymethyl)-4-pyridazine-carboxylate, 23
Ethyl 5-[(phenylthio)methyl]-4-pyridazinecarboxylate, 23
Ethyl 4-pyridazinecarboxylate, 23
Evodimine, 218

Febrifugine, 208
Fiscalin B, 209
Fiscalins, 214–216
Flavazoles, 149, 150
Fluorobenzo[*c*]cinnolines, 51
Fluoroionophores, 159
3-Fluoro-6-phenylpyridazine, 22
2-Fluorophenyl 3-pyridazinyl ketone, 22
Fluoropyrazine, lithiation, 107
Flutimide, 128
4-Formyl-2*H*-1,2-thiazine 1,1-dioxide, 78
5-Formyl-1,2,4-triazine oximes, 243
Friedel–Crafts alkylation, 167

Fullerenes, 191
Fumiquinazoline A, 214
Fumiquinazolines, 219–221, 224, 231
Fumiquinazolinones, 219–221
Fumitremorgin B, 222
Furazan-*N*-methanides, 269
Furo[2,3-*b*]quinoxalines, fluorinated, 146
Furo[2,3-*b*]quinoxalinones, 149, 150
4-(2-Furyl)-2-methyl-2-methylaminopyrimidines, 72

Gallocyanine, 184, 185, 188
Gliadins, 191
Glyantrypine, 214–216

4-Halogenocinnolines, 42
3-(4-Halophenyl)pyridazinium ylides, reactions with aryl isocyanates, 25
Halopyrazines, synthesis from tetrahydropyrazine-2,5-diones, 106
3-Halopyridazines, palladium coupling reactions, 19
Haloquinoxalines, 145–147
4-Heteroarylphthalazinones, 53
Hexafluoroquinoxalines, 146, 147
Hexahydrohexamethylbenzo[*b*]-phenazines, 171
Hexahydrophenazines, synthesis from 1,2-phenylenediamines/nitrosochlorides, 165
Hexahydropyrazines, 133
Hexahydropyridazines, 5
Hexahydroquinoxalines, 161
Hexazine radical anion, 264
Hexazines, 263–265
–, existence, 264, 265
Hinckdentine A, 230, 231
Hinsberg synthesis, 139
Hydrazaline, 69
3-Hydrazino-5,6-dihydrobenzo[*h*]-cinnolines, 52
5-Hydrazino-2-phenyl-3(2*H*)-pyridazinones, reactions with Pyrazolo[3,4-*d*]pyridazinones, 17
1-Hydrazinophthalazine, reactions with 1,3-diketones, 62
3-Hydrazinopyridazines, as antiinflammmatory agents, 33
5-Hydrazinopyridazinones, reactions with ethyl bromopyruvate/phenacyl bromide, 2
5-Hydrazinopyridazin-3(2*H*)-ones, 17, 18
6-Hydrazinopyridazin-3(2*H*)-ones, 15
3-Hydrazono-1,1,1-trifluoroalkan-2-ones, reactions with aldehydes, 239
3-Hydroperoxyindolines, 227
7-Hydroxechiozolinone, 210
3-Hydroxyanisotine, 210
1-Hydroxy-1,2-dihydrophthalazines, 63
6-(Hydroxyiminomethyl)-4-methyl-1-phenylpyridazinium-3-sulfate, 32
2-Hydroxyimino-4-methylpentanoic acid, 114
2-Hydroxy-6-methoxypyrazine, 112
2-Hydroxynaphthoquinones, reactions with 1,2-phenylenediamines, 170, 171
2-Hydroxyphenazine-1,5-dioxides, synthesis from nitroanilines/phenols, 165, 166
6-(2-Hydroxyphenyl)-1,2,4-triazines, 247
N-Hydroxypyrazines, metal complexes, 122
5-Hydroxypyrazinoic acids, 116
Hydroxypyridazinones, *O*-alkylation, 18
2-Hydroxypyrimidines, allylation with cyclopentadiene monoepoxide, 84
4-Hydroxyquinazoline, 206
3-Hydroxyquinazolines, synthesis from 2-aminobenzohydroxamic acid/adehydes, 95
1-Hydroxyrutaecarpine, 217, 218
7-Hydroxyrutaecarpine, 216, 217
4a-Hydroxytetrahydropterins, dehydration, 89

Imidazoles, synthesis from 1,3,5-thiadiazines, 265, 266
Imidazolol[1′,2′:1,2]pyrrolo-[3,4-*d*]-pyridazines, 39
Imidazolo[1,2-*b*]pyrazines, 111
Imidazolopyrazines, 111
Imidazo[1,2-*a*]pyrazolo[4,3-*b*]indoles, 261
Imidazo[1,2-*b*]pyridazine-2-acetic acids, as analgesics, 37

Imidazo[1,2-*b*]pyridazines as CNS agents, 39
6-Imino-2-(tetrazol-5-yl)-1,2,3,4,5-pentazine, 263, 264
Indenopiperazines, 137
Indolopyridoquinazoline alkaloids, 214–223
Indoloquinazoline alkaloids, 214–223
Indoloquinazolines, 230
Indolylmethylpyrazine, *N*-oxides, 113, 114
5-Iodo-2,4-dimethoxypyrimidine, 83
2-Iodo-6-methoxypyrazine, 108
4-Iodo-2-methylthiopyrimidine, 86
Iodo-1,2,4-triazines, cross-couplings with alkenes, 244
Isocytidine isosteres, 119
Isoxazoles, 87
2-Isoxazolines, 87
Isoxazolo[3′,4′:3,4]pyrido[2,3-*d*]pyrimidines, 88
Isoxazolo[2,3-*a*]quinoxalines, 140

Janus green, 186, 189
Janus green B, 188
Janus green D, 188
Janus green V, 188

Ketopiperazines, 133–136

Lamotrigine, analogues, 238
Lavanducyanin, 168, 169
Lucigenin, 190
Luminol, 67, 68

Mappicine, 245
Meldola blue, 179–181
5-Mercaptophthalazin-1(2*H*)-one, 56
2-Mercaptoquinazolin-4(3*H*)-one, reaction with hydrazonyl halides, 97
Merocyanines, 192
Metflurazon, 40
1,6-Methano[10]annulenes, synthesis from 4a,8a-methanophthalazines, 60
4a,8a-Methanophthalazine, 55
4a,8a-Methanophthalazines, 60
3-Methoxycarbonylmethylene-quinoxalin-2-one, 149
6-Methoxycarbonyl-4-phenyl-1,2,3-triazine, X-ray crystallography, 234
3-Methoxy-6-methylthio-1,2,4,5-tetrazine, 258
4-Methoxy-2-phenylquinazoline, metalation, 96
1-Methoxyrutaecarpine, 217
2-Methoxythiopyrimidin-4(3*H*)-ones, 81
3-Methoxy-1,2,4-triazine, reactions with secondary amines, 243
7-Methoxyvasicinone, 210
S-Methylalkylidenehydrazinecarbo-dithiolates, 265
Methyl 3-aminopyrazine-2-carboxylate, 110
Methyl 3-aminopyrazinoate, 115
2-Methylcinnolin-3-one, X-ray crystallography, 47
6-Methyl-2,3-diphenylquinoxaline, metalation/alkylation, 158
Methylene blue, 174, 175, 177, 179–186, 190–192
Methylene green, 182, 183, 185
Methylenequinoxalines, 149
Methylene violet, 186
Methyl 3-ethoxy-2-nitroacrylate, Michael additions, 126
3-Methylidene-2,3-dihydrothiazolo[2,3-*c*][1,2,4]benzotriazine, 244
N-Methylindole, 117
Methyl 3-iodopyrazine-2-carboxylate, 109, 110
3-Methyl-5-nitropyrimidin-4(3*H*)-one, 87
Methylpyrazine, 102
N-Methylpyridazines, NMR spectra, 30
2-Methylpyridazin-3(2*H*)-ones, Diels–Alder reactions, 23–25
2-Methylquinoxalines, hydrogenation, 143
–, reaction with ethyl bromopyruvate, 144
–, – with ninhydrin, 143
N-Methylquinoxalines, synthesis from 2-nitroanilines, 158, 159
1-Methylquinoxalin-3(1*H*)-one, 143
2-Methylsulfinylpyrazine, 112
2-(4-Methylsulfonylphenyl)-6-phenyl-hexahydro-4-pyridazinol, 2

2-Methylsulfonylpyrazine, 112
2-Methylthiopyrimidin-4(3*H*)-ones, 81
2-Methylthio-4,6-pyrimidinyl ditriflate, 86
4-Methylthio-1,3,5-triazin-2(3*H*)-ones, 251
4-Methyl-1,2,3-triazine, reactions with enamines, 237
2-Methyl-1,2,3-triazinium iodide salts, nucleophilic substitution, 236
N-Methyl-1,3,5,2-trioxazinane, 271
Michael–Stetter reaction, 5
Minaprine, synthesis, 34
Minisci reaction, 117
Mitomycin A, reaction with *N*-methyl-1,2-phenylenediamine, 171
Monodontamide F, 205, 206
5-Morpholinopyridazin-3(2*H*)-one, 37

3-*β*-Naphthylamino-6-^{i}propylamino-pyridazine, 41
Neutral red, 181, 186
New methylene blue, 184
New methylene blue N, 182, 183
Nile blue, 179, 180, 186, 189
Nile red, 173
Nitrobenzo[*c*]cinnolines, 51
3-Nitropyridinones, 87
5-Nitroquinoxaline-2,3-diones, 160
Norfluazon, 40
Nortryptoquivaline, 209

Octahydropyrazines, 137
Octahydropyridopyrazines, 137
1,2,3-Oxadiazines, 268, 269
1,3,5-Oxadiazines, 269, 270
Oxadiazino[5,6-*b*]quinoxaline, 152
Oxadiazolo[2,3-a]quinoxaline 5-oxides, deoxygenation, 154
Oxadiazoylquinoxalines, 156
1,2,3,5-Oxathiadiazine 2,2-dioxides, 272
Oxazine dyestuffs, 192
Oxazolo[3′,2′:1,2]pyrrolo[3,4-*d*]pyridazines, 39
Oxazolo[4,5-d][1,2,3]triazoles, ring-expansion, 270

Peganine, 211–213
Peganine *N*-oxide, 211
Peganol, 213
Pelagiomicin A, 162
Pentazines, 263–265
Perhydropyrimidin-2-ones (thiones), synthesis from 1,3-diaza-1,3-butadienes/enamines, 76
Perhydroquinazolin-2-ones (thiones), synthesis from 1,3-diaza-1,3-butadienes/enamines, 76
Phenacylpyrazines, enolisation, 101
Phenacylquinoxalines, enolisation, 139
1,10-Phenanthroline, 187
Phenanthro[9,1-*e*][1,2,3]triazines, 235
Phenanthro[9,10-*e*][1,2,4]triazines, 242
Phenazine antibiotics, 162, 163
Phenazine-1-carbonyhydrazide, reactions with pyrilium salts, 168
Phenazine-1-carboxylic acids, 162, 163
Phenazine-1,6-dicarboxylic acids, 162, 163
Phenazine-1,6-diol, 162
Phenazine-1,6-diol 5,10-dioxide, 162
Phenazine 5,10-dioxide, 164
Phenazine-fused [60]fullerenes, 171
Phenazines, 162–171
–, methods of synthesis, 164
–, physical and spectroscopic properties, 164
–, reduced, 168
Phenazine-tethered oligonucleotides, 167
Phenazinium salts, 174
Phenazinomycin, 168, 169
Phenazinone dyestuffs, 173, 174
Phenazinones, 168, 169
Phenazostatins, 162
Phenosafranine, 174, 186, 192
Phenothiazine dyestuffs, electro-chemistry, 179
Phenothiazinium salts, 175, 176, 178, 180, 183, 187, 190
4-[4-(Phenoxyethyl)-1-piperazinyl]-pyridazin-3(2*H*)-ones, 33
1-Phenoxyphthalazines, 60
N-Phenylbarbituric acid, 82
2-Phenylpyridazin-3-(2*H*)-ones, 36
6-Phenylpyridazin-3-(2*H*)-ones, antinociceptive activity, 37, 38

6-Phenyl-3-(2*H*)-pyridazinones as hypotensive agents, 35
2-Phenylquinazolines, 93
3-Phenylquinazoline-4(3*H*)-thiones, 93
Phenylseleno-1,4,2-dioxazines, synthesis from *O*-allyl hydroxamic acids, 270, 271
5-Phenylsulfonylmethyl-1,2,3-triazines, 236
2-Phenylsulfonylpyrazine, 112
2-Phenyl-2*H*-1,2-thiazines, nitrosation, 12, 13
2-Phenylthiopyrazine, 112
Phthalazin-1,4-diones, antineoplastic activity, 69
Phthalazin-1,4(2*H*,3*H*)-diones, 66
Phthalazine *N*-oxide, 60
Phthalazine polymers, 53
Phthalazine quinones, 66, 67
Phthalazines, acylated, 59
–, [*f*]-annulated, 55
–, as chiral auxiliaries, 61
–, Cinchona derivatives, 61
–, electronic absorption spectrum, 58, 59
–, heat of formation, 58
–, hydrazino derivatives, 16, 17
–, methods of synthesis, 53–59
–, physical properties, 58–60
–, polyphenylated, 63
–, Raman spectra, 58
–, reactions, 59–63
–, – with 2-chloroimidazoline, 59
–, – with ynamines, 61, 62
–, synthesis from cycloaddition reactions, 54–57
Phthalazine ylides, 63, 64
Phthalazinium ylides, 56, 57
Phthalazinones, 23, 24
–, methods of synthesis, 53, 54
Phthalazin-1(2*H*)-ones, aldose reductase inhibitors, 69
–, phenylation, 65, 66
Phthalazinophane macrocycles, 67
Phthalizines, synthesis from amidines, 57
Phthalocyanines, 191
Piperazin-2-carboxamides, kinetic resolution, 132
Piperazinediones, 134–136
Piperazine-2,5-diones, alkylidene derivatives, 123
Piperazines, 133
–, conformationally restricted, 137
–, fused-ring systems, 137
–, monoacetyl, 129, 130
–, synthesis from 2-oxazolidinones, 129
2-Piperazinoic acids, 131
Piperazino-*β*-lactams, 131
Piperazin-2-ones, use in stereo-controlled reactions, 133
6-Piperazinyl-3(2*H*)-pyridazinones, bronchodilating activity, 33
5-(4-Piperazinyl)pyridazin-3(2*H*)-ones, as alpha-blocking agents, 33
Podophyllotoxin A, pyridazine analogues, 40
Poly(imidepyridazinamide)s, 41
Prizidlol, 32
Propellanes, 55
N-2-Propenyl-5,6-dimethyl-1,2,4-triazine-3-carboxamide, cycloaddition reactions, 245, 246
N-(2-Propenyl)-*N*-(4-methylsulfonyl-phenyl)hydrazone, reaction with benzaldehyde, 2
6-(2-Propynylaminomethyl)pyrazinones, 125
3-(Prop-2-ynylsulfanyl)-1,2,4-benzotriazines, 244
Protoporphyrin IX, 183
Pteridinones, 110
Pyranopyrimidinecarboxamides, 83
Pyrazinamides, 116
2,3-Pyrazinedioic acid, 115
Pyrazine 1,4-dioxides, 115
Pyrazine *C*-nucleosides, 100, 119, 121
Pyrazine *N*-oxides, 113, 115
Pyrazines, 99–137
–, bicyclic, 125
–, 1,3-dipolar cycloaddition reactions, 103
–, fused, 137
–, general chemistry, 100
–, metalation, 103
–, physical properties, 101
–, Suzuki couplings, 108, 109

Pyrazines (*cont'd*)
–, synthesis from chloromethyl dipeptidyl keontes, 102
–, – from dialkynyl-1,2-diones, 101
–, – from enolates/*O*-(4-toluenesulfonyl)-isonitrosomaleonitrile, 101, 102
–, – from 1,5-hex-1,5-diync-3,4-diones/ diaminomaleonitrile, 101
Pyrazinium dicyanomethylide, cyloadditions, 105, 106
Pyrazinoic acid, hydroxylation, 112
Pyrazinoic acids, 115, 116
Pyrazinols, 111
Pyrazinones, 111
–, atropisomeric, 123
–, photochemistry, 112
–, synthesis from aryl diazonium salts/ enones, 10
–, – from phosphonium ylides, 3
–, – *via* cycloaddition reactions, 10–12
Pyrazin-2(1*H*)-ones, cycloadditions with 1,2,4-triazoline-2,5-dione, 122
Pyrazinopsoralens, 109
Pyrazolo[4,3-*c*]cinnolines, 48
Pyrazolopyrimidines, 254
Pyrazolylquinoxaline oxides, 152
4-Pyridazinecarbonitriles, Diels–Alder reactions, 25
Pyridazine-3-carboxylic acids, synthesis from hydrazinopentanoates, 5
Pyridazine diones, synthesis from anhydrides, 14
Pyridazine fatty acids, 15
Pyridazine macrocycles, 29
Pyridazine metal complexes, 28, 29
Pyridazines, biological properties, 32–41
–, ethylation by bis(tributylstannyl)-acetylene, 28
–, hydrazino derivatives, 16, 17
–, physical properties and spectra, 30–32
–, reactions with chloroformate and allyltributyltin, 28
–, side-chain metalation, 22
–, synthesis from aryl hydrazones, 1–3
–, – from benzil monohydrazone/ enolates, 3
–, – from chloroazadienes, 11
–, – from furans, 12–14
–, – from isoxazoles, 12–14
–, – from pyrazoles, 12–14
–, – from 1,2,4,5-tetrazine, 11, 12
–, – from tetrazines, 12–14
Pyridazinium ylides, [3 + 2]-cycloadditons, 25, 26
–, spectra, 32
Pyridazino[3,4-*d*][1,5]benzodiazepines, 15
Pyridazino[1,5]benzothiazepines, 260
Pyridazino[1,4]benzothiazines, 260
Pyridazino[1,2-*a*][1,2,4]benzotriazin-6-ones, 15, 16
4-Pyridazinols, synthesis from 1,5-diarylpent-1-yne-3,5-diones/ hydrazine, 3, 4
Pyridazinomorphinans, 12
Pyridazinone ligands, 28
Pyridazinones, physical properties and spectra, 30–32
–, ring-fused, 125
–, synthesis from (arylethyl)furanones, 8
–, – from 4,5-dihydropyridazinones, 26
3-(2*H*)-Pyridazinones, synthesis from ketones, 4
Pyridazin-3(2*H*)-ones, 5
–, synthesis from 1,2-dicarbonyl compounds, 14, 15
–, – from 4,5-dihydroisoxazoles, 9
Pyridazino[4,5-*d*]pyridazine, 55
Pyridazino[3,4-*b*]quinoxalines, 149, 150
Pyridazinyl oestradiols, 36
Pyridazin-3-yl triflates, methoxycarbonylation, 20
Pyridazocidin, 42
Pyridimidines, physical properties, 72
Pyridines, synthesis from 1,2,4-triazines, 245
Pyrido[1′,2′:1,2]imidazolo[4,5-*b*]-quinoxalines, 147
Pyrido[2,3-*b*]pyrazines, 120, 121
Pyrido[2,3-*c*]pyridazines, 21
Pyrido[3,4-*d*]pyridazines as CNS agents, 39
Pyrido[1′,2′:3,4]pyrimidino[4,5-*d*]-pyrimidines, 83
Pyridoquinazoline alkaloids, 214–223
Pyridoquinoxalines, 156

Pyrido[1,2-*b*][1,2,4]triazinium salts, 240
Pyridylpyrimidines, 77
Pyridyl-1,2,4-triazines, 240
Pyrimidine-5-carboxylic acids, synthesis from bromopyrimidines, 84
Pyrimidine orthoquinodimethanes, 88
Pyrimidines, alkylated, 85
–, as agrochemicals, 91
–, fluorinated, 73
–, general literature, 71
–, metalation, 84
–, methods of synthesis, 72–79
–, "Principle Synthesis", 72
–, reactions, 79–91
–, – with cyclic ketones, 83
–, stannylated, 85
–, synthesis from acetylpyrroles, 75
–, – from amidines, 73
–, – from anilines, 73, 74
–, – from cyanothioacetamides, 73, 74
–, – from 1,3-diaza-1,3-butadienes/chloroketenes, 75
–, – from enamines, 73
–, – from enaminones, 73, 74
–, – from ketene dithioacetals, 74
–, – from ketones/cyanamides, 77
–, – from oxazoles, 78
–, – from phenyltrichloromethane/nitriles, 76
–, – from *N*-tosyl-2-pyrrolinine, 75
–, – from *N,N,N*-tris(trimethylsilyl)-amidines/vinamidinium salts, 78
–, – from ureas, 72
–, – from vinylamines/aldehydes/ammonia, 73
–, use in non-linear optics, 90
Pyrimidin-2(1*H*)-one, bromination, 80
Pyrimidinones, synthesis from 3-amino-1-ethoxycarbonylmaleimides, 78
–, – from 1,3-diaza-1,3-butadienes/chloroketenes, 76
–, – from 1,1-dicyano-2-(*N*-phenyl-amino)ethene, 75
Pyrimidinyl triflates, reactions with stannanes, 85
Pyrimido[4,5-*b*]azepines, 88
Pyrimido[4′,5′:4,5]pyrido[6,1-*a*]isoquinoline, 83
Pyrimido[4′,5′:4,5]pyrido[6,1-*a*]-phthalazine, 83
Pyrimido[4,5-*b*]quinoxalin-4(3*H*)-one, 156
Pyrrolo[2,1-*a*]phthalazines, 63, 64
Pyrrolo[2,3-*b*]pyrazines, 117
Pyrrolopyridazines, 26
Pyrrolo[2,3-*d*]pyridazinones, 2, 3
Pyrrolo[2,3-*d*]pyridazin-4(5*H*)-ones, 17, 18
Pyrroloquinazolines, 210–214
Pyrrolo[1,2-*a*]quinoxalines, 144
Pyrrolo[1,2-*d*][1,2,4]triazines, 247

Quinazoline alkaloids, 203
–, synthesis, 206–208
Quinazoline-2,4-dione, 203
Quinazolines, bicyclic, 203
–, biological properties, 98
–, general literature, 71
–, methods of synthesis, 91–95
–, physical properties, 91
–, reactions, 95–98
–, synthesis from nitrobenzaldehydes and ketones, 93
–, – from phthalazines, 63
Quinazolinethiones, 96
Quinazolinium salts, 94
Quinazolinocaboline alkaloids, 216, 217
Quinazolin-4-ones, 227–229
Quinazolin-4(1*H*)-ones, 92
Quinazolin-4(3*H*)-ones, 92, 96, 203–210
Quinolines, pyridazinyl, 22
Quinoxaline, standard enthapy of formation, 138
Quinoxaline carboxyureides, 161
Quinoxaline charge-transfer complexes, 142, 143
Quinoxalinediones, 150, 151
Quinoxaline-2(1*H*),3(4*H*)-diones, tautomerism, 150, 151
Quinoxaline 1,4-dioxides, 137, 138
Quinoxaline-1,4-dioxides, 165, 166
Quinoxaline 1,4-dioxides, heats of formation, 155
Quinoxaline *N*-oxides, 151, 155
Quinoxaline-2,3-quinodimethane, 144
Quinoxalines, 137–161
–, hydrogenation, 160

Quinoxalines (*cont'd*)
–, methods of synthesis, 139–142
–, physical and spectroscopic properties, 138, 139
–, protium–deuterium exchange, 143
–, reduced, 158–161
–, synthesis from 1,2-phenylenediamine/fluorobenzils, 144
–, – from 1,2-phenylenediamines/α-dicarbonyl compounds, 139, 140
–, – from 1,2-phenylenediamines/oxoaldehydes, 138, 139
–, tetrasubstituted, 157
Quinoxalinones, 143
–, tautomerism, 150
Quinoxalin-2-(1*H*)-ones, 148–151
Quinoxalinoporphyrazines, 156
Quinoxalinylpyrazoles, 149
Quinoxazlin-2-yl palladium (II)complexes, 141

Resazurin, 186
Rhodamine 6G, 187
4-(β-D-Ribofuranosyl)pyridazines, 11
5-(β-D-Ribofuranosyl)-1,2,4-triazines, 239, 240
Ritterazines, 99
Rutaecarpine, 217, 218

Saccharin, 189
Safranine T, 186, 187
Sanfranamycin analogues, 136
Secretagogues, 128
Sevron blue 5G, 189
1-Silyloxymethyluracils, 84
2-Silyloxypyrazines, coupling reactions with sugars, 122
Silyloxypyrimidines, cleavage, 90
SNC 80, 130
Spirodihydro[indole-2,4′-piperidines], 195
Spirooxazines, 192
Spirooxetanes, 197
Spiropyrans, 192, 196
Spiroquinazolines, 223, 224, 231
Spirotetrahydropyridazinones, 19, 20
Streptnigrin, analogues, 157
ω-Sulfamoylalkoxylimidazo[1,2-*b*]-pyridazines, as antiasthmatics, 39
5-(2-Sulfamoylvinyl)pyrimidines, 78
Sulfoxonium ylides, 274
Sulfur dyestuffs, 200

5,5′,6,6′-Tetraaryl-3,3′-bi(1,2,4-triazines), 241
1,2,9,10-Tetraazaphenanthrenes, 42, 43
Tetraazaspiro compounds, 261
3,4,5,6-Tetrachloropyridazine, 20
Tetrachloroquinoxaline quinones, 158
Tetradotoxins, 230, 231
3,4,5,6-Tetrafluoro-1,2-phenylene-diamine, 146
5,6,7,8-Tetrafluoroquinoxalines, 146
Tetrahydrocinnolines, synthesis from phosphonium ylides, 3
1,2,3,4-Tetrahydrocinnolines, oxidation, 45
5,6,7,8-Tetrahydrocinnolines, synthesis by cycloaddition reactions, 45
1,2,3,4-Tetrahydrocinnolin-4-one, acylation, 49
2-(1,2,3,4-Tetrahydrocinnolin-4-onyl)-acetic acid, 49
5,6,7,8-Tetrahydro 3,4-diphenylcinnolin-5-ones, 46
Tetrahydro-1,4,3-oxathiazine 4,4-dioxides, synthesis from sulfonamides, 271
1,2,3,4-Tetrahydrophenazine, 165
Tetrahydrophenazines, 171
1,2,3,4-Tetrahydropyrazines, 103, 104
–, asymmetric hydrogenation, 127
Tetrahydropyrazinoindoles, 117
Tetrahydropyrazino[2,3-*b*]-indoloprophyrazines, 118
1,2,3,6-Tetrahydropyridazine, 15
2,3,4,5-Tetrahydropyridazine-3-carboxylic acid, 12
Tetrahydropyridazine-3-carboxylic acids, synthesis from γ-formyl-(α-hydrazinobutyric) acids, 10
Tetrahydro-3,6-pyridazinedione-3-hydrazones, 7
Tetrahydropyridazine herbicides, 40
1,4,5,6-Tetrahydropyridazines, 13
–, synthesis from hydrazines/2,5-dimethoxytetrahydrofuran, 5
Tetrahydropyridazin-3-one, synthesis from 6-(4-biphenylyl)-4-(arylamino)-

2,3,4,5-tetrahydropyridazin-3-one, 5
1,4,5,6-Tetrahydropyridazin-3(2*H*)-ones, 6
Tetrahydropyrido[3,2-*f*]quinoxalines, 148
1,2,3,4-Tetrahydroquinazolines, synthesis from *N*-(chloroalkyl)pyridinium chlorides, 92
1,2,3,4-Tetrahydroquinoxalines, 160, 161
5,6,7,8-Tetrahydroquinoxalines, 141, 161
1,4,5,6-Tetrahydro-1,2,4-triazin-3(2*H*)-ones, azacyclic peptide mimics, 242
1,1′,5,5′-Tetramethyl-6,6′dioxobisverdazyl, 262, 263
1,2,3,4-Tetramethylpyrazine *N*-oxides, 115
1,4,5,8-Tetraphenylphthalazine, 63
2,4,5,8-Tetraphenylquinazoline, 63
3,3,6,6-Tetraphenyl-1,2,4,5-tetrathiane, thermolysis, 274
1,2,3,4-Tetraquinoxalines, synthesis from quinoxalines, 143
1,2,3-5-Tetrathiane, 273, 274
Tetrathianes, 273, 274
1,2,4,5-Tetrathianes, synthesis from alkane dithioic acids, 275
Tetrazalophthalazines, 66
Tetrazines, equilibrium geometry, 255
–, methods of synthesis, 255
–, physical data, 255
1,2,3,4-Tetrazines, cyclopenteno, 255
–, synthesis and structure, 254–256
1,2,3,5-Tetrazines, synthesis and structure, 254–256
1,2,4,5-Tetrazines, as azadienes, 258–262
–, cycloadditions with azulenes, 260
–, – with 1,5-benzothiazepine and 1,4-benzothiazines, 260
–, – with cyclopenta[*e*]thiopyran, 260
–, – with cyclopropanobenzene, 260
–, – with dienamines, 260
–, – with 2,3-dioxabicyclo[2.2.2]oct-5-ene, 260
–, – with [60]fullerene alkenes, 260
–, – with heterodienophiles, 260, 261
–, – with indoles, 260
–, – with 1-methyl-2-methylthio-2-pyrroline, 260
–, – with nucleophilic singlet carbenes, 261
–, – with thiopyranocyclopenta[1,2-*a*]pyridazines, 260
–, physical data, 256, 257
–, reactions, 258–262
–, solid phase synthesis, 259
–, synthesis and structure, 256–258
–, – from iminium chlorides/*S*-methylisothiocarbonohydrazide iodide, 257
Tetrazolopyridazines, 16
1,2,5-Thiadiazines, synthesis from 2,4,6-trichloro-1,3,5,2,4,6-trithiatriazine, 268
1,3,5-Thiadiazines, synthesis from amino(thiocarbonyl)amidines, 265
1,3,4,5-Thiatriazines, desulfurisation, 266
–, methods of synthesis, 266
Thiazolo[2,3-*a*]phthalazines, 64
Thiazolo[4,3-*a*]phthalazines, 64
Thiazolo[4,5-*d*]pyridazines, as antibacterials, 37
Thienocycloheptapyridazines, as muscarinic affinity labels, 34
Thienopyridazinium, mesoionic species, 56, 57
Thieno[2,3-*d*]pyrimidines, 81
5-(2-Thienyl)-2,4-dimethoxypyrimidine, 83
Thienylpyridazinones, 5
Thiobarbiturates, photocyclisation, 89
Thiobarbituric acid, 71
5-Thioformyluracils, 89
Thionine, 176, 177, 184, 191
Thionine blue, 184, 190
Thiophenes, synthesis from dithines, 267
Thiopyronine, 177, 192
Thymine, alkylation by 4-bromobutyl acetate, 84
Titanocene chelate complexes, 273
Toluidine blue, 183, 184
Toluidine blue O, 176, 179–181, 187
4-(4-Tolyl)hydrazino-2-phenyl-2-oxazolin-5-one, 44

Trialkylpyrazines, 104
Trialkylsilyloxyphenazines, 168, 169
2,4,6-Triallyl-1,3,5-triazacyclohexanes, 254
Triaryldihydroquinazolines, 93
2,4,6-Triaryl-1,3,4,5-oxatriazines, 270
2,4,5-Triaryl-1,2,3-triazolium-N^1-arylimides, 266
Triarylverdazylium salts, reduction by ascorbic acid, 262
1,3,5-Triazine-2,6-diones, synthesis from isocyanates/*N*-trimethylsilylimines, 252
1,2,4-Triazine nucleosides, 239
1,3,5-Triazine-2-one-6-thiones, synthesis from isocyanates/*N*-trimethylsilylimines, 252
1,2,4-Triazine 1-oxides, reactions with benzyne, 246, 247
1,2,3-Triazine pyrolysis, 237
1,2,3-Triazines, electronic absorption spectrum, 234
–, inverse-electron demand cycloadditions, 236, 237
–, nucleophilic substitution, 235
–, – with chloromethylphenyl sulfone, 236
–, optimised structure, 233, 234
–, photolysis, 237, 238
–, reactions, 235–238
–, – with silyl enol ethers/1-chloroethyl chloroformate, 236
–, – with tetracyanoethylene oxide, 236
–, synthesis and structure, 233–235
–, – from 1-aminopyrazoles, 234
–, – from cyclopropenyl azides, 234
1,2,4-Triazines, cycloaddition reactions with indoles, 246
–, cycloadditions with cyclic vinyl ethers, 245
–, metalation, 244
–, NMR spectra, 238
–, physical data, 238
–, polymeric, 240
–, reactions, 242–248
–, synthesis and structure, 238–242
1,3,5-Triazines, 88
–, acidic hydrolysis, 253
–, allylation, 254
–, cycloadditions with amidines, 253, 254
–, photo-disassociation, 253
–, physical data, 248, 249
–, reactions, 253
–, synthesis and structure, 248–252
–, – from *N*′-acyl-*N*,*N*-dimethylamidines, 250
1,3,5-Triazine-2,4,6-triones, synthesis from 1,3,4-oxadiazol-2(3*H*)-ones, 251
1,2,3-Triazinium-2-dicyanomethylides, synthesis, 235, 236
1,2,4-Triazino[3,2-*a*]isoquinolinium salts, 242
1,2,4-Triazino[2,3-*b*]isoquinolinium salts, 242
1,3,5-Triazinones, synthesis from ethoxycarbonylhydrazones/aromatic aldehydes/aryl or alkyl isocyanates, 252
2,4-Triazoline-2,5-dione, 122
1,2,3-Triazolium-N^1-ylides, 266
Triazolopyridazine, anticonvulsant properties, 34
Triazolopyridazines, 16
1,2,4-Triazolo[4,3-*a*]quinazolin-5-ones, 97
Triazolylquinoxalines, 156
2,3,5-Tri *O*-benzyl-D-ribo-1,4-lactone, 120
Tribromoindolo[1,2-*c*]quinazolines, 230
2,4,6-Trichlorophenol, reaction with sodium sulfide, 201
3,4,5-Trichloropyridazin-6-one, 20
2,4,6-Trichloropyrimidine, reaction with pyridines, 82
2,4,6-Trichloro-1,3,5,2,4,6-trithiatriazine, 268
Tricyclic pyridazinones, aldose reductase inhibitors, 33, 34
1,2,4-Tri(ethoxycarbonyl)-hexahydropyridazine, 27
5-Trifluoromethyl-2,3-dihydro-1,2,4-triazines, 239
5-Trifluoromethyl-2,5-dihydroxy-2-methyl-1,2,4-triazines, 239
4-Trifluoromethyl-2-methylthiopyrimidine, lithiation, 86

3-(Trifluoromethyl)piperazine, synthesis from 3,3,3-trifluoro-2-oxoPropanal/ *N*-benzylethylenediamine, 130
2-Trifluoromethylpyrazine, synthesis from 2-iodopyrazines, 107
Trifluoromethyl-4-pyridazinones, 3
Trifluoromethyl-1,2,4-triazines, 238
5-Trifluoromethyl-1,2,4-triazines, 239
2,3,4-Trihydro-1,2,4,5-tetrazin-1-yl radicals, 262, 263
1,2,5-Trimethylpyrrole, 117
2-(Trimethylsilylmethyl)phthalazinium triflate, 64
Trimethylstannylpyrimidines, 85, 86
2,4,6-Trinitro-1,3,5-triazine, theoretical prediction, 248, 249
1,2,4-Trioxanes, 272
1,2,4,5-Trioxazines, 271
Triphenodioxazine dyestuffs, 196–200
2,4,6-Triphenyl-1,3,5-triazine radical anion, 250
4,5,6-Tris(dimethylamino)-1,2,3-triazine, X-ray crystallography, 234
Trisheterenium salts, 82
1,2,4-Trithiane, 273
1,3,5-Trithiane-1,1,3,3,5,5-hexaoxide, silylation, 273, 274
Trithianes, 273, 274
Tryptanthrin, 218, 219, 230
Tryptoquivalines, 209

Ugi condensation, 132
Uracils, bromination, 80
Ureidopyridazines, anticonvulsant properties, 34

Vascinolone, 212
1-Vasicine, 210–212
Vasicinol, 212
Vasicinone, 211–213
Vasicinone *N*-oxide, 211
Vasnetine, 210, 211
Verdazyls, 262, 263
Vinblastine, 223
6-Vinylcytosines, 81
5-Vinylpyrimidines, 83

Wipf dehydration, 215, 216
Wool fast blue BL, 186

Zolpidem, 111